MACMILLAN INTEGRATED SCIENCE Book 1

John Allen
Jenny Crocker-Michell
Maggie Hannon
Richard Page-Jones
Tony Thornley

Editor: Mick Michell

First published 1986

Published by
MACMILLAN EDUCATION LTD
Houndmills, Basingstoke, Hampshire RG21 2XS
and London
Companies and representatives
throughout the world

Printed in Hong Kong

British Library Cataloguing in Publication Data
Science 14-16.
Bk. 1
1. Science
I. Michell, Michael
500 Q161.2
ISBN 0-333-36224-1

10 9 8 7 6 5
00 99 98 97 96 95 94 93 92 91

MACMILLAN INTEGRATED SCIENCE Book 1

Contents

Introduction for students and teachers

Although there are many good traditional text-books on the market which concern themselves with the separate parts of science such as Chemistry and Biology we feel that those of you who have been wise enough to continue to study science as broadly as possible have been poorly served. The textbooks that exist are either designed to back up one particular course or are mixtures of bits taken from the separate Biology, Chemistry and Physics syllabuses. We set out with the idea of trying to write as modern a book as possible which would combine the best features of a traditional type of textbook, but which would also make your study of science broader and more interesting. In doing this our main aims are:

1. to provide you with a textbook which covers all that is required for GCSE, O-level and CSE science syllabuses;
2. to write for you, as far as we can, in an interesting and up-to-date way which includes many relevant examples;
3. to cut out any unnecessary scientific 'jargon' and to explain things in simple language so that you understand the difficult, as well as the simple, parts of science;
4. to help you to understand what a scientist does and what science is - and to emphasise that science is not just about remembering a lot of facts;
5. to help you to learn how best to approach your study of science so as to make it as interesting and relevant as possible;
6. and, most importantly of all, to help you to enjoy the study of science and to want to continue finding out about it.

The two books are divided into ten sections which we call *units*. Each unit contains a number of *topics*.

In each topic there are:

summaries of what you need to know at the end of each topic;
questions at the end of each unit to help you test your understanding;
assignments at regular intervals for you to do in school or at home. These are designed to extend your understanding or to reinforce it;
investigations which are aimed at getting you to look at science in a practical way.

Whilst we believe that the experiments can be carried out in safety and comply with the latest informed opinion, advice does change with time and it is the professional responsibility of the teacher to judge what is safe in his/her own circumstances.

Eye protection should be worn wherever chemicals are handled. A reminder about this is included in those experiments where the hazard is particularly great.

Mick Michell

Acknowledgements

The authors and publishers wish to thank the following who have kindly given permission for the use of copyright material:

The British Petroleum Company p.l.c. for extract from '*B.P. Briefing Paper: Energy for Development, September 1982*'.
The Campaign for Nuclear Disarmament for extract from *Questions and Answers about Nuclear Weapons* (1981)
The Guardian and Manchester Evening News for extract and illustration from the *Guardian* 10.1.83
William Heinemann Ltd for extract from *The Grapes of Wrath* by John Steinbeck
Heinemann Educational Books Ltd for Table 3.7 in '*Mineral Resources*', Reader G, Science in Society series.
The Controller of Her Majesty's Stationery Office for extract from *Defence Fact Sheet 3*, February 1983
The Observer Limited for extract, 'US Plans to build a Village in Space' the *Observer*, 24.10.82
Shell Times Educational for map on the proposed Severn Barrage from 'Alternatively Speaking', Shell Times Educational Pull-out Supplement No.1
Every effort has been made to trace all the copyright holders but if any have been inadvertently overlooked the publishers will be pleased to make the necessary arrangement at the first opportunity.

The authors and publishers wish to acknowledge the following photograph sources:

Argos Ltd p.195; Austin Rover Group p. 187; British Aircraft Corporation fig. 3.51(b); Reproduced by courtesy of the Trustees of the British Natural History Museum figs 1.62, 5.4(a); British Petroleum plc fig. 3.30; Jim Brownbill fig. 3.17(a); Camera Press figs. 2.34(d), 3.1(c), 3.17(c), 3.33, 4.1; Bryn Campbell fig. 5.7(b); CENCO fig. 5.3; Central Electricity Generating Board p.216, p.165; Central Office of Information fig. 5.35; Central Press Photos fig. 3.1(b); CERN p.246; Bruce Coleman Ltd fig. 2.2(b), 2.6, 2.24(a), 2.24(b), 2.34(b), 2.34(c), 2.34(e), 2.34(f), 2.35(b); Colorsport fig. 3.1; Daily Telegraph Colour Library figs. 1.20, 1.22, 1.26(a); Department of the Environment (Crown Copyright) figs. 1.74(c), 3.16(b); K.I. Dobson fig. 4.9; FAO Photo Library fig. 3.16(a); Philip Harris Ltd fig. 1.29; Alison Hart figs. 1.53, 1.54(c), 2.35(a), 2.29; p.115; Griffin and George Ltd figs. 4.5, 5.27; I.C.I. Mond Division fig. 5.22; Reproduced by courtesy of the Trustees of the Institute of Geological Sciences figs. 1.49(a), 1.49(b), 1.50(a), 1.50(b), 1.52, 1.56(d); Rodney Jennings fig. 2.8(a), 2.8(b), 2.10, 2.32(a), 2.32(b), 3.20(a), 3.24(a), 3.24(b), 3.25, 4.57, 4.58, 5.26(b); Dr Leedale fig. 2.19(b); Mullard Ltd figs. 4.38; p.225; NASA figs. 1.1, 1.12; NEI Parsons Ltd fig. 4.75; Novosti Press Agency figs. 1.44; Photo Source fig. 3.51(c); Popperfoto figs. 1.58, 3.56(b), 5.7(a); Post Office Telephones fig. 3.67; Raleigh Ltd fig. 3.14(b); Mervyn Rees fig.1.54(d); Dave Richardson fig. 3.51(b); R. S. Components Ltd fig. 4.34; Reproduced by courtesy of the Trustees of the Science Museum figs. 3.66; p.243; Science Photo Library Ltd figs. 3.1d, 3.58 (Ray Ellis), 3.68; p.20; Soil Survey of England and Wales fig. 1.71; T. I. Machine Tools p.187; TOA Electronics Ltd fig. 3.57(a); Topham Picture Library 1.26(b), 1.72, 2.33, 3.17(b), 5.26(a); Transport and Road Laboratory fig. 3.2(a); Jim Turner figs. 3.1(e) 3.20(b), 3.36; UAC International fig. 3.56(a); UKEA 5.32; p.263; Zefa (UK) Ltd figs. 1.54(b), 2.31, 3.41; p.263.

The publishers have made every effort to trace the copyright holders, but where they have failed to do so they will be pleased to make the necessary arrangements at the first opportunity.

The authors and publishers would like to thank Dr T. P. Borrows, Chairman of the ASE Laboratory Safeguards Sub-committee, who offered advice on safety.

1. The Earth and its Place in the Universe

THE EARTH AND BEYOND

THE EARTH

We live on the planet Earth. The earth is a sphere 12 750 kilometres in diameter.

Like all planets the earth spins or ROTATES on its axis. It also orbits or REVOLVES around the sun. The earth is active. Not only does it move through space, but its surface is also constantly changing. From space we can see the thin crust or surface of the earth. This is called the LITHOSPHERE. On average it is about 100 km deep. The lithosphere has a wrinkled appearance. It has been folded by slow but steady movements over many thousands of years. Where the lithosphere is thin the earth is covered with oceans of water.

Cloaking the earth is a mixture of gases called the ATMOSPHERE. These gases move across the earth in great swirling patterns carrying with them water vapour. The vapour is clearly seen as clouds in the photograph. The atmosphere is held in place by the earth's gravity. Without the atmosphere the earth would be lifeless. The atmosphere keeps the temperature of the surface of the earth roughly constant at about 5°C on average. It shields the earth from the dangerous ultraviolet light and X-rays from the sun. It also traps part of the sun's energy that is essential for life.

Around the earth there is a thin layer of air, water and soil in which all life exists. This is called the BIOSPHERE. On land the biosphere goes down as far as the deepest tree roots. In the sea most life is in the top 500 metres. Some birds and insects fly high into the sky but most animals could not even live on the earth's highest mountain. The earth's biosphere supports a great richness and diversity of life that has evolved and adapted over millions of years. Visits to our close neighbours in space and a careful listening ear in the form of giant radio telescopes have yet to find evidence that even the most basic life form exists outside the earth's biosphere.

Fig. 1.1

THE EARTH ON THE MOVE

The earth, like all objects in the universe, is moving. There are two important parts to its motion (Fig. 1.2).

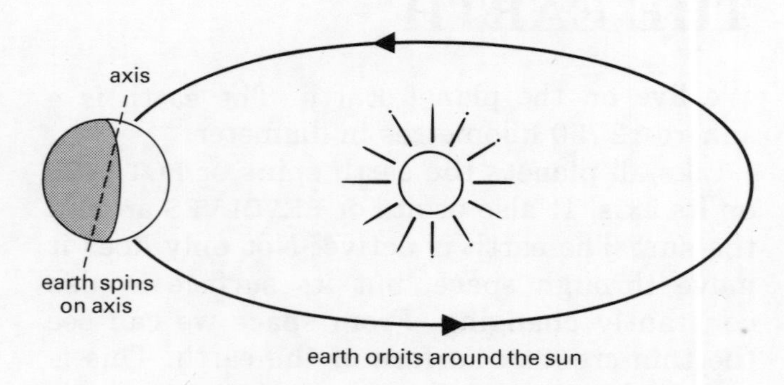

Fig. 1.2 The motion of the earth

(a) It is spinning or rotating on its axis.
(b) It is orbiting or revolving around the sun.

It is only recently that we have been able to get away from the earth so that we can see the movement. The motion of the earth has been known about for hundreds of years because of how it affects the way that we see other objects.

THE EARTH'S ROTATION

If you look at the position of the sun during the day you can see that the sun appears to rise, move through the sky and set once every 24 hours. This could be explained by saying that the earth rotates once every 24 hours while the sun remains still. It could also be explained by saying that the sun orbits the earth every 24 hours.

You get more evidence by looking at the night sky. The star patterns change during the night. The diagrams in Fig. 1.3 are drawn from photographs of star patterns. Photograph (a) was taken with the camera pointed at the North Star. The exposure time was short. In (b) the camera pointed the same way but the exposure was 4 hours. Photograph (c) was taken with a 4-hour exposure by a camera on the equator pointing straight upwards.

Again there are two possible explanations for this. Either the earth is spinning around on an axis that points north *or* all the stars are revolving around the North Star.

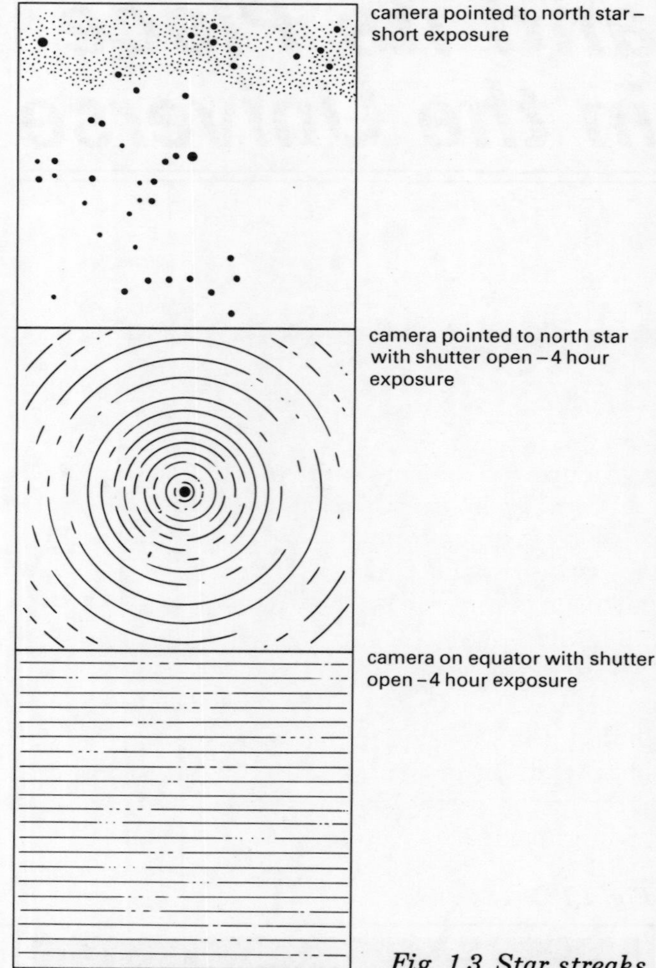

Fig. 1.3 Star streaks

The evidence gained by looking at the sun and stars can be explained in two ways. Either the earth rotates on its axis every 24 hours or the sun revolves around the earth and the stars revolve around the North Star every twenty-four hours.

The idea of the earth spinning is the simpler explanation, and it is accepted by everyone today. About 350 years ago these ideas produced a tremendous argument, among scientists, that also involved the Church. You can find out more about this in the assignment on page 4. The satellite evidence shown in Fig. 1.4 would have been very useful in 1616. A satellite in orbit over the poles sends back different pictures of the earth as the earth rotates beneath it.

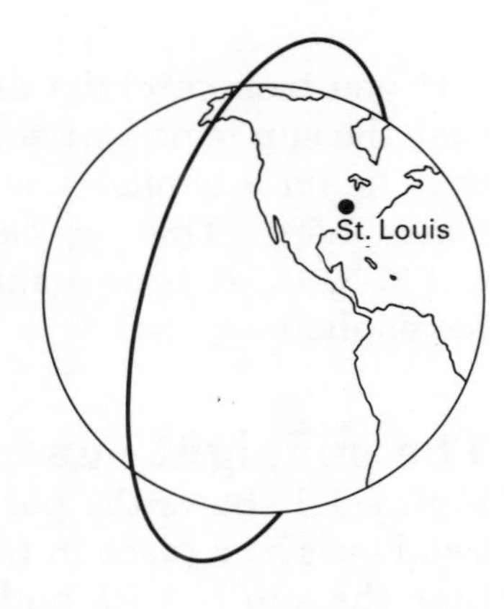

Fig. 1.4 A satellite in orbit over the poles shows that the earth rotates

THE EARTH'S REVOLUTION

Evidence for the earth's revolution around the sun can also be seen from the way that our view of distant objects changes. If you look at the sky at different times of the year, the star patterns change. Figure 1.5 shows why.

The fact that the earth revolves around the sun, together with the fact that the earth spins on a tilted axis also explains the SEASONS. Why should the weather in Britain be warmer in summer than in winter? Why should there be parts of the earth where people experience the midnight sun?

The seasons

The earth is warmed by heat rays from the sun. The amount of heat the earth picks up depends on the angle at which the heat rays hit it. Look carefully at Fig. 1.6.

The horizontal lines show heat rays from the sun. The right hand section shows the different parts of the earth. The three diagrams on the left show heat rays hitting these three areas, which are the same size. Count the rays hitting each area. Which has the most rays hitting it? From this we can see that the more a surface is tilted away from the sun, the fewer rays hit it and therefore the cooler it will be.

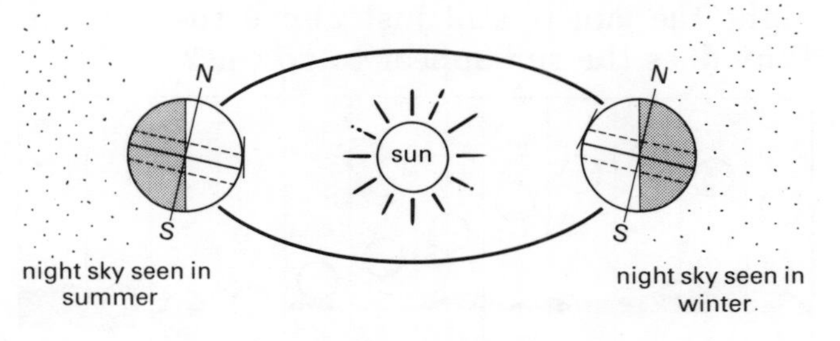

Fig. 1.5 The earth's position in winter and summer

As the earth moves around the sun the angle of the surface to the sun at a particular place changes because of the tilting of the earth's axis.

The places where the sun is directly overhead in June lie on a circle around the earth. This circle is called the TROPIC OF CANCER. Half a year later the surface of the earth at the Tropic of Cancer is tilted away from the sun, so not so much heat hits the earth here. It is then winter in the northern part of the earth.

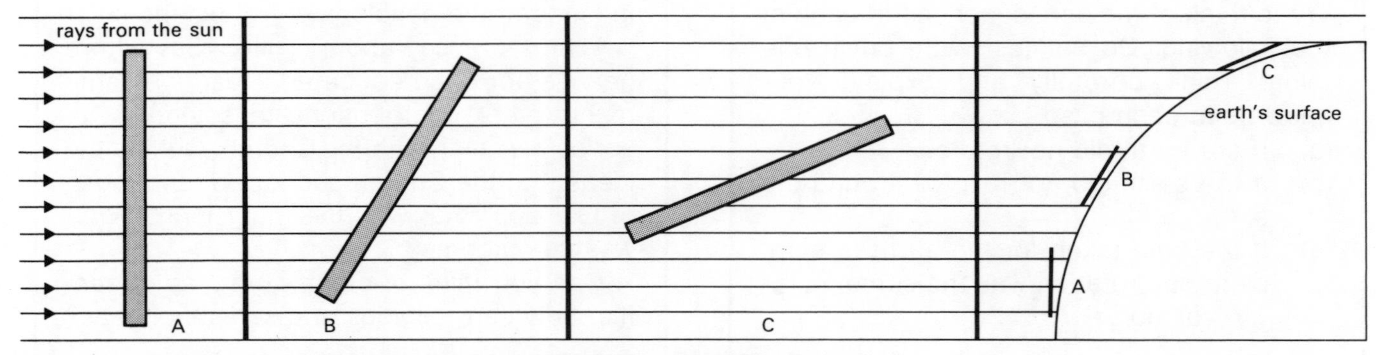

Fig. 1.6 The more a surface is tilted, the fewer the rays that hit it

If you look carefully at Fig. 1.5 you can see that the sun is now directly overhead at places that lie on a similar line around the southern hemisphere. This is called the TROPIC OF CAPRICORN. It is now summer in the southern hemisphere.

The midnight sun

Figure 1.7 shows the position of the sun every two hours at a place in the Arctic. You can see that the sun has its highest point at midday, but it never sets. It does not go dark. At midnight the sun is still just above the horizon. Why does the sun appear to do this?

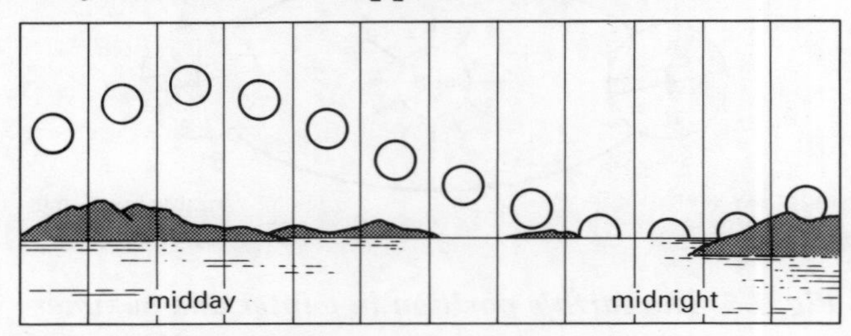

Fig. 1.7 The midnight sun

It is another consequence of the fact that the earth spins on a tilted axis. Look carefully at Fig. 1.8. At midday the sun would be seen at its highest point. As the earth rotates the sun appears to get lower, but even twelve hours later at midnight, because of the tilt of the earth, the sun can still just be seen. Notice that for a similar place in the southern hemisphere the sun would not be seen at all. It would be dark for the whole twenty-four hours. Six months later the position would reverse. It would be dark in the northern Arctic and the midnight sun would be seen in the south.

Assignment – The moving earth

Try to think of the consequences of each of the following. Do not just write down one thing. Think carefully and explain your answers in each case.

(a) If the earth did not rotate.
(b) If the earth did not revolve around the sun.
(c) If the time taken for the earth to complete one rotation was the same as for 1 revolution.

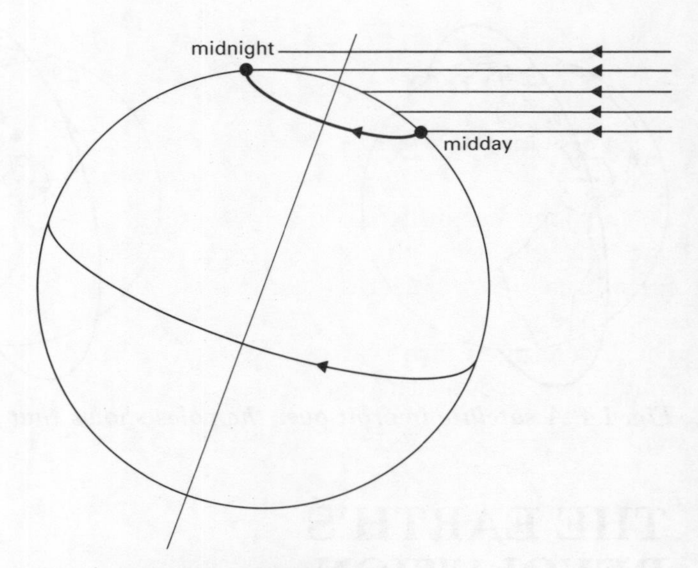

Fig. 1.8 Regions of midnight sun

Assignment – Galileo

The great thinker Aristotle died in 322 BC. For almost 2000 years afterwards people slavishly stuck to his ideas. It was not until the time of Galileo (1564–1642) that science broke out of these 'Dark Ages'.

It was Galileo who supported Copernicus' ideas about the position of the earth in the universe and who did more than anyone to convince people that the earth was not the centre of the universe, but was just one of a number of planets orbiting the sun.

He paid a heavy price for his beliefs, since they were in opposition to the main teachings of his day that were supported by the Church. He was persecuted and imprisoned. His writings were banned.

Find out as much as you can about Galileo. It is important that you realise that science is not just words in books but is the result of the activities of men and women, and that all sorts of outside factors have influenced the development of science.

Write an essay about Galileo. Say where he was born and educated. Describe some of his other contributions to science. Describe his clash with the Church. Outline the ideas that the Church supported, and those of Galileo. What was the Inquisition? What was its outcome?

Can you think of other examples of conflict between religious and scientific ideas?

OUR NEAREST NEIGHBOUR — THE MOON

If you look carefully at the moon each night you can see two changes taking place. The moon rises at a different time each night. It also appears to have a different shape. Both of these changes are caused by the moon's movement around the earth.

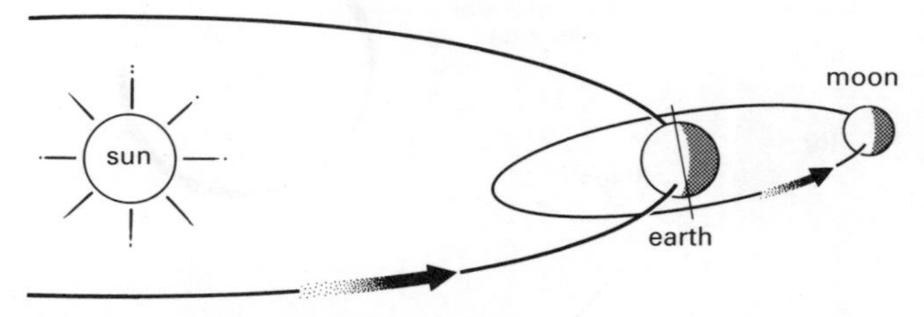

Fig. 1.9 The orbit of the moon

You can see from Fig. 1.9 that the moon orbits the earth on a path that is tilted with respect to the earth's orbit around the sun. We see the moon because sunlight is reflected off the moon's surface. As the moon orbits the earth we see different amounts of the surface hit by the sun. This causes us to see the moon with a different apparent shape or different PHASE. Look carefully at Fig. 1.10

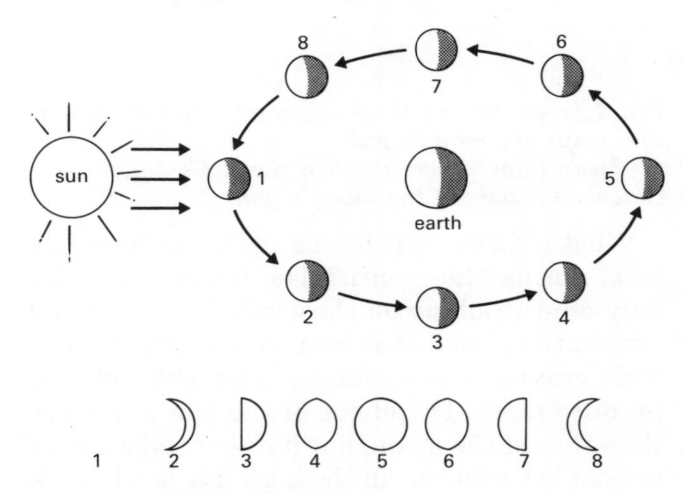

Fig. 1.10 Phases of the moon

When the moon is in position 1 the lit face points away from the earth. The moon cannot be seen. When the moon is in position 2 a small amount of the lit surface can be seen. To help decide on the shape it may be useful to turn the book around and look directly along a line from the earth to the moon. You can see that the right hand side of the moon is lit. If you looked into the night sky when the moon is in this position you would see only the right hand side of the moon in a crescent shape. The moon takes $27\frac{1}{3}$ days to orbit the earth.

TIME

Before looking more closely at the moon we should consider the three ways we have of measuring time.

(a) The earth spins on its axis causing the sun to rise and set every day.
(b) The earth revolves around the sun once a year.
(c) The moon revolves around the earth once in $27\frac{1}{3}$ days.

Each of these events causes changes that are important to humans. The earth's spin causes day and night. Its revolution causes the seasons. The moon's revolution causes changing tides.

To give us more useful chunks or units of time, the day is divided into smaller bits. These are 24 hours in a day, with 60 minutes in each hour and 60 seconds in each minute.

The problem arises when fitting days into months and years on a calender. There are $365\frac{1}{4}$ days in a year. There are 29½ days in a month.

To fit days into the year we have 365 days each year with 366 in a leap year which is every four years.

To fit days, months and years together we have twelve months in the year with 30 or 31 days in a month (except for February which has 28, and 29 in a leap year). Do you know the rhyme which helps us all to remember this?

Standard time

Having decided that the time taken for the earth to spin once on its axis should be the basis of time, people built more and more accurate clocks for marking the time off. There was still a problem. Everyone could agree that each day would begin at midnight and end 24 hours later on the following midnight. The problem was agreeing when midnight was.

Each town or city would set the clocks to its own apparent time so that noon came when the sun was most nearly overhead. This was acceptable before the days of travel, but in a large country like the USA communications became increasingly difficult in the 19th century. In the 1880s railroad timetables were in hopeless confusion caused by the hundreds of different local times. The railways set up four standard time zones across the country. This was the beginning of the modern system of standard time.

The earth's surface is now divided into twenty-four standard time zones. The time in each zone is one hour earlier than in the zone to the east and one hour later than in the zone to the west.

THE TIDES

The earth's gravity pulls on the moon and keeps it in orbit around the earth. The moon also pulls on the earth. The most noticeable effect of this is the tides. If you watch the water level in the sea carefully you can see two main patterns. The level is constantly changing. Twice a day there are high tides and twice a day low tides. If you watch these over a period of several months you will see that twice a month there are very high tides called *'spring' tides* - nothing to do with seasons. Also twice a month there are tides that are much lower than usual, called *'neap' tides*. (See Fig. 1.11.)

The gravity of the moon pulls everything on earth towards it. It is this pull of gravity on the earth's oceans that causes tides.

A tidal bulge appears on the side of the earth facing the moon. A similar bulge is formed on the opposite side of the earth. As the earth rotates, two high tides and two low tides pass each point during the 24 hours. In fact because the moon is in orbit around the earth, the earth does slightly more than one revolution before the second high tide. It takes 24 hours and 50 minutes for both high tides to pass a given spot. This means that high tide is at different times each day - an important fact to know if you work or live near the coast, or even if you are visiting the coast for a holiday!

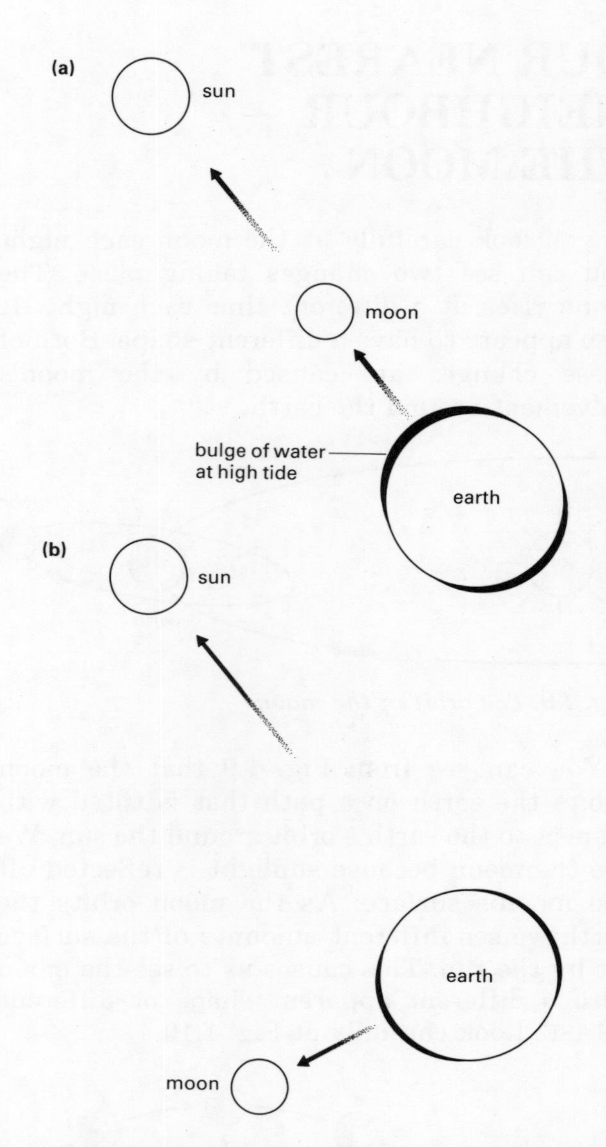

Fig. 1.11 (a) Spring tides - formed when the sun's pull helps the moon's pull
(b) Neap tides - formed when the sun's pull cancels out part of the moon's pull

What then causes the change in the high tide level during the month? The moon is not the only object pulling on the earth. The sun has a gravitational pull that keeps the earth in orbit. This gravity also combines with the moon to produce tides. The effect of the sun is weaker than that of the moon, but it does produce very noticable variations in the high tide level. Look carefully at Fig. 1.11.

Diagram (a) shows a spring tide. Here the sun's gravitational pull adds to the moon's, so a very high tide is produced. Diagram (b) shows a neap tide. Here the gravitational pull of the sun is cancelling part of the moon's pull. The high tide is lower than usual.

Assignment – Tides

Look at Table 1.1, which shows variations in the high tide level at Dover. Plot a graph of depth of water at high tide against date. When plotting graphs, don't forget to label the axes carefully, showing what you are measuring and the units you are using. Choose the scale so that you use up as much of the graph paper as possible. Don't make it so small that your graph is all crammed up in the corner. Likewise don't make it so large that you can't fit all the information onto it.

Plot the points carefully. Put the best line you can through the points. Drawing the best line is a difficult problem. In science different graphs do different jobs and we use different techniques. When we are looking for a pattern between two things that we can write as a mathematical formula, we put the best smooth line we can between the points. Often this is a straight line and it does not necessarily pass through all the points.

On other occasions, when we just want to see how something is changing, we join up dot to dot - the temperature graphs that you see at the bottom of a hospital bed are good examples of this.

As you do more and more science you will learn to judge which type of graph to draw. For this one you should make sure your line passes through all the points.

When you have drawn your graph, answer these questions.

1 What does a peak in the graph mean?
2 What does a trough mean?
3 On your graph mark 'spring tides' and 'neap tides'.
4 How many spring tides are there each month?
5 Draw diagrams to show how the earth, moon and sun are positioned for these spring tides.
6 What phases would the moon be in during these spring tides?

Table 1.1 Variation in high tide level at Dover

Date in month	1	2	3	4	5	6	7	8	9	10	11	12	13	14	15	16	17	18	19	20	21	22	23	24	25	26	27	28	29	30	31
January																															
Lunar phase		full moon								last quarter							new moon							first quarter							
Height of tide/m	6.4	6.5	6.6	6.4	6.6	6.5	6.4	6.3	6.0	5.8	5.5	5.4	5.4	5.7	6.0	6.4	6.6	6.8	6.8	7.0	7.0	6.9	6.7	6.4	6.1	5.9	5.7	5.7	6.0	6.2	6.4
February																															
Lunar phase	full moon								last quarter								new moon						first quarter								
Height of tide/m	6.5	6.3	6.6	6.5	6.5	6.3	6.1	5.9	5.6	5.4	5.3	5.4	5.8	6.3	6.6	6.9	7.0	7.0	6.9	7.0	6.8	6.5	6.1	5.7	5.4	5.5	5.8	6.1	6.3	—	—
March																															
Lunar phase	full moon								last quarter							new moon							first quarter								full moon
Height of tide/m	6.4	6.5	6.3	6.5	6.4	6.3	6.2	6.0	5.7	5.5	5.2	5.4	5.8	6.2	6.6	6.9	7.1	7.0	7.1	7.0	6.7	6.4	6.0	5.5	5.2	5.4	5.7	6.0	6.2	6.3	6.4

THE MOON'S SURFACE '. . . ONE GIANT LEAP FOR MANKIND'

Even before the momentous step by Neil Armstrong on 20 June 1969, the moon was quite well known. It had been viewed by telescopes for hundreds of years and accurate drawings made of the cratered surface. Only one side of the moon can be seen from earth. As the moon orbits the earth, so it rotates keeping the same side facing the earth. This meant that half the moon was unseen and unknown until recent satellite visits. The unmanned trips of the early 1960s and then the voyages of Armstrong and others have given us very full information about the nature of the moon.

Fig. 1.12 The surface of the moon

Look carefully at the photograph (Fig. 1.12). You can see a large number of craters that vary in size from 80 km across to a few hundred metres across. You can also see that part of the moon is in brilliant sunlight whilst other parts are in deep dark shadow. What has caused these craters and why does the earth not have a similar appearance?

Look at the information below. Remember DENSITY just means the mass of a one metre cube of the material.

	Diameter in metres	Density in kilograms per cubic metre	Pull of gravity on 1 kg at surface in newtons
Earth	12 750 000	5500	10
Moon	3 476 000	3300	1.7

As you can see, the earth is more dense and larger than the moon. It therefore has a greater gravitational pull.

This is one reason why the earth has an atmosphere. As we will see later in the book, gases are made of molecules in very rapid motion. The earth's gravity is large enough to pull these fast moving molecules back to earth. Any gas molecules on the moon would escape from the moon's gravity. So the moon has no atmosphere. This lack of atmosphere has a number of serious consequences that had to be carefully considered before any moon landing was considered.

Temperature

Because there is no atmosphere to absorb the sun's rays, the surface of the moon gets very hot: during the moon day temperatures reach over 100° C. During the moon night there is no atmosphere to keep in the heat and temperatures fall as low as −100°C. This large change in temperature has caused rocks on the moon to break up and cover the surface with fine dust.

Sunlight and shadows

If you looked up into the sky on the moon it would look very different to an earth sky. You would see a brilliant sun in a black sky. Any shadow cast on the moon would be very dark.

On earth the sky is blue because the atmosphere bounces or scatters light down to earth. Sunlight is a mixture of colours, but blue light is scattered more than red so that we have a blue sky. There is no atmosphere to scatter sunlight on the moon so the sky is black. Shadows are not completely dark on the earth because, although direct rays from the sun are cut off, light is scattered into the shadow from the rest of the sky. This does not happen on the moon.

Craters

The craters on the moon are formed by the impact of meteorites - fragments of rock or metal from outer space. These collide with the moon's surface and throw large amounts of debris upwards in spectacular explosions, leaving craters behind.

These meteorites also head towards the earth but as they enter the atmosphere the friction generates so much heat that they burn up. You may be lucky enough to see this happen: the event is commonly called a shooting star.

THE SOLAR SYSTEM

The sun is the star at the centre of our solar system. It is orbited by nine PLANETS. The word 'planet' is an ancient Greek word meaning wanderer. The planets were so called because they seemed to wander through the night sky appearing in different places on different nights. They are shown in Fig. 1.13.

The planets do not give off light of their own. We can see them only because they reflect light from the sun. Some planets are orbited by moons. There are also many ASTEROIDS in the solar system. These are chunks of rock varying in diameter from about a metre to 1000 km. A belt of these appears between Mars and

Assignment – Moonwalk

The place: Tranquility base - a stark but beautiful boulder-strewn wasteland on the right hand side of the moon in the Mare Tranquilitatis.

The time: First quarter (moon) 3.52 a.m. 21st June 1969 (BST) 109.19.16 (Apollo mission time) and the begining of the moon age.

MISSION CONTROL: 'Stand by. Okay Neil, we can see you coming down the ladder now.'

ARMSTRONG: 'Okay. I just checked-getting back to that first step Buzz. It's not even collapsed too far but it's adequate to get back up. It takes a pretty good little jump. I'm at the foot of the ladder. The LM* foot pads are only depressed in the surface about one or two inches. Although the surface appears to be very finely grained as you get close to it it's almost like a powder. Now and then it's very fine. I'm going to step off the LM now. That's one small step for a man, one giant leap for mankind. As the surface is fine and powdery I can pick up loosely with my toe . . . .

'There seems to be no difficulty in moving around as we suspected. It's perhaps even easier than the simulations at one-sixth that we performed in the simulations on ground. It's actually no trouble to walk around . . . . .

'It's quite dark here in the shadow and a little hard to see if I have good footing. I'll work my way over into the sunlight here without directly looking into the sun.'

(*LM = lunar module)

1 Why should the moon's surface be covered with dust? Suggest two ways in which it could have been formed.
2 Why did Armstrong say 'one giant leap for mankind'?
3 What did Armstrong mean by 'even easier than the simulations at one-sixth'?
4 It's actually no trouble to walk around.' In what sense would walking on the moon be easier than walking on earth?
5 'It's quite dark here in shadow.' Explain why this should be so.
6 Why should Armstrong have been very careful about 'directly looking into the sun'?
7 Armstrong and Aldrin were completely dependent on their suits whilst on the moon. Think carefully about the suits as life support systems. Find as much about them as you can. Describe their structure and how this fits the jobs they had to do.

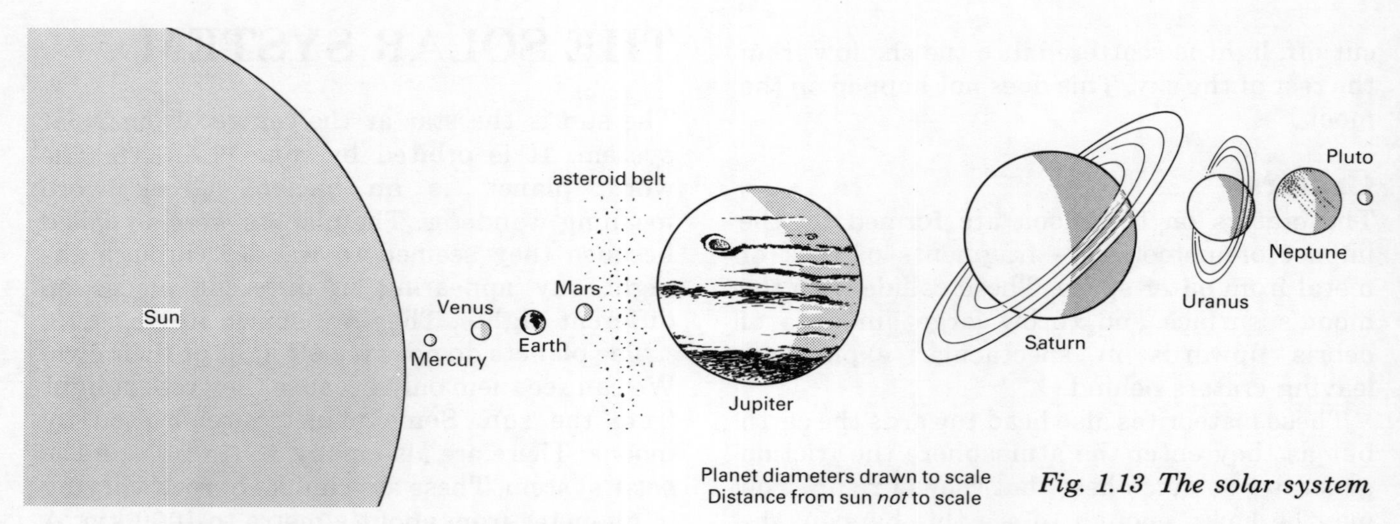

Fig. 1.13 The solar system

Jupiter. Despite this, the solar system is an empty place. Imagine an empty football ground such as Wembley Stadium. A puff of smoke spread through the stadium's entire volume would represent a much greater concentration than the matter in the solar system.

Assignment – Planet patterns

Scientists spend much of their time doing experiments and collecting measurements and results. They then need to look very closely at the results to see if there are any patterns. If they see clear patterns they think up theories to explain the patterns. They then use these theories to make predictions which they test out with more experiments.

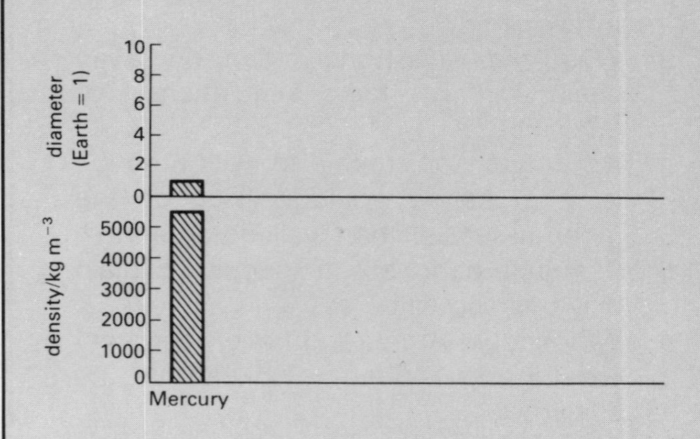

Look carefully at Table 1.2. Can you see any clear patterns? Look carefully at the column headed Diameter and the column headed Density – remember density is just the mass of a 1 metre cube of the material.

It is often difficult to spot patterns from numbers, so draw a graph. For this sort of information a BAR GRAPH would be best. Draw two bars for each planet, one to represent diameter, the other to represent density. Draw the bar graphs directly above one another as shown. The bars for Mercury are done for you.

Use the graphs to answer these questions.

1 What pattern can you see from the graphs?
2 Use the graphs to divide the planets into two groups. Write down the planets in each group and say how the groups differ.
3 Which planet does not fit into the pattern?
4 Now look back at the table. Into what other patterns do these groups fit?
5 Again, which planet does not fit into the pattern?
6 Why do you think that Jupiter, Saturn, Uranus and Neptune are called 'gas giants'?
7 Look back at Fig. 1.13 and Table 1.2. In what way does Mars not fit in with general trends?
8 Try to suggest reasons for this.

Table 1.2 The main members of the solar system

Body	1 Diameter (Earth=1)	2 Mass (Earth=1)	3 Surface gravity (Earth=1)	4 Density/ kg m^{-3}	5 Period of spin			6 Angle of tilt between axis and orbit	7 Average distance from Sun (Sun–Earth=1)	8 Period of orbit/ years	9 No. of moons (* = plus rings
					days	hours	minutes				
Sun	109.00	333 000.00	28.00	1400	25	9		97°			
Mercury	0.40	0.06	0.40	5400	58	16		90°	0.4	0.2	0
Venus	0.95	0.80	0.90	5200	244	7		267°	0.7	0.6	0
Earth	1.00	1.00	1.00	5500		23	56	113°	1.0	1.0	1
Mars	0.53	0.10	0.40	4000		24	37	114°	1.5	1.9	2
Jupiter	11.18	317.00	2.60	1300		9	50	93°	5.2	11.9	16*
Saturn	9.42	95.00	1.10	700		10	14	116°	9.5	29.5	15*
Uranus	3.84	14.50	0.90	1600		10	49	187°	19.2	84.0	5*
Neptune	3.93	17.20	1.20	2300		15	48	118°	30.1	164.8	2
Pluto	0.31	0.0025	0.20	400	6	9	17	?	39.4	247.7	1

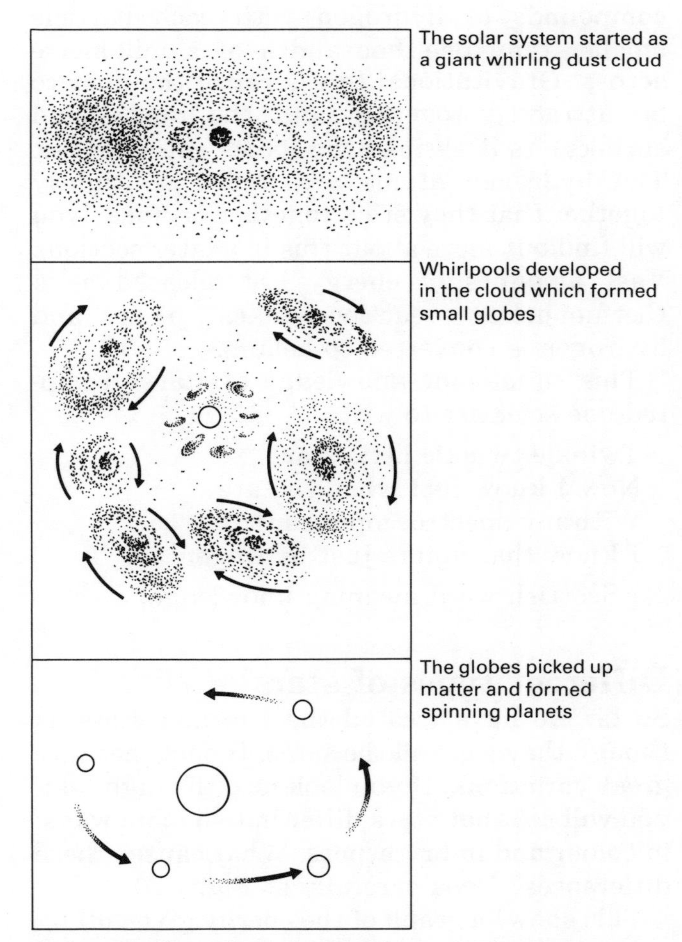

Fig. 1.14 How the solar system was formed

The origin of the solar system

As more and more space probes have visited other planets in the solar system, vital information has been gathered about the structure of these planets. Using basic laws of gravity and motion and very complex mathematics scientists have worked out how the solar system may have been formed. The most favoured theory at the moment is the one shown in Fig. 1.14.

This theory is able to explain the two different types of planet. The four near the sun were heated and any gases on the surface escaped. The gas giants, being further from the sun, were less affected by the sun's energy so the gas remained trapped, forming massive but less dense planets. The planet Pluto may be an escaped moon from another planet.

The asteroid belt may have been formed by planet explosion. The problem with this theory is explaining why a planet should explode. It may be material that has not been collected by Mars which could explain why Mars is smaller than we might expect.

THE SUN AND OTHER STARS

Energy from the sun

At the centre of the solar system lies the sun. The sun is a huge ball of gas that is giving out a constant supply of energy. This energy has warmed the surface of the earth and has made life possible.

Whilst we can find out more about planets in the solar system by sending probes to fly very close or soft-land, this is not possible in the case of the sun. We must carefully collect the radiation coming from the sun. By measuring how much there is, and what kind of radiation it is, we are able to put forward ideas about what exactly the sun is and how it manages to pour out energy constantly.

So far we have considered only visible light coming from the moon and other planets. As you will read in a later chapter, light is a wave motion and part of what we call the ELECTROMAGNETIC SPECTRUM (Fig. 1.15). In addition to these radiations the sun is also firing out streams of charged particles (see Unit 5).

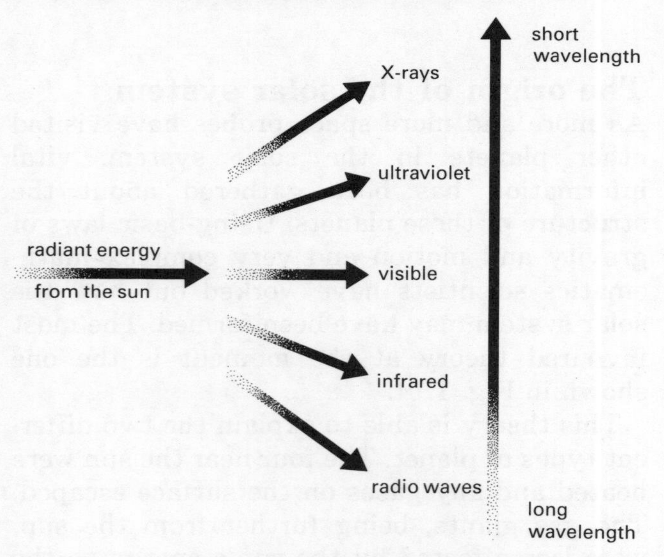

Fig. 1.15 Different types of radiation from the sun

Astronomers use detectors to pick up as much of the radiation as possible, which is then carefully analysed, normally using computers. You can try an experiment now where you catch some visible light from the sun and use it to measure the sun's diameter.

The stars – just other suns

By examining radiation from the sun and radiation from stars we can see that the sun is in fact just another star. Gathering radiation from stars is difficult because of the great distances involved. Telescopes are used to gather the radiation.

The Mount Palomar telescope is the world's largest light-gathering telescope. It has a curved mirror 5 m across. The one at Jodrell Bank is a radio telescope. It collects radio waves in a dish that is 76.2 metres across. With large telescopes like this people don't look down them. The radiation from light-gathering telescopes is usually fed into a SPECTROMETER which sorts it out into the different wavelengths.

The evidence that we get from these telescopes tells us that all stars are made of very hot gas. In the case of the sun the temperature is about 14 000 000 °C at the centre. The stars were formed from dust that litters the universe. By 'dust' astronomers mean frozen compounds of hydrogen with each particle smaller than one-thousandth of a millimetre across. Gravitational forces cause the dust to be attracted together and the dust cloud shrinks. As it shrinks the forces get stronger. The hydrogen atoms are pulled so tightly together that they stick together or 'fuse' (you will find out more about this in a later section). Vast amounts of energy are released as a thermonuclear reaction takes place and hydrogen is converted to helium.

This significant knowledge of stars prompted one scientist to write

Twinkle twinkle little star,
Now I know just what you are
With my spectroscopic ken*,
I know that you're just hydrogen.

*(a Scottish word meaning knowledge)

Different types of stars

So far we have treated the sun and stars as though they were all the same. In fact there are great variations. If you look into the night sky you will see that stars differ in two main ways: in colour and in brightness. What causes these differences? Look carefully at Fig. 1.16.

This shows a graph of the energy given off by a hot object plotted against its wavelength.

Investigation 1.1 The sun's diameter

In this investigation you will measure the diameter of the sun.

WARNING - never look directly into the sun either with your naked eye or through sun-glasses or any instrument like a telescope or binoculars.

Collect

metre rule
2 pieces of card
scissors
pin
pair of compasses

What to do

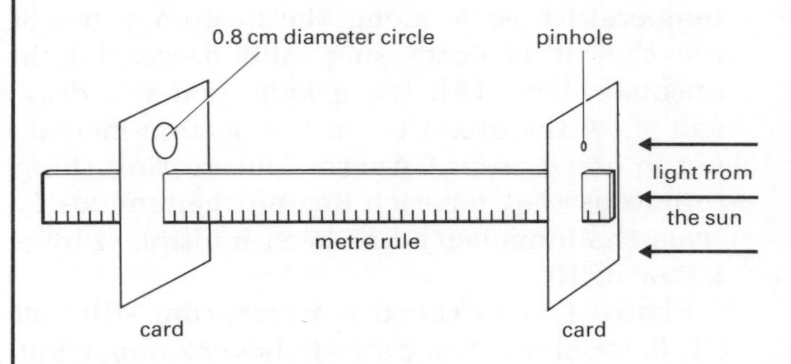

1 Cut slots in the cards so that they will fit on the ruler as shown.
2 Make a pinhole 2 cm above the top of the slot on the first card.
3 Draw a circle 0.8 cm in diameter with the centre 2 cm above the slot on the second card.
4 Mount the cards as shown in the diagram and point the metre stick at the sun. The second card should be in the shadow of the first card.
5 Adjust the position of the second card until the image of the sun just fills the 0.8 cm circle (the image of the sun will look like a bright circular patch).
6 Measure the distance between the cards.

Doing the calculation

The diagram below shows how the image of the sun is formed.

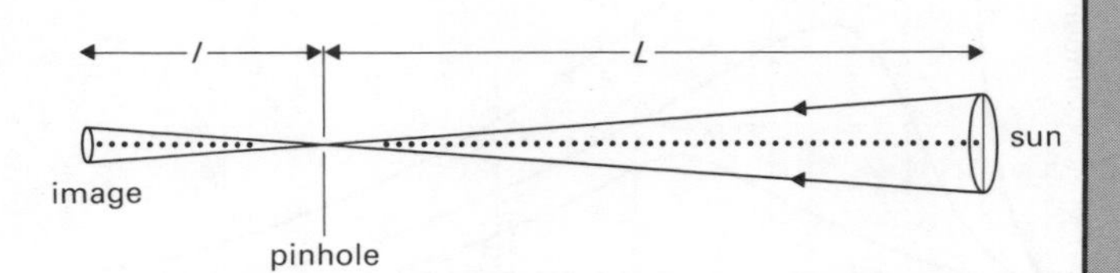

The diagram shows rays of light going through the pinhole and forming an image on the second card. Rays of light from the top and bottom of the sun are shown.

If you look carefully you can see that the two triangles shown are scale models. This means

$$\frac{\text{diameter of sun}}{\text{diameter of image}} = \frac{\text{distance from sun to pinhole}}{\text{distance from image to pinhole}}$$

The distance from sun to pinhole is 150 000 000 km.
The diameter of the image is 0.8 cm.
Both of these must be converted to metres.

$$\frac{\text{diameter of sun}}{0.008} = \frac{150\,000\,000\,000}{\text{distance from image to pinhole}}$$

So, diameter of sun =

$$\frac{150\,000\,000\,000}{\text{distance from image to pinhole}} \times 0.008.$$

You can put your measurement in and calculate the diameter of the sun. The distance from image to pinhole is just the distance between the cards - it must be in metres.

Do you think that you could use this method to measure the size of the planet Mars? Explain your answer.

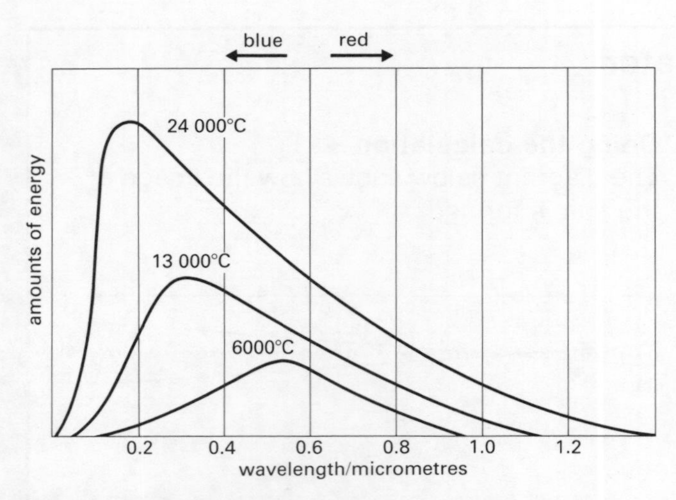

Fig. 1.16 Graph of amount of energy against wavelength for a hot object

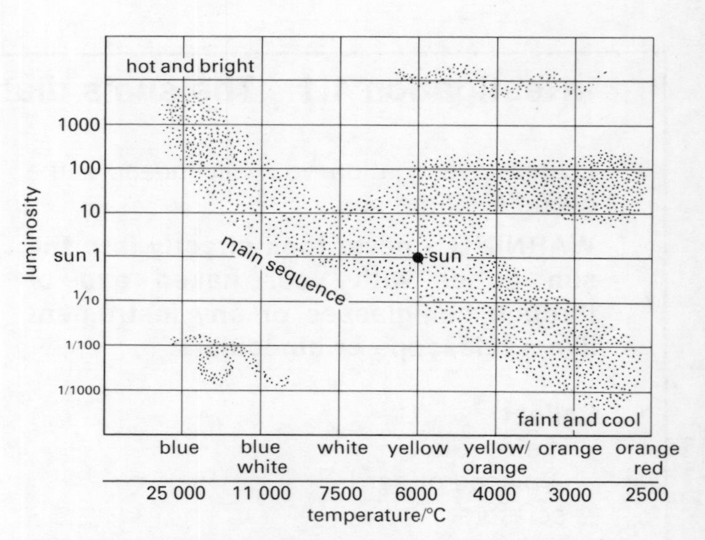

Fig. 1.17 Stars plotted on a Herzsprung-Russell diagram

The wavelength tells us the colour of light given off. These colours are given along the top of the graph. Graphs are given for objects at three different temperatures. There are two main patterns shown.

(a) The hotter an object gets the bluer it looks. You can see this by the way the peaks move to the left as the temperature increases. This gives us important information about stars. Blue stars are very hot and red stars are much cooler. Our sun would look yellow from outer space. It has a surface temperature of about 6000 °C. (b) The hotter an object gets the more energy it gives off. This can be seen by the way the peaks get bigger as the temperature increases. This gives us another important piece of information about stars. Hot blue stars give off more energy than cooler red stars of the same size.

When we compare the brightness of stars seen from earth we cannot say that those giving off the most energy are necessarily the brightest. It depends how far away the stars are. A nearby star giving off a small amount of energy could look brighter than a very distant bright star. Astronomers get around this problem by working out how far away each star is. They then work out how bright the stars would be if they were all at the same distance. They call this the LUMINOSITY of the star.

The life cycle of a star

When the temperature and luminosity of many stars are measured, important patterns emerge. These patterns can be used to understand the changing fortunes of stars as they are born, grow old and eventually die. Look at Fig. 1.17.

This is not really a graph. If you look at the temperature scale along the bottom you will see that it is decreasing, and decreasing in unequal steps. (All the graphs you will draw will show the quantity on the bottom increasing in *equal* steps.) Another interesting thing to note is that for each line on the luminosity scale the luminosity has been multiplied by a factor of 10.

Figure 1.17 is called a Herzsprung-Russell (H–R for short) Diagram. It is very important to astronomers. It shows there is considerable variation in the luminosity and temperature of stars. However, most stars form a pattern. They lie on a line called the MAIN SEQUENCE. As we would expect, the hotter the star the more luminous it is. You can see that our sun lies right in the middle of the main sequence.

Look carefully at the diagram. You can see a distinct group of stars in the top right corner. These stars are cool but they give off a good deal of light. How can they do this? They must be larger than the other stars. They are called RED GIANTS.

Now look at the bottom left. There is another group of stars that are very hot, yet appear faint. They must be very small. These are called WHITE DWARFS.

From the H–R diagram we can work out a star's life cycle. This is shown on Fig. 1.18. The star is born from gas or dust. This is shown at point A on the diagram. The star becomes hotter until it joins the main sequence at B.

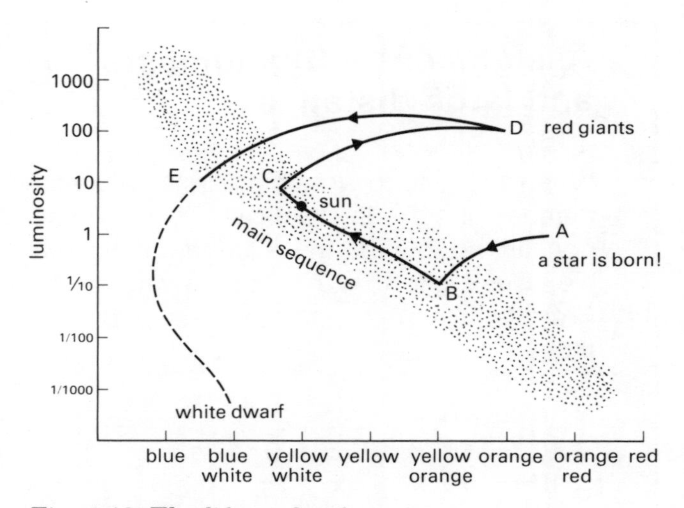

Fig. 1.18 The life cycle of a star

The more massive the star is the further up the Main Sequence it will join. A large star will be brighter and hotter than a small star.

Stars spend most of their lives as main sequence stars. It has been estimated that a star the size of the sun would spend ten thousand million years in the main sequence. During this time that star gets hotter and brighter.

At some stage, marked C on the diagram, a dramatic change occurs in the star. The hydrogen in the star's core has all been transformed into helium. The star starts to burn the hydrogen outside the core and suddenly increases in size, forming a red giant. If our sun expands like this, the position that you are now in will be inside the sun! Not all stars follow this course: some continue up the main sequence.

The star at D on the diagram then undergoes another equally dramatic change. It collapses, giving off a great deal of energy, and passes the main sequence to become a hot white star at E. It then goes on collapsing, becoming a white dwarf. It has used all its nuclear fuel and will gradually cool to form a dark, very dense corpse that would appear lifeless and invisible.

This may sound like guesswork, but it is backed up by good scientific evidence. Look at Fig. 1.19. It shows two H-R diagrams. They are for star clusters, which are groups of stars of the same age.

Galactic clusters (Fig. 1.19(b)) are groups of stars that have formed relatively recently and so it is difficult to work out their ages. You can see that all stars lie on the main sequence.

Now look at the globular cluster (Fig. 1.19(a)). This is an immense group of stars formed when the galaxy was young. Here the stars clearly show their age. There is an obvious absence of hot blue stars. All the stars in the cluster are about the same age but the blue stars have matured into red giants.

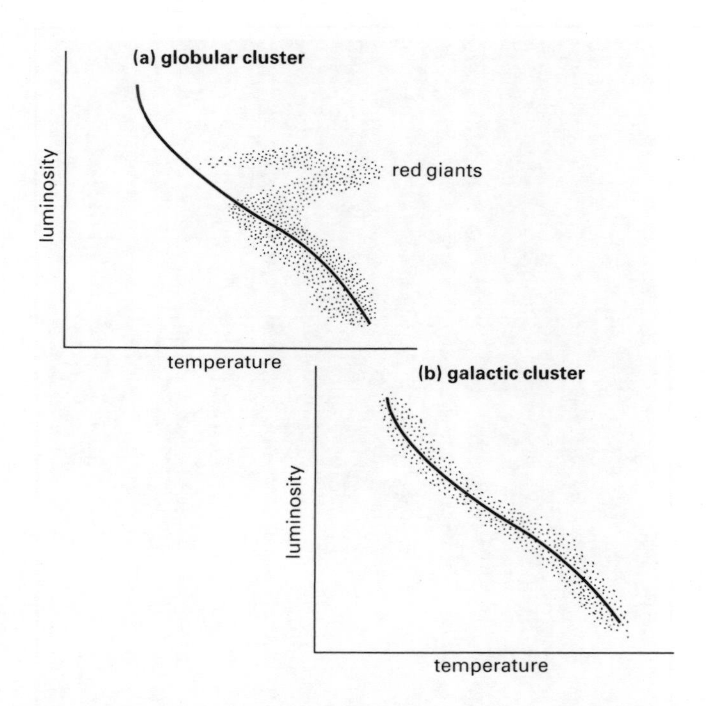

Fig. 1.19 H-R diagrams for (a) a globular cluster and (b) a galactic cluster

THE GALAXY AND BEYOND

Our sun is one of the millions of stars that form a GALAXY. If you look into the sky on a summer evening there is a broad band of faint light that stretches across the sky. If you look at it through binoculars you can see that it is made of millions of stars. A good telescope shows these stars, together with dark clouds of dust that form the birthplace of new stars. When you look at the Milky Way you are looking towards the centre of the galaxy. From the outside it would look like Fig. 1.21(a) overleaf.

The Milky Way Galaxy has a spiral shape, with the sun towards the edge of the galaxy in a spiral arm. (Fig. 1.21(b)). There are about 100 000 000 000 stars in the galaxy. Some of them are grouped in GLOBULAR CLUSTERS.

Fig. 1.20 The Milky Way - a telescope shows up many more stars than you can see with your naked eye

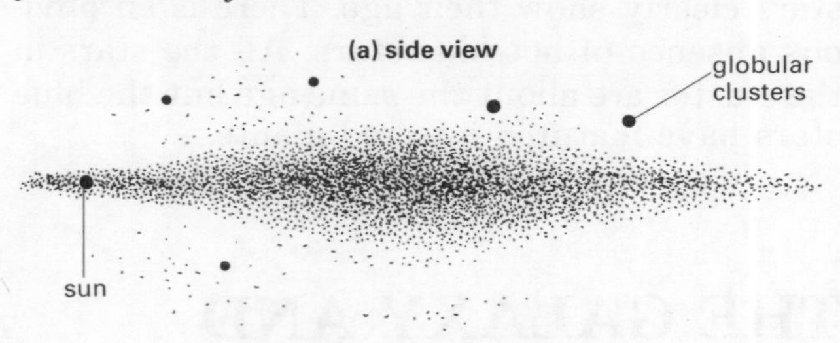

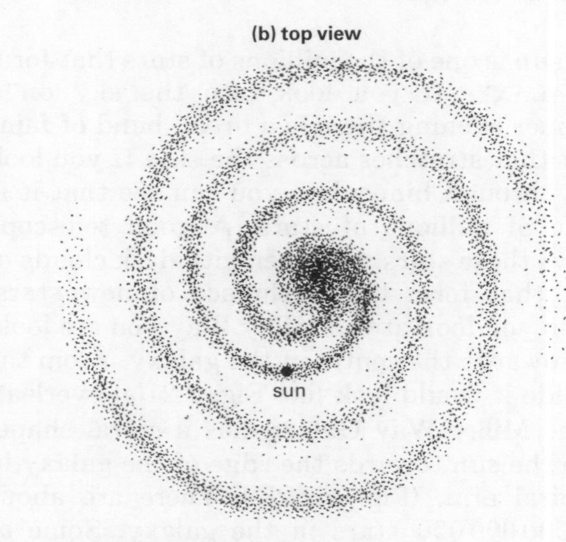

Fig. 1.21 The Milky Way galaxy

Assignment – Big numbers and large distances

A. Big numbers

When we need to write big numbers like the number of stars in a galaxy, it is very time-consuming to write them out in full. Instead we use a shorthand using powers of 10. From your maths lessons you should know that

$$10^1 = 10$$
$$10^2 = 10 \times 10 = 100$$
$$10^3 = 10 \times 10 \times 10 = 1000$$
$$10^4 = 10 \times 10 \times 10 \times 10 = 10\,000.$$

You can see the pattern. The small number above the 10 is called the INDEX. It tells us the number of noughts there are after the 1. *For example*: There are 100 000 000 000 stars in the galaxy. This can be written 10^{11} for short.

Suppose the number is not so simple. How could we write 176 000 000 in our shorthand?

All you do is write it in two parts:

1.76	×	100 000 000
↑		↑
this part is always between 1 and 10		this is what you have to multiply by to get your number

We then write this as 1.76×10^8.
This shorthand is called STANDARD FORM.

1 Put these numbers into standard form.
(a) 100 (b) 1 000 000 (c) 10 000 000
(d) 200 (e) 22 000 (f) 3 200 000

2 Work out in standard form
(a) the number of seconds in an hour.
(b) the number of metres in 1000 kilometres.

THEORIES OF THE UNIVERSE

The Milky Way is just one amongst millions of galaxies. Astronomers have noticed that galaxies tend to form groups. There are seventeen galaxies in the Local Group, the nearest of

B. Large distances

When we start to measure the distance between stars and the size of galaxies we are involved in large distances.

The nearest star, Alpha Centauri, is 40 400 000 000 000 kilometres away. We could write this as 4.04×10^{13} kilometres. Astronomers prefer to measure these large distances in LIGHT YEARS. Although the term 'light year' seems to be a measure of time it is a measure of distance. It is the distance that light travels in a year. Light travels very quickly - at about 3×10^8 metres per second, so it goes a long way in a year.

1 light year = 9 400 000 000 000 kilometres
= 9.4×10^{12} kilometres

The nearest star - Alpha Centauri - is 4.3 light years away. The Milky Way is 100 000 light years across.

The sun's light takes 8.5 minutes to reach the earth.

1 How far does light travel in 1 second?
2 How far does light travel in 1 minute?
3 How far does light travel in 8.5 minutes?
4 How far is the sun from the earth?
5 The Great Nebula in Andromeda is 2 200 000 light years away. How far is this in kilometres?
6 The star Vega is 26.5 light years away. If you look at Vega in the night sky, light that comes from Vega enters your eye. What was happening on Earth when the light left Vega? Give an answer like 'my mum and dad were watching the Beatles' or 'Mount Everest had just been climbed'.

which is Andromeda about 2 million light years away. If we look further with our powerful telescopes we can see other groups of galaxies. Figure 1.22 shows a group that is 300 million light years away.

The question is: how far can we see out into the universe? With optical telescopes we can see things that are up to 6 billion light years

Fig. 1.22 A globular cluster

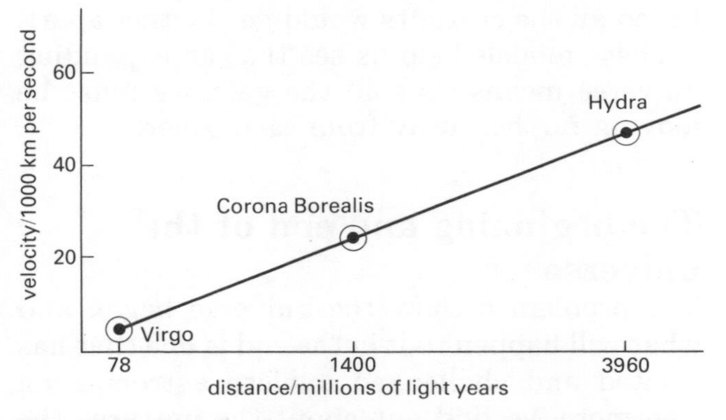

Fig. 1.23 Graph of distance against velocity for galaxies shows that the further away a galaxy is the faster it is moving

away. One very important theory that can be worked out from studying galaxies is that they are all moving away from us. Also, as Fig. 1.23 shows, the further away the galaxy is the faster it is moving. The graph shows the galaxies called Virgo, Corona Borealis and Hydra.

The expanding universe

The fact that all other galaxies are moving away from us means that the universe must be expanding. It does not mean, though, that the Milky Way Galaxy is at the centre of the universe. Figure 1.24 shows a simple model.

Fig. 1.24 Balloon model of the expanding universe

Imagine a half inflated balloon. This represents the universe. Dots painted on the surface represent galaxies. Look what happens when the balloon is inflated. All of the dots move further apart, although none of them is at the centre of the 'model' universe. This is only a two-dimensional model. We have put all our galaxies on the surface of the balloon. For a three-dimensional model we would have to imagine an inflatable currant bun. The currants represent the galaxies. If the bun was inflated all the currants would get further apart.

These models help us see that an expanding universe means that all the galaxies must be moving further away from each other.

The beginning and end of the universe

The problem of how the universe began and what will happen to it in the end is one that has excited and challenged modern astronomers. The more we find out about the universe the more we realise that it is in a constant state of change. Even stars, which most people think of as being always there, are born, grow old and eventually die.

When did the universe begin?
What is happening to it now?
What will happen to it in the future?

These are the questions that the astronomer is faced with.

The Big Bang

The theory that most astronomers accept at the moment is the BIG BANG. This says that the universe began with a gigantic explosion. Matter was thrown outwards. Stars were born from gas and grouped together as galaxies. This theory is able to explain why all the galaxies should be moving apart at high speeds.

The theory also predicts that there should be radiation - a sort of afterglow from the big bang - spread through the universe. Astronomers have detected this radiation using radio telescopes. This provides powerful evidence for the big bang theory.

Most astronomers agree that the universe began with a gigantic explosion 10 000 000 000 000 years ago. What about the future? There are two main ideas:

(a) universe continues to expand (Fig. 1.25(a));
(b) universe reaches a certain size then all the galaxies move back together. The universe *implodes* (Fig. 1.25(b)).

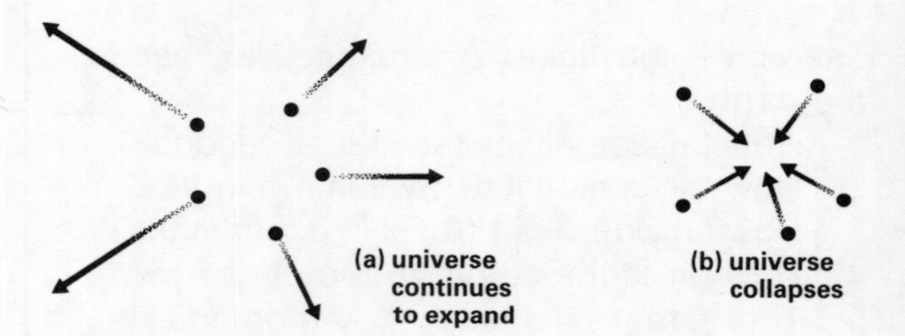

Fig. 1.25 The fate of the universe

More evidence is needed before astronomers can decide which theory is best. You need not worry too much about the collapsing universe - even if the theory was right it would take longer than the age of the universe to happen!

The earth and beyond

WHAT YOU SHOULD KNOW

1 The earth has (a) a lithosphere - a thin crust about 100 km deep; (b) an atmosphere of gases which keeps the earth at a roughly constant temperature; (c) a biosphere or thin layer near the surface where life is found.

2 The earth rotates on a tilted axies. This rotation causes the sun to rise and set every day and causes the star patterns to appear to move through the night sky.
3 The earth revolves around the sun every $365\frac{1}{4}$ days. This revolution and the fact that the earth's axis is tilted causes the seasons, changes in the length of the day and changes in the night sky during the year.
4 The moon rotates on its axis every $27\frac{1}{3}$ days and orbits the earth every $27\frac{1}{3}$ days.
5 The gravitational pull of the moon causes the oceans on the earth to bulge, producing tides.
6 As the moon orbits the earth we see different amounts of the surface illuminated by the sun. This causes the phases of the moon.
7 Because the moon is small it has no atmosphere. This means that the surface has extremes of temperature, the sky is black and shadows are very dark, and the surface is pitted with impact craters.
8 Nine planets orbit the sun. These, together with asteroids and comets, form the solar system.
9 The four inner planets tend to be small and dense. The next four planets are large, but have low density.
10 The sun, like all other stars, was formed by gas being pulled together by gravitational force. The energy comes from a nuclear reaction where hydrogen is converted to helium.
11 Hot stars are blue and bright. Cooler stars are red and more faint.
12 Stars have a distinct life cycle. They are born from dust and gas. They grow old and eventually burn out.
13 Stars form groups called galaxies. Our sun is in the Milky Way Galaxy.
14 A light year is the distance light travels in one year.
15 Galaxies are moving apart. The further away galaxies are, the faster they seem to be moving.
16 Astronomers think that the universe is expanding. They think that the universe began with a big bang and that stars and galaxies have formed from the flying fragments.

QUESTIONS

1 This diagram shows a photograph that was taken with a camera pointed into the sky with the shutter open.

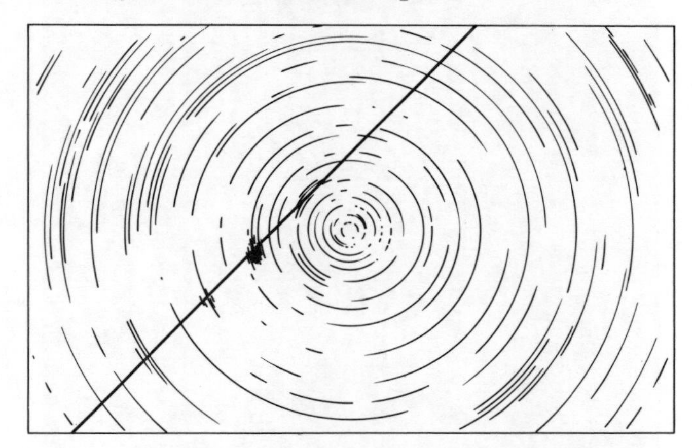

(a) Explain what the curved lines are.
(b) In which direction was the camera pointed?
(c) What do you think caused the straight line?
(d) How long was the shutter open for?
(e) Draw a diagram to show what you would see if a similar photograph was taken from the equator with a camera pointed straight up?

2 Look carefully at this diagram.

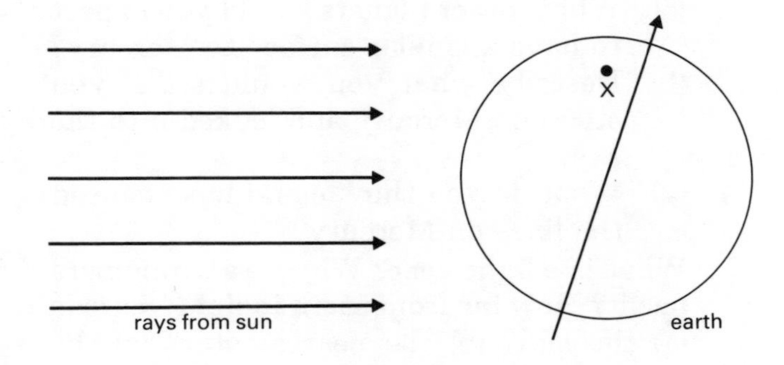

(a) Would you say that the place marked X was experiencing winter or summer?
(b) Do you think point X is in day or night? Explain your answer.
(c) Shade in the area of the earth that would be dark for 24 hours. Give a reason for this.
(d) Shade in the area that experiences the midnight sun.

3 Explain with diagrams how the moon influences the seas on earth.

4

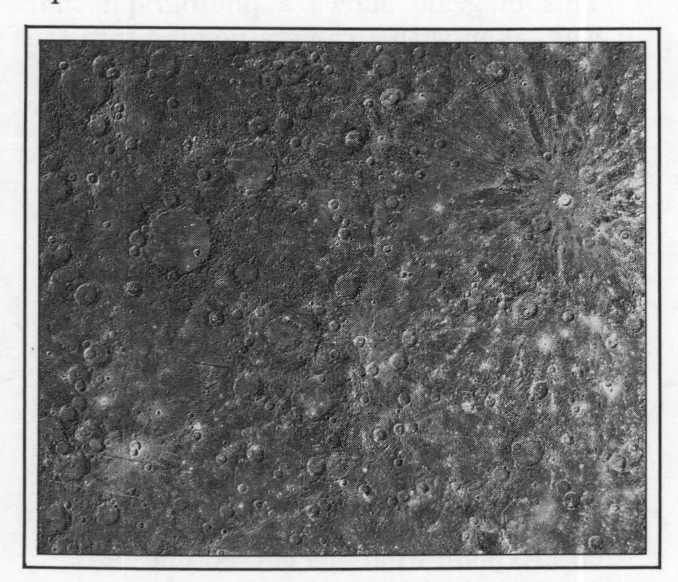

Look carefully at the photograph above. It shows the surface of Mercury.

(a) What do you think the circular shapes are?
(b) What has caused them?
(c) Do you think that Mercury has an atmosphere? Explain your answer.
(d) Look carefully at the table on page 11 (Table 1.2) What other information could you use to answer part (c)?
(e) Which other planets would you expect to have a similar surface to Mercury?
(f) Describe what you would see if you stood on Mercury and looked into the sky.
(g) What do you think could have caused the lines on Mercury?

5 What is a light year? Why do astronomers use it? How far from earth in light years is (a) the sun? (b) the nearest star? (c) the centre of the Milky Way? (d) the nearest galaxy? (e) the furthest galaxies that we can see well with our largest telescopes?

6 If some stars are moving as rapidly as 20 km per second across our line of sight, why do the star patterns remain constant from night to night?

7 (a) State the difference between a star and a planet.
(b) List the planets in the solar system in order of increasing distance from the Sun.
(c) (i) Why do Venus and Mercury show 'crescent phases' (similar to the moon) at certain times, whereas other planets in the solar system do not?
(ii) Mercury and Venus are bright planets. Explain why Venus is brighter than Mercury even though Mercury is closer to the Sun.
(d) (i) Whereabouts in the solar system are the asteroids?
(ii) What are the asteroids?
(iii) How are they thought to have originated?
(iv) Why could they be a hazard in any space journey?

8 Look at the H–R diagram on page 14 (Fig. 1.17). What is the relationship between star colour and temperature? How does the temperature of the sun compare with other stars? With the help of the H–R diagram on page 15 (Fig. 1.18) describe the likely future of the sun.

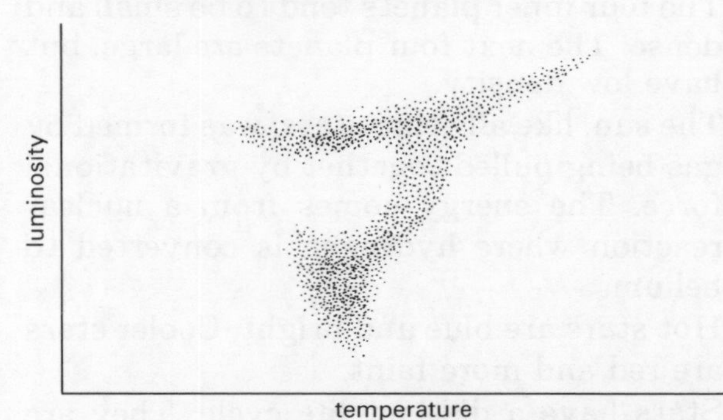

The diagram above is an H–R diagram of a globular cluster called MS, in Canus Venatici. All the stars in the cluster are the same age, although they vary in size. Copy the diagram out and sketch in the position of the main sequence. Which stars are missing from the main sequence? What has happened to them?

Sketch an H–R diagram for a similar but very much younger cluster.

The Earth and its Place in the Universe

TRAVELLING IN SPACE

Figure 1.26 shows two important spacecraft. *Pioneer 10* was launched in 1972. After a flight of nearly two years it passed close to Jupiter and sent pictures and scientific data back to earth. It then travelled towards the edge of the solar system to begin its journey towards other stars. In case it is ever found by an alien civilisation it carries a plaque giving vital information about our civilisation. The plaque is made of gold-coated aluminium and is designed to last for 100 million years. How is it that *Pioneer* can travel so far? Does it need to keep using fuel to keep it going? What sort of motion does it have in outer space away from any planet or star?

The space shuttle was first launched in 1981. It is the first reuseable spacecraft. It enables important experiments to be carried out and important tests to be made with satellites and orbiting space stations. What sort of motion does the shuttle have? How can it stay in orbit? What does it feel like to be inside an orbiting spacecraft? These are important questions. To find out the answers we need to do careful experiments and understand the basic laws of moving objects. These laws are not just for explaining the motion of spacecraft, they also help us to understand the motion of our bicycles and cars and, of course, ourselves.

Fig. 1.26 (a) (left) Pioneer 10, (b) (right) the Space Shuttle, Challenger

SPEED

We all have some idea of what is meant by SPEED. We know that a fast train will get us from London to Birmingham in less time than a slow train. Descriptions like 'fast' and 'slow' are useful for comparing trains but not precise enough for us now. What is 'slow' for a train is very fast for you riding a bicycle. We need to define speed carefully. Look at Fig. 1.27.

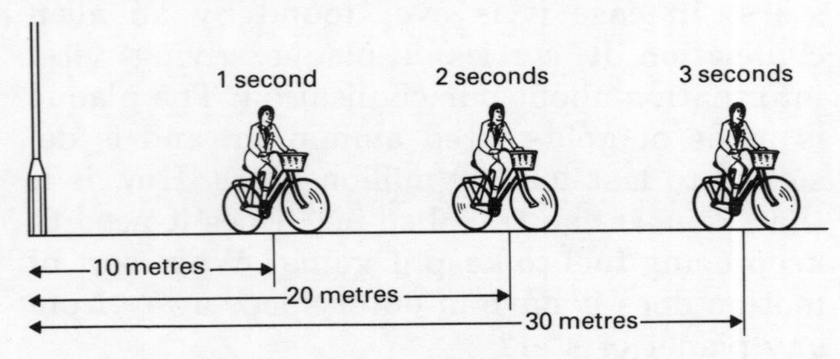

Fig. 1.27 Travelling at a steady speed

A person is pedalling steadily on a bicycle. When she passes a lamp post a clock starts and her progress each second is shown on the diagram. The cyclist is going at a steady speed. We define speed as the distance travelled each second. The cyclist is going 10 metres each second.

$$\text{speed} = \text{distance travelled each second}$$
$$= 10 \text{ metres per second}$$

Important note - units

Whenever we measure things our answer has two parts. It has a NUMBER which tells us how much and a UNIT which tells us what we are measuring. For example, if you measure your body temperature you will probably get an answer of 37 degrees Celsius.

37 is the number
degrees Celsius is the unit

We write ° C for short.

The unit of distance is the metre.
The unit of time is the second.
The unit of speed is the metre per second.

Working out speed

We often need to find the speed of an object but we do not know how far it travels each second but only in, say, 10 seconds or 20 seconds. How do we then work out the speed? Look at the diagram of the cyclist again. If we did not have the two middle pictures, only the end ones, could we work out the speed? Yes, easily! We could say that 30 metres in 3 seconds means 10 metres each second.

$$\text{speed} = \frac{\text{distance travelled}}{\text{time taken}}$$

You must realise that this applies only to things that travel at a steady speed. Figure 1.28 shows a different sort of journey.

What do you get if you work out the speed for the journey?

$$\text{speed} = \frac{\text{distance travelled}}{\text{time taken}} = \frac{1000 \text{ metres}}{200 \text{ seconds}}$$
$$= 5 \text{ metres per second.}$$

The cyclist was obviously not going at 5 metres per second for the whole journey. There was a time when the speed was zero. The cyclist must also have been going faster than 5 metres per second for some time. Is the calculation useless? Not really, because it tells us the *average* speed of the journey, which is very useful.

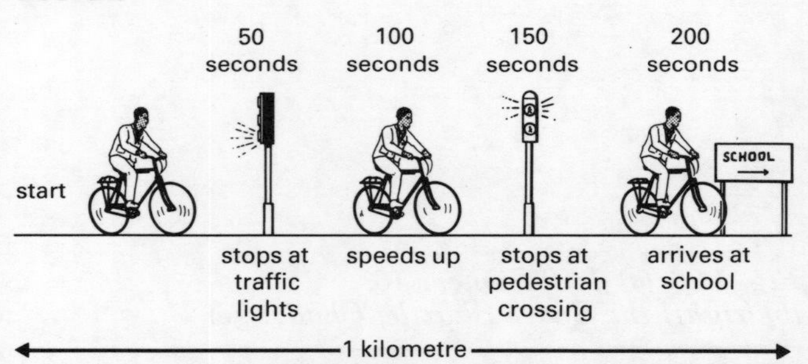

Fig. 1.28 A journey with different speeds

From now on we should say

$$\text{average speed} = \frac{\text{distance travelled}}{\text{time taken}}$$

Measuring speed in other units

Measuring speed in metres per second makes sense for scientists but it is not so useful for car drivers. In Britain distances are measured in miles and we often time our journeys in hours, not seconds! We measure car speeds in miles per hour. In many other countries distances are measured in kilometres and speed in kilometres per hour.

Assignment - Working out speeds

Copy the table below into your book. Fill in the blanks. Make sure that you put in the right units.

Distance	Time	Speed
100 metres	10 seconds	—
50 kilometres	$\frac{1}{2}$ hour	—
10 metres	2 seconds	—
—	2 seconds	20 metres per second
—	10 seconds	10 metres per second
—	2 hours	50 kilometres per hour
1000 metres	—	10 metres per second
50 kilometres	—	25 kilometres per hour

Investigation 1.2 Measuring speeds

In this experiment you will work out the speed at which you can run.

Collect
tape-measure
stop watch

What to do
1 Measure out a 20 metre distance in the school yard or playing field.
2 Get your partner to measure the time taken for you to run 20 metres.
3 Use the formula,
$$\text{average speed} = \frac{\text{distance travelled}}{\text{time taken}},$$
to calculate your average speed.

Questions
1 Do you think that the answer you get tells you your fastest speed? Explain your answer.
2 Explain how you could easily measure your fastest speed using the same apparatus.
3 An Olympic sprinter runs 100 metres in 10 seconds. What is the sprinter's average speed for the race? Have a guess at the fastest speed that he or she runs at.

Measuring small time intervals

A stop watch is fine for measuring the time taken for things that travel large distances. If we want to look at objects whose speed changes we need to be able to measure very small time intervals. Figure 1.29 shows a TICKER-TIMER.

Fig. 1.29 A ticker-timer

When the ticker-timer is connected up to a power supply the arm vibrates up and down 50 times each second. If paper tape is pulled through, dots are made in the tape every 1/50 second. Here is what you might see:

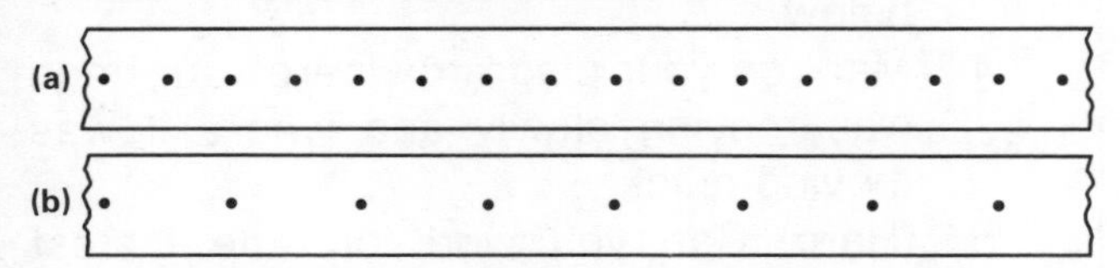

Fig. 1.30 Ticker-tapes: both were pulled at a steady speed, but (b) was pulled faster than (a)

In each case the dots are equally spaced. This means that the speed of the tape did not change. We say that it has a CONSTANT SPEED.

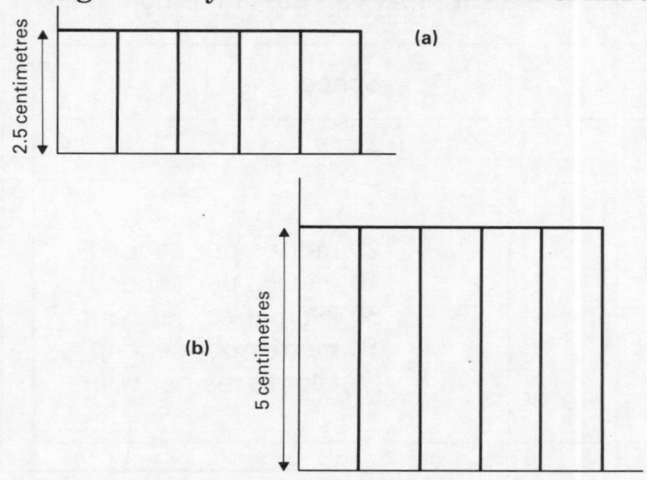

Fig. 1.31 Tape charts. Each length of tape has 5 dots; it shows the distance travelled in $\frac{1}{10}$ second

Tape charts

We usually analyse motion by cutting the tapes into 5-dot lengths and sticking them together to form a chart.

Figure 1.31 shows what the tape charts produced from tapes (a) and (b) would look like. Notice that the height of each strip tells us the distance moved in $\frac{1}{10}$ second.

CHANGING SPEEDS

When an object gets faster we say that it ACCELERATES. As with speed, it is important that we can measure acceleration.

Figure 1.32 shows the position of a cyclist each second. The speed of the cyclist is also shown. The cyclist's speed is increasing steadily. We say that there is a constant *acceleration*.

Investigation 1.3 Analysing motion

In this investigation you will pull tape through the ticker-timer at different speeds. You will then make up tape charts and analyse the motion of your hand.

Collect

ticker-timer
ticker-tape
power pack
connecting leads

What to do

1 Connect up the ticker-timer to the power supply. Make sure that you use the correct connections and the correct voltage.
2 Thread the tape through the ticker-timer. Turn on the power. Pull the tape through slowly at first then quickly, then slowly again.
3 Cut the tape into 5-dot lengths and stick them into your book. Stick them side by side with the bottom level as shown below.
4 Mark on your diagram where your hand was moving slowly and where it was moving quickly.
5 (Hard) Can you work out the fastest speed of your hand?

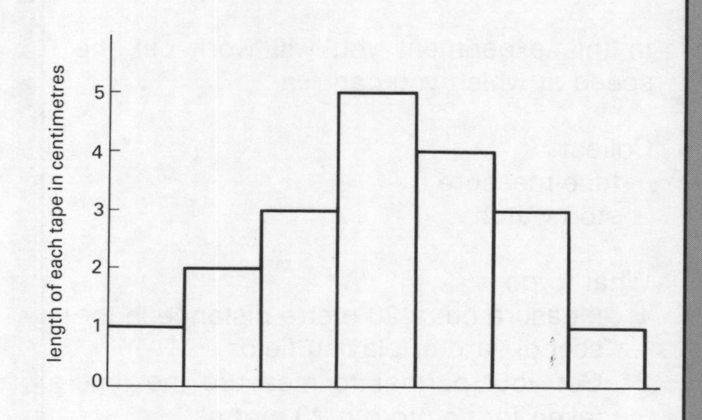

Questions

This tape chart shows the results that a pupil obtained for this investigation. Sketch them in your book.

1 Mark on the graph where the pupil's hand was moving the fastest.
2 Mark where it was going the slowest.
3 Now look at the longest tape. It is 5 centimetres long. This means that it went 5 centimetres in $\frac{1}{10}$ second. Suppose that it kept going at this speed. How far would it go in 1 second? What is the speed?
4 How far did the pupil's hand move altogether?

Fig. 1.32 Getting faster - a constant acceleration

We define acceleration as the increase in speed each second. The cyclist accelerates at 2 metres per second each second.

acceleration = 2 metres per second per second

Acceleration – a strange unit

The unit of acceleration is the metre per second per second. This may seem very strange at first. It may help you to understand why the unit should be like this if you think of a similar example. Think of your pocket money. Suppose you get £3 per week. You are being paid at a steady rate. Now suppose that each week your parents increase your pocket money by 10 pence. Now your pocket money is accelerating. The acceleration is the increase per week.

pocket money acceleration = 10 pence per week per week

Investigation 1.4 Diluted gravity

Objects fall towards the earth because of gravity. In this investigation you will 'dilute' gravity by letting a trolley roll down a slope. You will then analyse the motion of the trolley.

Collect

runway
ticker-timer
ticker-tape
trolley
power supply
2 connecting leads

What to do

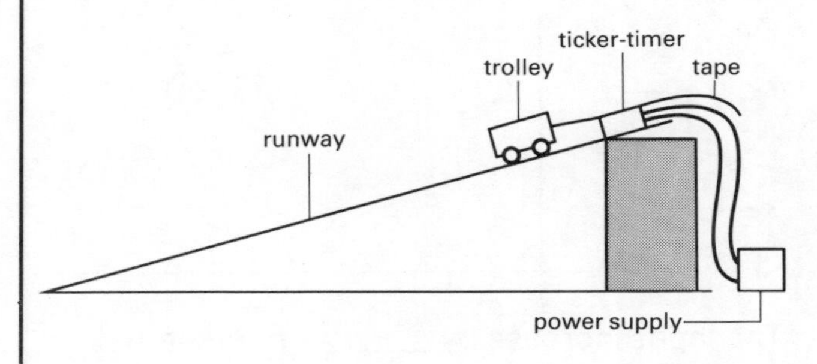

1 Slope the runway by putting books or some other support under one end.
2 Set up the apparatus as shown. Place the ticker-timer on the top of the runway and connect up as shown in the diagram.
3 Thread the tape through the ticker-timer and attach the tape to the trolley.
4 Switch on the power. Release the trolley. Catch the trolley at the bottom of the runway. Make sure the trolley does not crash. It could damage an expensive trolley.
5 Cut the tape into 5-dot lengths and make up a tape chart.
6 Use the tape chart to describe the motion of the trolley as carefully as you can.

Questions

The diagram below shows the results from a pupil's diluted gravity experiment.

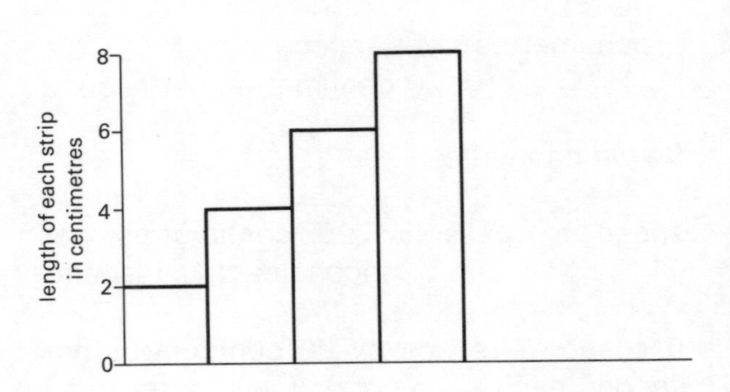

1 What can you say about the amount by which the steps are increasing?
2 What can you say about the acceleration of the trolley?

Working out acceleration

Look again at Fig. 1.32. Suppose you had only the first and the last diagrams, could you work out the acceleration? Yes; if the speed increases by 6 metres per second in 3 seconds, in 1 second it would increase by 2 metres per second. We can say

$$\text{average acceleration} = \frac{\text{change in speed}}{\text{time taken}}$$

Important note - shorthand for units

So far in this book the units of distance and time have been written out fully. There is a shorthand that can be used.

metres can be written m
seconds can be written s

It is very important to realise that s means seconds and it must not be added onto other units to form plurals.

for example 5 metres is written 5 m
never 5 ms

speed has the units of metres per second. The shorthand way of writing this is $m\,s^{-1}$

acceleration has units of metres per second per second. The shorthand way of writing this is $m\,s^{-2}$.

Do not worry if you find these strange looking shorthands a little difficult - everyone has problems with them at first. In this book, all units will be written out fully when they are first used. After this we will use shorthand. You can give answers fully or in shorthand, but be careful.

Assignment – Accelerations

Later on in the course you will need to work out accelerations. This assignment shows you how to do it. Look at the diagram in Investigation 1.4, which shows a tape chart for an accelerating object. You can see that each strip is 2 centimetres longer than the one before it. This means that the speed is increasing by the same amount each $\frac{1}{10}$ second: the object has a constant acceleration.

speed increase = 2 centimetres per $\frac{1}{10}$ second each $\frac{1}{10}$ second

2 centimetres per $\frac{1}{10}$ second = 20 centimetres per second

So we can write:

speed increase = 20 centimetres per second each $\frac{1}{10}$ second.

If speed increases by 20 centimetres per second each $\frac{1}{10}$ second, it will increase by 200 centimetres per second each second. So we can write:

speed increase = 200 centimetres per second each second.

So

acceleration = 2 metres per second each second.

Now try this example. Look carefully at the tape chart below. Write out the account below, filling in the blanks.

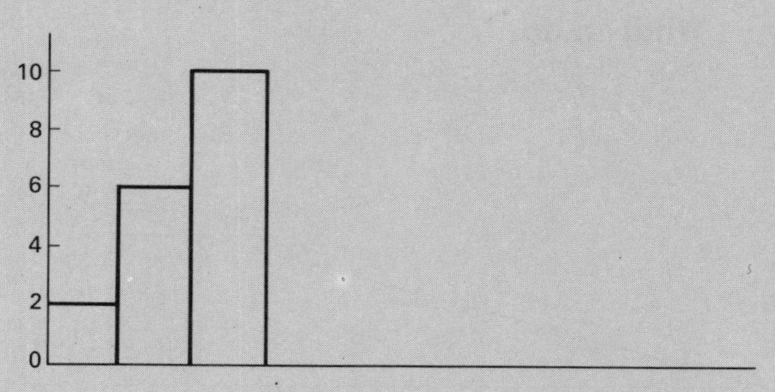

The tapes increase by ________ centimetres each $\frac{1}{10}$ second.
The speed increase = ________ centimetres per $\frac{1}{10}$ second each $\frac{1}{10}$ second.
The speed increase = ________ centimetres per second each $\frac{1}{10}$ second.
The speed increase = ________ centimetres per second each second.
Acceleration = ________ metres per second each second.

MOTION WITHOUT FORCE

At the beginning of this section we asked a very important question - what happens to a spacecraft far away from any planet if its rockets are turned off? Does it slow down and stop or does it keep on going? This question of how objects move when no forces act worried people long before the days of space travel.

In the 4th century BC Aristotle thought that if you wanted to keep something moving you must keep pushing it. This idea was thought to be true up until the 17th century. It was Galileo who changed people's minds. He said that if no force acted on an object it would go on at the same speed in the same straight line. The reason why most things slow down is because of FRICTION.

Frictionless motion

We can get rid of friction in the laboratory by floating objects on a cushion of gas. Figure 1.33 shows how this can be done. Where does the gas come from in each example?

In each case a small push will cause the 'hovercraft' to coast along at a steady speed. This shows that Galileo was right: if moving objects are left alone they keep going with a steady speed in a straight line.

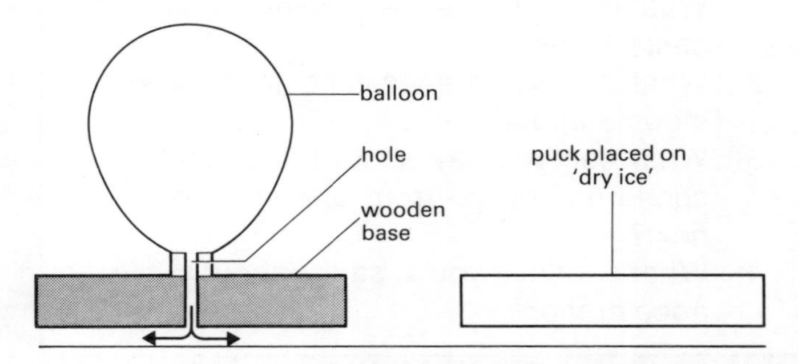

Fig. 1.33 Getting rid of friction - model hovercraft

Newton's Laws

The work of Galileo was carried on by Newton. In 1687 he published his famous Laws of Motion. These laws tell us what happens when forces are applied to objects. They are still very important today. They allow us to explain and to predict the forces involved in such things as aeroplane flights and car crashes. Newton's Laws have also enabled space scientists to soft-land a lunar module on the moon and bring it safely to earth again. You will meet one of Newton's famous laws now, and the other two in Unit 8 of this course.

> NEWTON'S FIRST LAW
> If no total force acts on an object it either stays at rest
> *or*:
> if it is moving it carries on with the same steady speed in the same straight line

So we have an answer to the question about *Pioneer 10* on page 21. It does not need to carry vast amounts of fuel. Once it is away from the gravitational pull of planets it will travel on and on without needing to be pushed. If there is no force on it, it will travel with a constant speed.

Balanced forces

Newton's First Law can also be written backwards. If an object stays at rest or if it moves in a straight line with a steady speed then no total force acts on it. The word 'total' is important. Look at Fig. 1.34.

Fig. 1.34 Balanced forces on a bicycle means steady speed

The bicycle is going at a steady speed. This means that, according to Newton's Law, no total force is acting on it. Yet the cyclist is pedalling and producing a pushing force. Does this mean that Newton's Law is wrong? Look at the diagram again. Although the pedalling force is pushing the bicycle forwards, there is another force acting. There is friction in the wheels. The force of friction tends to pull the bicycle back. The friction force and pedalling force are in opposite directions. They cancel out. The total force is zero.

Balanced forces give steady speed. What happens if the forces are unbalanced? The speed must change. An acceleration must be produced.

Investigation 1.5 Motion with a steady force

What sort of acceleration does a force produce? In this investigation you will apply a steady force to a trolley and analyse its motion. This is a difficult experiment that calls for great care and patience.

Collect

runway
trolley
ticker-timer
power supply
connecting leads
elastic band

What to do

1 Friction would confuse this experiment. We need to get rid of it. We cannot make it disappear so we compensate for it. Tilt the runway slightly.

2 Check that the runway has the right slope. Connect up the ticker-timer and thread the tape through. Stick the tape to the trolley. Start the ticker-timer and give the trolley a small push. If the slope exactly compensates for friction the trolley should move at a steady speed. The dots will be equally spaced.

3 You are now ready to pull the trolley with a constant force. This needs a bit of practice. Try it without any ticker-tape attached at first. Hook one end of the elastic band onto the trolley as shown in the diagram.

Pull the trolley with the elastic kept stretched by the same amount. In the diagram the elastic is stretched so that it is level with the front of the trolley. Practice doing this a few times.

4 Now thread some ticker-tape through the ticker-timer and stick the tape to the trolley. Switch on the power and pull the trolley down the friction-compensated runway with a steady force.

5 Cut the tape into 5-dot lengths and make a tape chart.

6 Describe the motion of the trolley as carefully as you can.

Questions

A pupil carries out the steady-force experiment and gets these results.

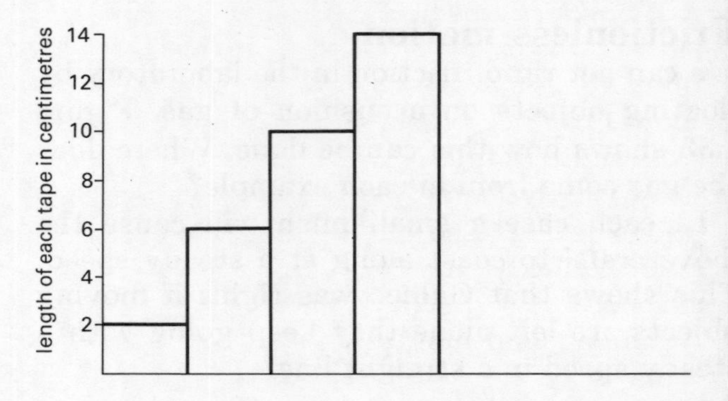

1 What is the difference in height between steps 1 and 2?

2 What is the difference in height between steps 3 and 4?

3 What can you say about the change in speed as you go from one strip to the next?

4 What can you say about the acceleration?

Motion with a steady force – an important result

Investigation 1.5 has been carried out many times by careful scientists. The results always form the same pattern.

A constant force gives a constant acceleration

This is an important result that will be useful to us in this section. We will also return to it in Unit 7.

The results that you got for your investigation probably do not fit exactly with this pattern. This does not mean that you have found an exception. We would need much stronger evidence before we thought that this pattern was not always true.

MASS

A constant force produces a constant acceleration. We also realise that the bigger the force the bigger the acceleration. There is something else that affects the acceleration.

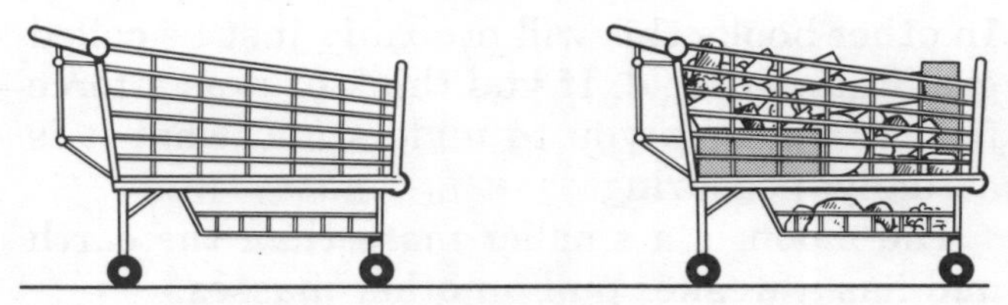

Fig. 1.35 Which trolley is easier to get going?

Look at Fig. 1.35. It shows two supermarket trolleys. The one on the left is empty. The one on the right is full. Now think about pushing them with the same force. What would their motions be like? The empty trolley would speed up more quickly. It would have a greater acceleration.

A similar (and more painful) way of looking at this is to ask what would happen if both trolleys were pushed towards you: which would be the easiest to stop? Yes, of course, the empty one.

We have a special name for how difficult it is to accelerate an object. We call it the MASS of the object. We say that the full trolley has more mass than the empty trolley. The unit of mass is the KILOGRAM (abbreviation: kg).

It is important to realise that mass is different from weight. If we took both trolleys into outer space on an intergalactic shopping trip they would have no weight. They would both be weightless. The full trolley would still be harder to accelerate: it would still have more mass.

The word 'inertia' is sometimes used to describe how difficult it is to accelerate an object. The word mass is shorter and means the same, so in this book the word inertia will not be used.

Assignment – Force, mass and acceleration

We know that the larger the acceleration, the larger the force needed to produce it. Also the larger the mass of an object the larger the force needed to accelerate the object. Use these ideas and Newton's First Law to explain *why* the following happen.

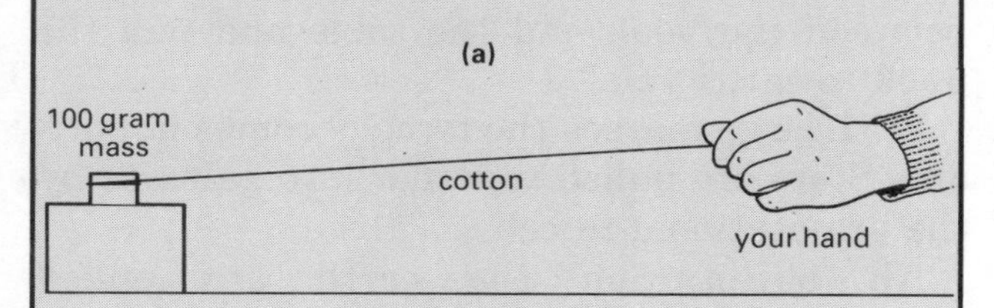

If you pull the cotton slowly, the mass glides along the table. If you pull it suddenly, the cotton breaks.

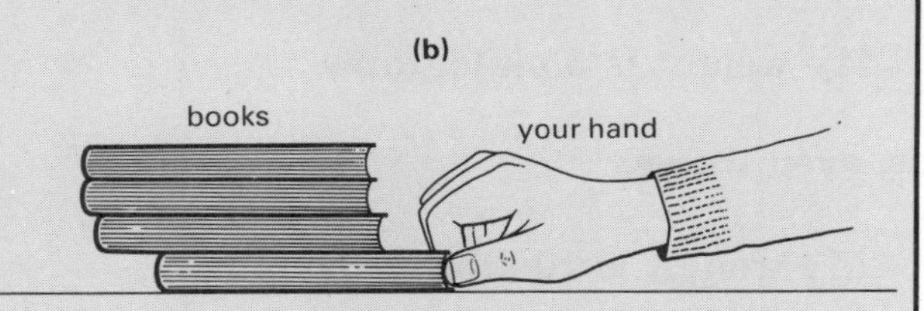

If you pull the bottom book slowly, all the books stay together and slide along the table. If you pull the book suddenly, the bottom book slides out and the rest drop to the table.

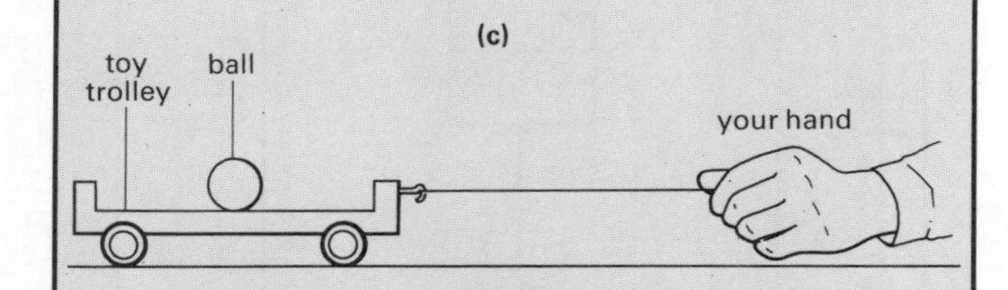

If you pull the truck forward the ball seems to go backwards and hits the back of the trolley.

Discuss these with your friends and your parents at home. They are quite difficult to explain properly. Try and write a paragraph on each.

Gravity

All objects in the universe are attracted together by a force called GRAVITY. The size of the force between two objects depends on their mass and how far apart they are. The greater the mass the larger the gravitational force.

The greater the distance apart, the smaller the gravitational force. There is a gravitational force between you and this book. It is very small - not enough to overcome the friction between the book and the table and pull the book towards you.

With large masses the force becomes important. Stars are pulled together into galaxies by the gravitational force.

All objects on the earth are pulled downwards by gravity. There is a special name for the gravitational pull of the earth on an object. It is called the WEIGHT. It is important to realise that weight is a force. Up until now you have probably said things like

'My weight is 55 kilograms'

or, even worse,

'My weight is 10 stone 7 pounds'.

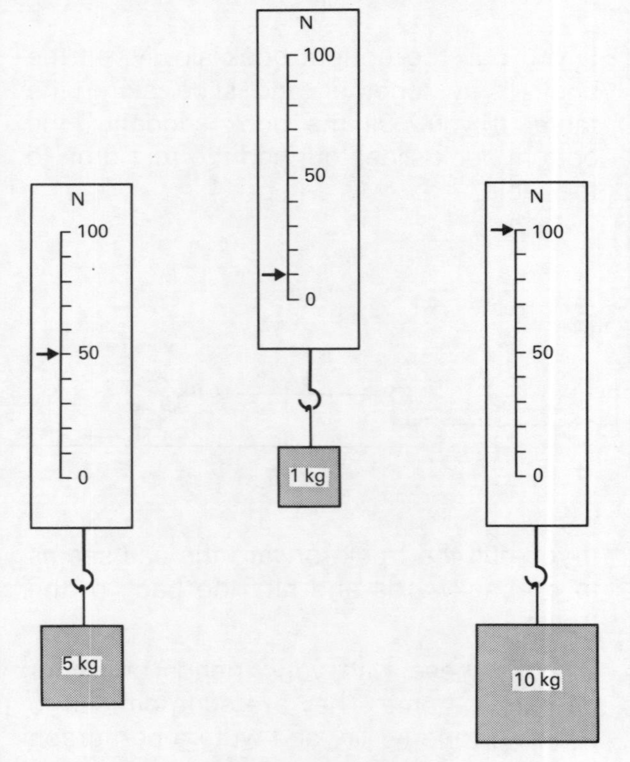

Fig. 1.36 The earth pulls each kilogram with a force of 10 newtons

This is fine for everyday use, but in science we must choose our words carefully. The unit of force is the NEWTON, which is abbreviated to N. The weight of an object must be in newtons.

Look at Fig.1.36. It shows three masses hanging on a spring balance. The earth pulls each kilogram down with a force of 10 newtons. There is a special name for the pull on each kilogram. It is called the GRAVITATIONAL FORCE FIELD and is measured in newtons per kilogram ($N\,kg^{-1}$).

earth's gravitational force field is
10 newtons per kilogram

In other books this will probably just be called gravitational field. If you think of it as a force field it will help you to understand what it is actually measuring.

The moon is a smaller mass than the earth and has a weaker pull on other masses.

The moon's gravitational force field is $1.7\,N\,kg^{-1}$.

You can see from Fig. 1.36 that the weight of an object can be worked out by multiplying the number of kilograms by the force on each kilogram.

weight = mass × gravitational force field

We can write this in shorthand, using these symbols:

weight = W
mass = m
gravitational force field = g

We can write

$$W = mg$$

For example, an astronaut has a mass of 65 kg. What is her weight on the moon?

mass = 65 kg
gravitational force field = $1.6\,N\,kg^{-1}$

Using
Weight = mass × gravitational force field,
weight = 65 × 1.6
= 104 N

Assignment – Measuring mass in space

Measuring mass on earth is easy. We can use a spring balance. We are actually measuring weight rather than mass. We know that on the earth, if we have twice as much mass, we will have twice the weight. Look at this diagram. You can see that for each extra kilogram the spring stretches the same extra length. We simply mark the balance off in kilograms and we can use it to measure an unknown mass.

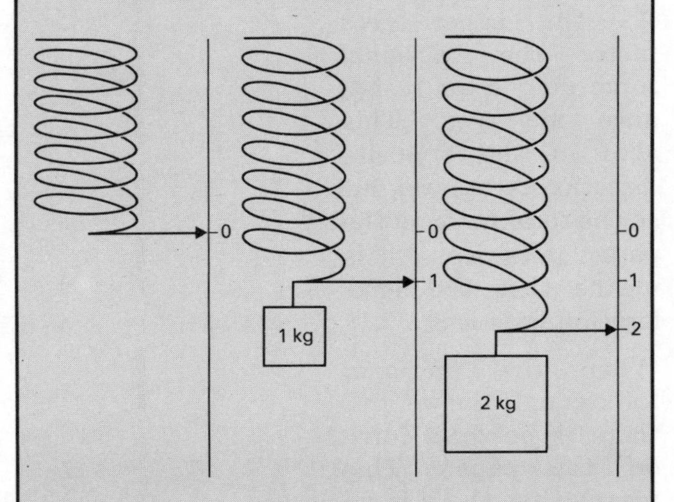

Would the mass balance work in outer space? Obviously not. The masses would have no weight and so would not stretch the spring at all.

How could you compare masses in space? Imagine you are in a spacecraft in outer space. Describe how you could compare masses. Design an instrument, using the equipment you have in your school laboratory, that would allow you to compare masses in outer space.

FALLING OBJECTS

What sort of motion does a falling object have? The earth pulls all objects down with a steady force. Investigation 1.5 has shown that a constant force gives a constant acceleration. This means that all objects should fall towards the earth with a constant acceleration.

Investigation 1.6 Free fall

In this investigation you will analyse the motion of a freely falling object.

Collect

ticker-timer
connecting leads
power supply
G-clamp
ticker-tape
100 gram mass

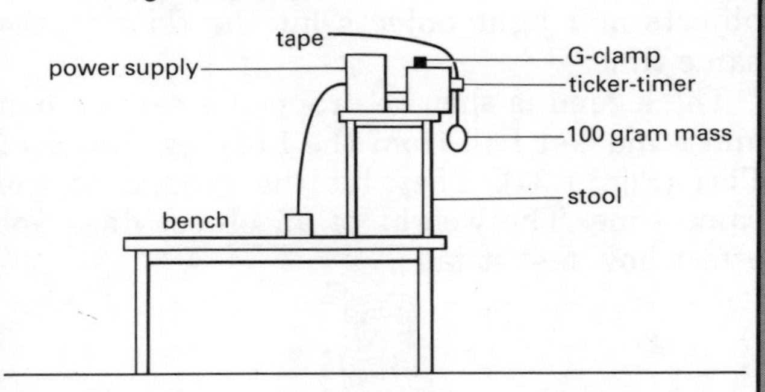

What to do

1. Set up the apparatus as shown.
2. Thread the tape through the ticker-timer and stick it to the 100 g mass. Turn on the power. Let the mass fall.
3. Cut the tape into 1-dot lengths and make up a tape chart.
4. Describe the motion of the 100 g mass as fully as you can.
5. If you have done the assignment on page 26, calculate the acceleration.

Questions

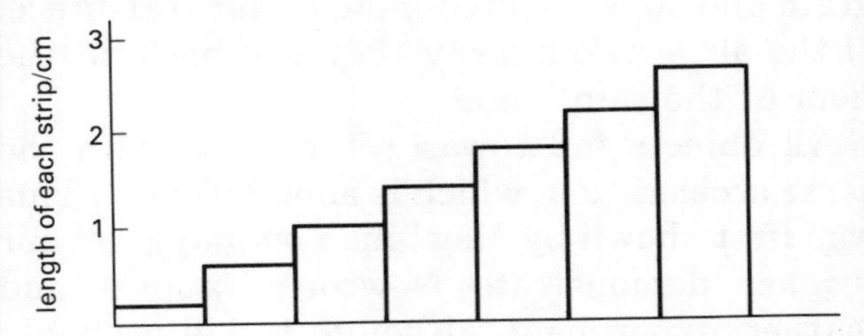

The diagram above shows the results from a pupil's book for the free fall investigation.

1. What can you say about the increase in height of each strip?
2. What is happening to the speed of the 100 g mass?
3. What can you say about the acceleration of the 100 g mass?

Free fall – different masses

Suppose you dropped two stones – a large stone and a small stone. If you dropped them at the same time from the same height, which would hit the floor first?

It is very tempting to say that the big stone is pulled with a bigger force and so will accelerate more. It should therefore hit the floor first. Tempting, but wrong – can you see why? Do not worry if you cannot. This problem fooled many people for hundreds of years. It was Galileo who convinced people that heavy objects and light objects hit the floor at the same time.

The legend is that he dropped a cannon ball and a musket ball from the Leaning Tower of Pisa (Fig. 1.37). They hit the ground at the same time. The weight of an object does not affect how fast it falls.

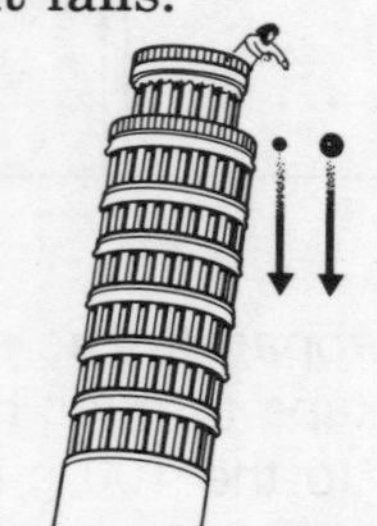

Fig. 1.37 Galileo on the Leaning Tower of Pisa: different weights hit the ground at the same time

Aristotle and others can perhaps be forgiven for thinking that heavy objects fall faster, because they were confused by AIR RESISTANCE. If a coin and a feather are dropped together the coin hits the floor first. This is because the feather has more surface area per gram and so is affected more by air resistance. If the air is taken away they will both hit the floor at the same time.

All objects fall towards the earth with the same acceleration, which is about 10 m s^{-2}. This was first shown by Newton. You may see your teacher demonstrate Newton's 'guinea and feather' experiment, although it will probably be a devalued version!

A dramatic demonstration of this experiment was given by Alan Sheppard when he dropped a hammer and a feather on the surface of the moon. They both hit the ground at the same time. Because the gravitational force field on the moon is less than on the earth, objects fall with a smaller acceleration – about 1.7 m s^{-2}.

Air resistance

What effect does air resistance have on the motion of a falling object? Does it actually slow it down or does it just stop it getting faster? Look at Fig. 1.38. It shows what happens when a small piece of tissue paper is dropped.

When the paper is released the only force acting is the gravitational pull of the earth – the weight of the paper. So the paper will accelerate.

As the paper gets faster the frictional force due to air resistance increases. This acts in the opposite direction to the weight, so the total force on the paper gets less. This means that the acceleration gets less.

When the frictional force equals the weight, there is no *total* force on the paper. This means that there is no acceleration. The paper falls with a constant speed.

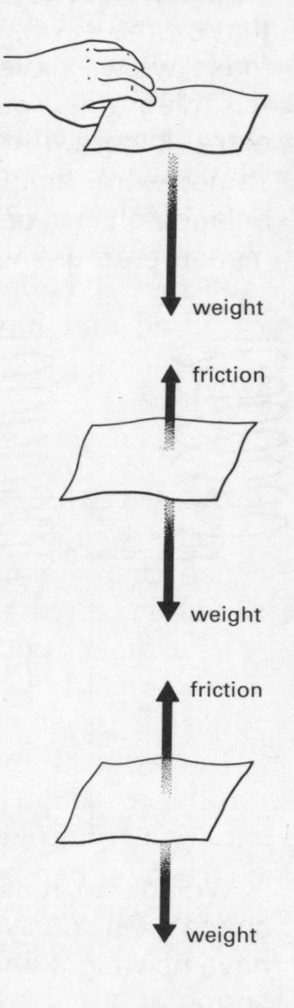

Fig. 1.38

There is a special name for this constant speed produced when air resistance balances the weight of an object. It is called the TERMINAL VELOCITY (velocity just means speed in a particular direction).

A piece of tissue paper has a very small terminal velocity. For most other objects this is larger and they fall a long way before they reach it. Sky divers fall with a terminal velocity of about 120 miles per hour.

The size of the terminal velocity depends on how much air resistance affects an object. It is the surface area of an object that determines the air resistance. So if two objects had the same weight, the one with the greater surface area would fall more slowly.

Does this mean that large objects fall more slowly than small objects? No, in fact the reverse is true. Large objects do have more surface area *but* small objects have more surface area per gram. This means that air resistance affects them more.

Very small particles of dust have a very low terminal velocity. They take a long time to fall and are easily blown about by air currents. This means that dangerous substances such as asbestos fibres, radioactive dust and coal dust can remain suspended in air for long periods and cause serious health problems. The low terminal velocity of small particles can also be a useful effect. Think of plants that rely on pollen being carried by the wind from one plant to another. It is because the pollen grains are very small and have a low terminal velocity that they can be carried a long way before they fall to earth.

Air resistance and space travel

In outer space air resistance is not a problem because there is no air. There is a problem however in getting out of and back into the earth's atmosphere. The air resistance makes it difficult to control fast moving crafts and also produces a great deal of heat. Air resistance can be reduced by designing the spaceship carefully so that the air flows over it easily. Look at the photograph of the space shuttle on page 21. You can see how it is streamlined to avoid air resistance. Compare this with *Pioneer 10* on the same page. Although this travels very fast, air resistance is not a problem.

THROWING THINGS SIDEWAYS

Figure 1.39 shows someone dropping a ball and throwing one sideways at the same time. Does the one that falls straight down hit the floor first? Try it. It seems as though they both hit the floor at the same time. The sideways motion of a falling object does not affect the time taken for it to fall downwards.

This can be seen clearly from Fig. 1.40, which shows the path taken by a ball that is dropped and one that is thrown sideways at the same time. The diagram shows the positions of the two balls each $\frac{1}{10}$ second. The positions were found by illuminating the balls with a flashing light called a stroboscope and photographing them. The light flashed every $\frac{1}{10}$ second so that the positions of the two balls each $\frac{1}{10}$ second were recorded on the photograph.

Look carefully at Fig. 1.40. You can see that the vertical distance fallen each $\frac{1}{10}$ second is the same for both balls. You can also see that these distances increase each $\frac{1}{10}$ second. Now look carefully at the ball moving sideways. The horizontal distance travelled each $\frac{1}{10}$ second is the same. The ball has a constant horizontal speed. This result shows clearly that an object thrown sideways has two parts to its motion that are independent of each other:

a constant vertical acceleration
a constant horizontal speed

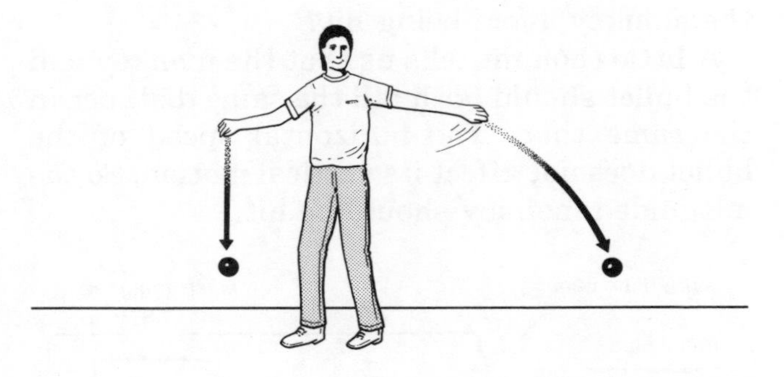

Fig. 1.39 An object thrown sideways hits the floor at the same time as one dropped

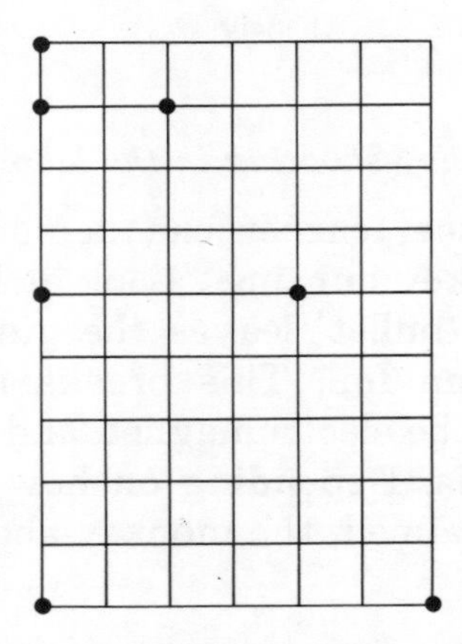

Fig. 1.40 Diagram showing the positions of a ball dropped vertically and one thrown sideways at the same time

The monkey and hunter problem

To end this section we can consider the amusing problem of the monkey and the hunter. This problem dates back to the cruel days of monkey shooting (now, fortunately, banned).

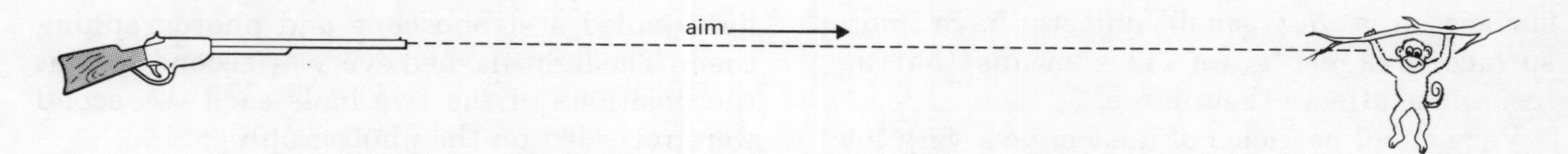

Fig. 1.41 The monkey and hunter problem – does the monkey avoid being hit?

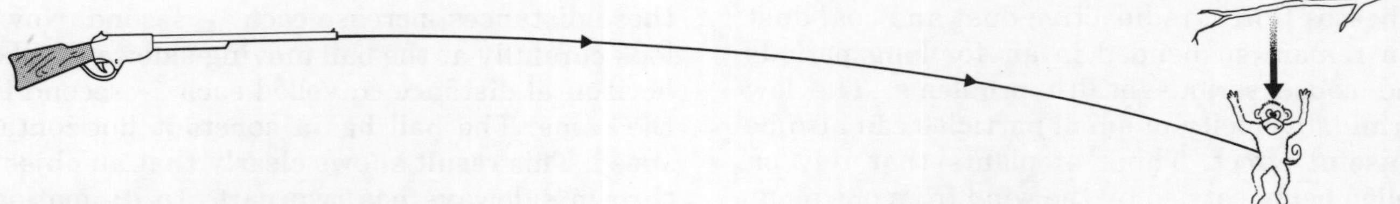

Fig. 1.42

A monkey is hanging from a tree. A hunter aims the gun straight at the monkey. The clever monkey realises the danger and lets go the instant that the bullet leaves the gun. Will the monkey avoid being hit?

A little thought tells us that the monkey and the bullet should both fall the same distance in the same time. The horizontal speed of the bullet does not affect its vertical motion. So the misguided monkey should be hit.

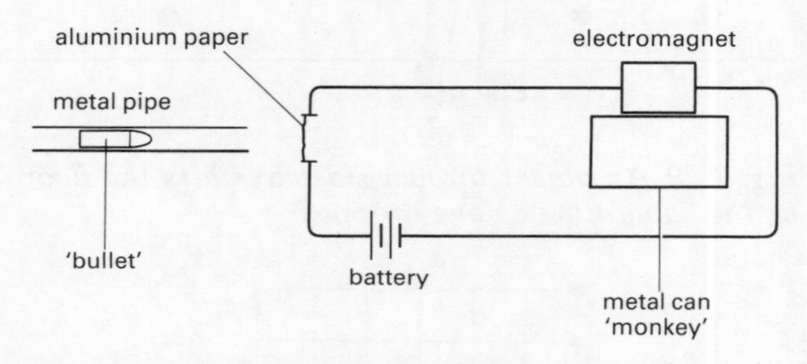

Fig. 1.43 Monkey shooting in the laboratory

Your science teacher can demonstrate her skill at 'monkey hunting'. Look at Fig. 1.43.

When the 'bullet' leaves the gun it breaks the aluminium foil. This breaks the circuit, switches off the electromagnet, and the tin can 'monkey' falls. Depending on how accurately the gun was aimed, the monkey should be hit!

SATELLITES

SATELLITES are objects that orbit the earth. The moon is a natural satellite. An increasing number of artificial satellites also orbit the earth. They are becoming more and more important.

Many of the satellites are communications satellites. Television signals are sent up to a satellite and then bounced back down to earth. Signals can be bounced around the curve of the earth. This means that events in Australia can be seen 'live' in Britain.

Fig. 1.44 Salyut 6

Figure 1.44 shows an orbiting laboratory called *Salyut 6*. This carries scientific instruments such as telescopes and allows astronauts to carry out experiments in weightless conditions.

Satellites provide the photographs that you see on the weather forecast. They can also be used for studying the natural resources of the earth. The cameras they carry are capable of showing great detail from a large distance and are being used increasingly for military purposes.

Assignment – Communications satellites

Find out more about comunications satellites. Write an essay making the following points.

(a) How they get into orbit.
(b) The importance of a geostationary orbit.
(c) The different sorts of signals that they can be used with.

By the time you read this book you may well be receiving your television signals direct from satellites. Explain how this works and what special equipment is needed to pick up the signals.

How satellites stay in orbit

Satellites do not stay in orbit because they escape from gravity. Far from it. They orbit the earth *because* gravity pulls them down. Why don't they hit the ground?

Imagine a cannon on a high hill firing a large shell. Figure 1.45 shows the paths of two shells. Shell B is fired out with a higher speed than shell A.

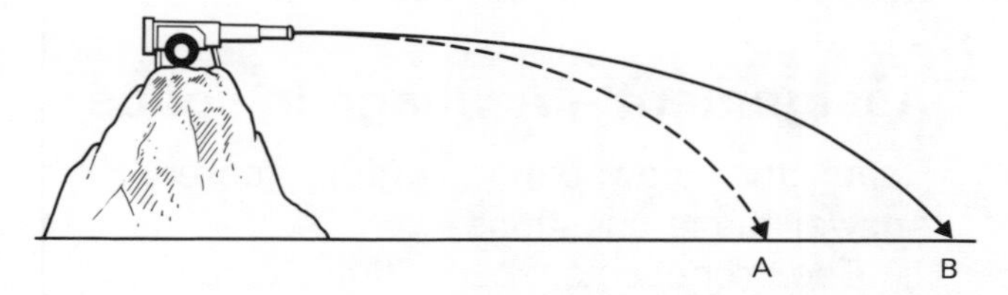

Fig. 1.45 Firing shells from a high hill. Shell B fired faster than shell A

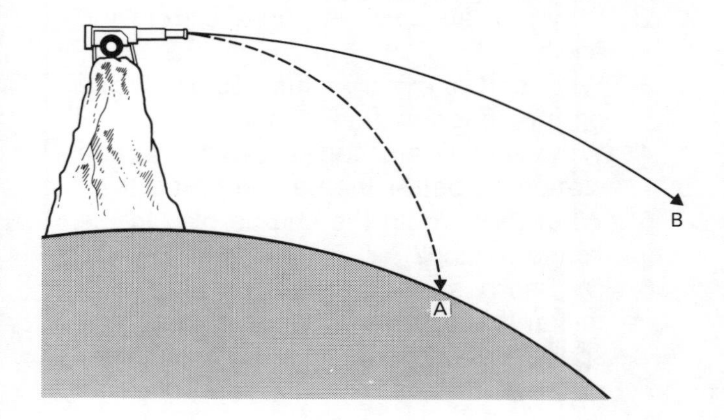

Fig. 1.46 Firing a shell into orbit

Now imagine a very high hill and a very powerful cannon (Fig. 1.46). Notice the earth is curved. Shell A travels a long way before it hits the ground. Look carefully at shell B. As it falls towards the ground the earth is curving away from beneath it. It never hits the earth. It always falls freely but the earth is curving from beneath it. It will orbit the earth at a constant height. It will be a satellite.

Weightlessness

If an object is in outer space far from any star or planet it will be weightless. There will be no gravitational force on it. Objects will also appear to be weightless if they are in free fall.

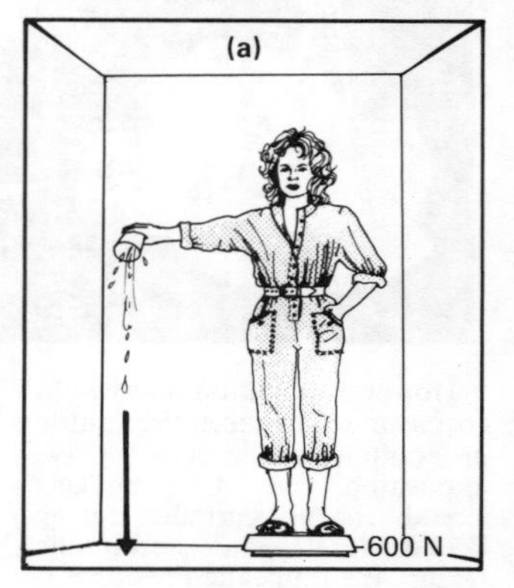

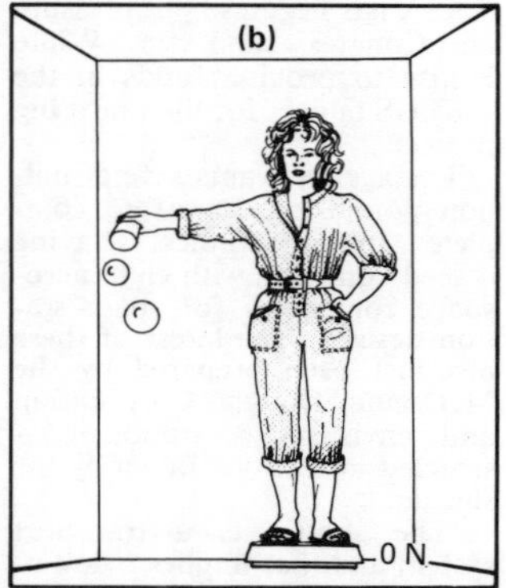

Fig. 1.47 (a) A lift moving at a steady speed.
(b) A lift falling freely

Figure 1.47 shows a person standing on a set of bathroom scales in a lift. The lift is not moving. The bathroom scales read 600 N. If the person tips a glass of water out the water falls to the floor. What would happen if the cable broke and the lift fell freely?

The lift would be falling, the person inside would also fall at the same rate. The person would no longer push down on the bathroom scales. The scales would read zero. If the person poured the water out of the glass it would not fall to the floor of the lift. All things inside the lift would appear 'weightless'.

A satellite or orbiting spacecraft is in free fall, which is why all things inside appear weightless. Astronauts in the space shuttle have not escaped from gravity. They are weightless because both they and the shuttle are falling freely towards the earth.

THE OBSERVER, Sunday 24 October 1982

US Plans to build a village in space

from ROBIN McKIE at Kennedy Space Center, Florida

AMERICAN scientists have begun detailed planning of a space station to be built 300 miles above Earth before the end of the decade.

The £7 billion station would house up to 12 astronauts and scientists and would act as a staging post for launching probes to the rest of the solar system.

The idea of a permanent space village — once the dream of science fiction writers — is being taken very seriously by National Aeronautics and Space Administration officials. Buoyed with new confidence after the success of the space shuttle flights, NASA has begun to put pressure on Congress and the White House to provide funds in the 1985 US budget for their next big project.

The agency wants several million pounds next year to complete initial studies, having placed contracts with eight aerospace companies for space station designs. The latest of these has just been prepared by the McDonnell Douglas Corporation and envisages a station constructed in sections flown by the shuttle.

'The shuttle is a transport system and that implies a destination,' said Mr Glenn Parker, a shuttle project manager. 'When building the reusable spaceship, we always had a space station in mind — and once we get that we really will have opened up the last frontier.'

Such a station would provide laboratories for growing plants, making crystals for electronics, perfecting new welding techniques and many other experiments requiring zero gravity or a perfect vacuum. Telescopes there would be free of Earth's atmosphere, allowing scientists to peer more deeply into the universe than is presently possible.

A space tug would carry satellites into high orbit, while the station would be used to build and launch probes to the Moon and planets. NASA officials are particularly keen to build an unmanned probe which would bring Martian soil back to Earth — to discover if water, and perhaps primitive life, exists deep below the planet's surface. Such an ambitious mission would need a space station for its construction.

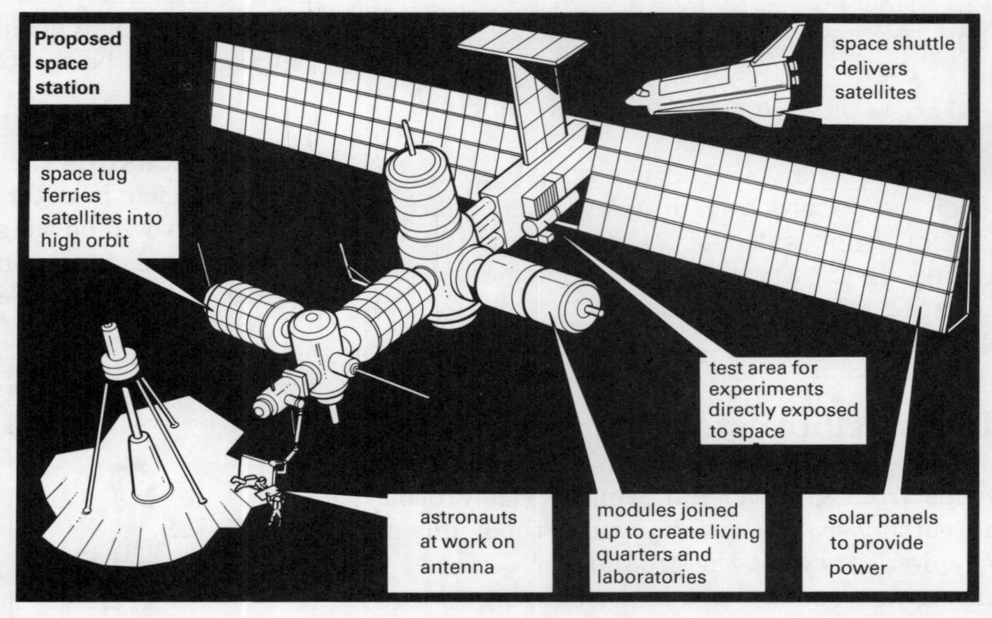

However, a station that would cost as much as the entire shuttle programme, would be a massive imposition on US budgets already sharply curtailed during President Reagan's economic reign. 'We think the President is sympathetic but our case will still have to be as economically attractive as possible,' said NASA future projects officer, Dr Alan Sharp.

One possibility is the involvement of the US Air Force which originally wanted its own station — a move blocked by Congress. 'We like the idea of a military presence,' added Dr Sharp. 'It would certainly help us get funds and approval.'

The agency also wants other nations — Canada, Japan and in particular Europe — to join the project. The European Space Agency — of which Britain is a member — has already spent £500 million building Spacelab, a small orbiting manned laboratory to be flown on board the shuttle next year.

Many designs, including the McDonnell plan, use Spacelab as a basic component — a move likely to be followed by ESA when it presents its own space station proposals to NASA.

A future space station may then have an encouraging international flavour about it — a co-operative spirit that would be somewhat offset by the presence of the military as well.

Assignment – A village in space

Read the newspaper article carefully. Answer these questions.

1 The proposed space station would allow probes to be launched to the rest of the solar system - where might these probes go and what could they do?
2 Why not launch these probes from the earth?
3 What sort of experiments could be done on the space station?
4 Why would telescopes on the space station be better than those on earth?
5 What part would the shuttle play in the space station?
6 Dr Sharp says 'We like the idea of a military presence'. Suggest how the space station could be used for military purposes.

WHAT YOU SHOULD KNOW

1 Average speed = $\frac{\text{distance travelled}}{\text{time taken}}$

2 Average acceleration = $\frac{\text{change in speed}}{\text{time taken}}$

3 If no *total* force acts on an object then it

either - stays at rest

or - if it is moving it carries on moving at the same speed in the same straight line.

4 If the forces acting on an object are equal and opposite they add up to give zero total force.

5 A constant force gives a constant acceleration.

6 The more mass an object has the more difficult it is to accelerate it.

7 All masses are pulled together by the force of gravity.

8 Weight is the gravitational force of an object.

9 The gravitational force field is the pull of gravity on each kilogram.

10 Weight = mass × gravitational force field.

11 If there is no friction, all objects fall towards the earth with an acceleration of 10 metres per second per second ($m\,s^{-2}$).

12 When air friction produces a force that is equal to the weight of an object, there will be no total force and the object will fall with a steady speed called the terminal velocity.

13 An object thrown sideways has two parts to its motion that are independent of each other:

a constant vertical acceleration
a constant horizontal speed

14 Objects that are a long way from planets are weightless.

15 Objects in a box that is in free fall appear weightless.

16 Satellites are in free fall but because the earth is curved they never hit the surface.

QUESTIONS

1 What is the average speed of a car if it travels (a) 10 metres in 5 seconds, (b) 100 metres in 5 seconds, (c) 1 kilometre in 100 seconds, (d) 600 metres in 1 minute?

2 A good sprinter can run at $10\,m\,s^{-1}$. How long will it take him to go
(a) 1 m, (b) 5 m, (c) 20 m, (d) 100 m, (e) 400 m, (f) 800 m? Why is the answer to (f) unlikely?

3 A cyclist travels at $20\,m\,s^{-1}$ How long will she take to go
(a) 10 m, (b) 40 m, (c) 100 m, (d) 1 km.

4 A car accelerates from 0 to $20\,m\,s^{-1}$ in 5 seconds.
(a) What is the average acceleration of the car?
(b) If it continued accelerating at this rate how fast would it go after
(i) 6 seconds?
(ii) 10 seconds?
(c) Why is the answer to (b) (ii) unlikely?

5 Copy and complete the following.
If no total . . . acts on an object it either stays at rest or . . . .
If it is moving it continues with the same . . . in the same . . . line.

6 A friend says, 'Our car needs a tank full of fuel to go just over a hundred miles. Just think how much *Voyager 2* would need to get to Saturn!' What would you reply?

7 A space craft is travelling in outer space, away from stars and planets. The rocket engines are turned off.
(a) What sort of motion will the space craft have?
(b) What will happen if the rocket engines are fired in the normal way?
(c) How could the rocket be slowed down?
(d) Use diagrams to explain how the direction of the space craft could be changed.

8 The diagram shows a passenger in a car.

The driver brakes and the car decelerates steadily.

(a) What provides the force that decelerates the car?

(b) If the deceleration is constant, what can you say about the force?

(c) If the passenger is to remain in his seat he must have the same deceleration as the car. What provides the force on the passenger under normal circumstances?

(d) If the car is stopped suddenly, the force that you described in (c) is unlikely to be large enough. What would provide the force needed to keep the passenger in his seat?

(e) The diagram below shows what might happen if a seat belt was not worn. Explain what is happening and why.

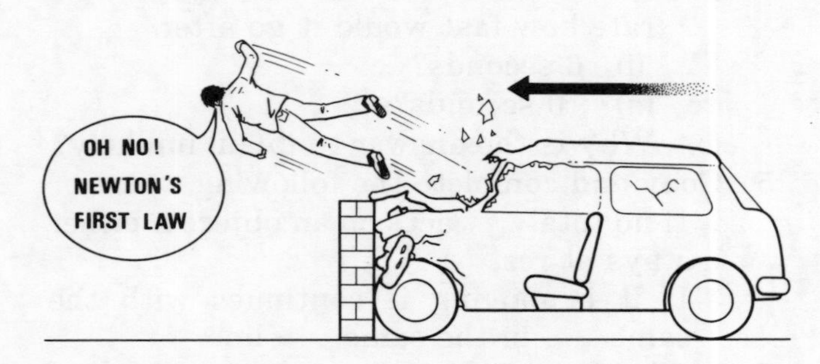

9 The earth's gravitational force field is $10\ N\ kg^{-1}$. What is the weight of the following masses?

(a) 1 kg, (b) 10 kg, (c) 25 kg, (d) yourself, (e) an astronaut of mass 80 kg, (f) a piece of paper of mass 1 g.

10 What would the weight of all these objects be in outer space - away from any planet or star?

11 What would the weight of each of them be on the moon where the gravitational force field is $1.7\ N\ kg^{-1}$?

12 If a snowflake falls gently down to earth, what can you say about air resistance? (Hard) Try and guess how big the air resistance might be.

13 A pupil drops the following objects from the top of a high building: a marble, a feather, a crumpled up piece of paper, a peanut.

(a) Write down the order in which they land.

(b) Why would this be a dangerous thing to do?

14 Copy and complete the following.
An object thrown sideways has ... parts to its motion: a ... horizontal ... and a ... vertical acceleration. These are ... of each other.

16 The diagram below shows a stroboscopic photograph of a ball rolling along a table top at a constant speed.

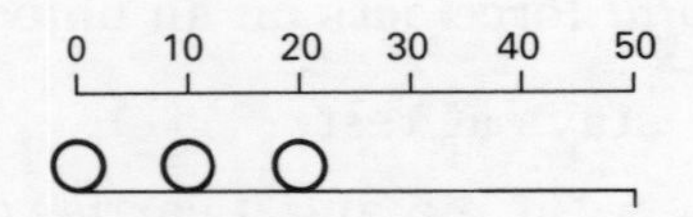

(a) Copy the diagram and draw the next three positions of the ball.

(b) If the stroboscope flashes 10 times per second, what is the speed of the ball?

17 The moment that the ball in question 16 reaches the edge of the table, another ball is dropped from the same height. This is shown below. Copy the diagram carefully into your book and draw the position of the sideways moving ball each $\frac{1}{10}$ second. Remember that it will have left the table top so it will be falling as well as moving sideways.

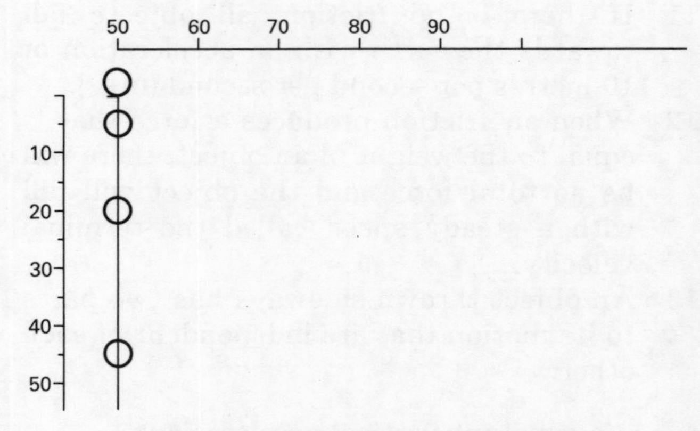

18 A cosmonaut in the Russian orbiting space station *Salyut 6* appears weightless.

(a) Is it true that no forces act on the cosmonaut?

(b) Why does the cosmonaut appear to be weightless?

(c) If *Salyut 6* is in free fall, why does it not crash to earth?

The Earth and its Place in the Universe

THE SURFACE AND BENEATH

Figure 1.48 shows the different layers that make up the earth. This is not just guesswork. The structure has been worked out by looking at the way that vibrations from earthquakes travel through the earth.

The earth's crust is made of ROCKS. The rocks themselves are made of naturally occurring crystals called MINERALS. In the same way that words are made up of one or more letters of the alphabet, so rocks are made up of one or more minerals.

ROCK TYPES

There are many different rocks in the earth's crust. They vary in the minerals they contain and in the shape and size of the mineral grains. Rocks are divided into three main groups.

Igneous rocks

These are formed from the cooling and hardening of hot molten material called MAGMA. The magma comes from within the earth. There are two main types of igneous rock. VOLCANIC igneous rock is formed when the magma comes to the surface and cools. Basalt is an example

Fig. 1.48 The layers of the earth

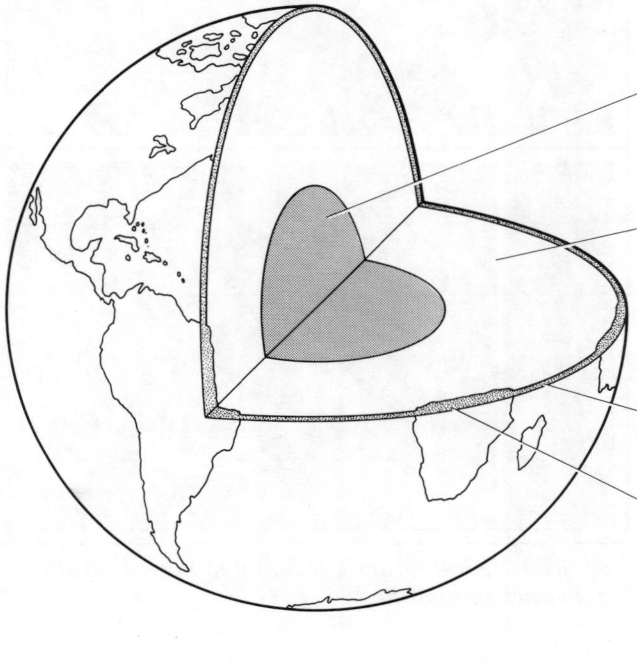
This part of the earth is called the CORE. It is very hot
This part is called the MANTLE. It is hot and part of it is made of molten rock called MAGMA
We live on the thin surface layer called the CRUST.
This is the OCEANIC CRUST. Here the crust is thin and covered with water.
This is the CONTINENTAL CRUST. Here it is much thicker and forms land.

of a volcanic igneous rock. PLUTONIC igneous rock is formed when the magma cools and forms into a solid beneath the surface. Granite is an example of a plutonic igneous rock. Notice that both types are made up of interlocking crystals. (Fig. 1.49)

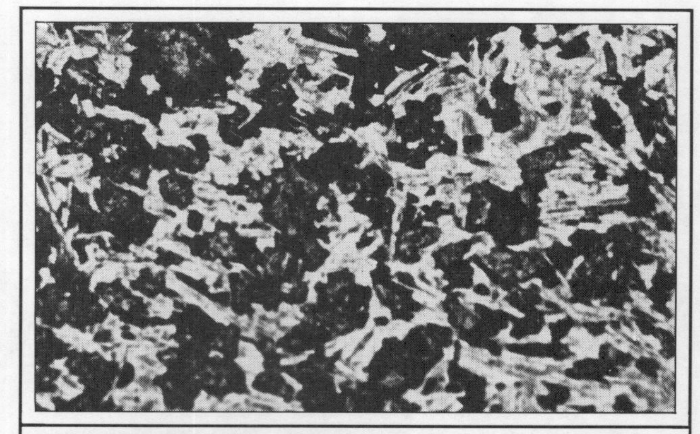

Fig. 1.49 Igneous rocks: (a) (above) basalt, (b) (below) granite

Sedimentary rocks

Two-thirds of the earth's surface is covered with sedimentary rocks. These are formed when solid particles carried in rivers and seas are deposited. Sedimentary rocks have definite layers and you can often see the STRATA or beds running through the rock. There is great variation in grain sizes and hardness, and sedimentary rocks often contain fossils. Limestone and sandstone are examples of sedimentary rocks which are made up largely of decomposed sea animals. They are made mainly of calcium carbonate and fizz when dilute hydrochloric acid is added to them.

Metamorphic rocks

These are formed below the surface of the earth. They are formed when older igneous and sedimentary rocks are transformed by a combination of heat, pressure and chemical liquids. Slate and marble are examples of metamorphic rocks.

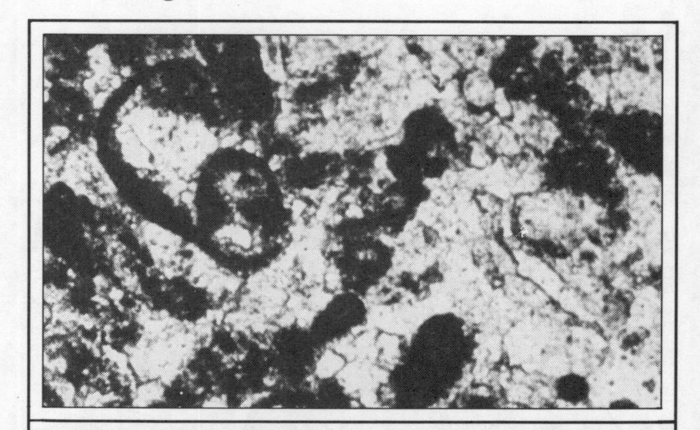

Fig. 1.50 Sedimentary rocks: (a) (above) limestone (b) (below) sandstone

THE CHANGING SURFACE

The most important point about this section is to realise that the earth's surface is forever changing. Where there are now grassy meadows with sheep grazing in South Wales, there was once a warm tropical sea full of corals rather like the Great Barrier Reef.

The earth's surface does not change overnight except in extreme cases such as a volcanic eruption, when a new island can suddenly appear out of the sea. It changes over millions and millions of years. This changing nature of the earth's surface can perhaps be best seen by considering how sedimentary rocks are deposited. Figure 1.51 shows how this happens.

The rocks on the higher land are broken up

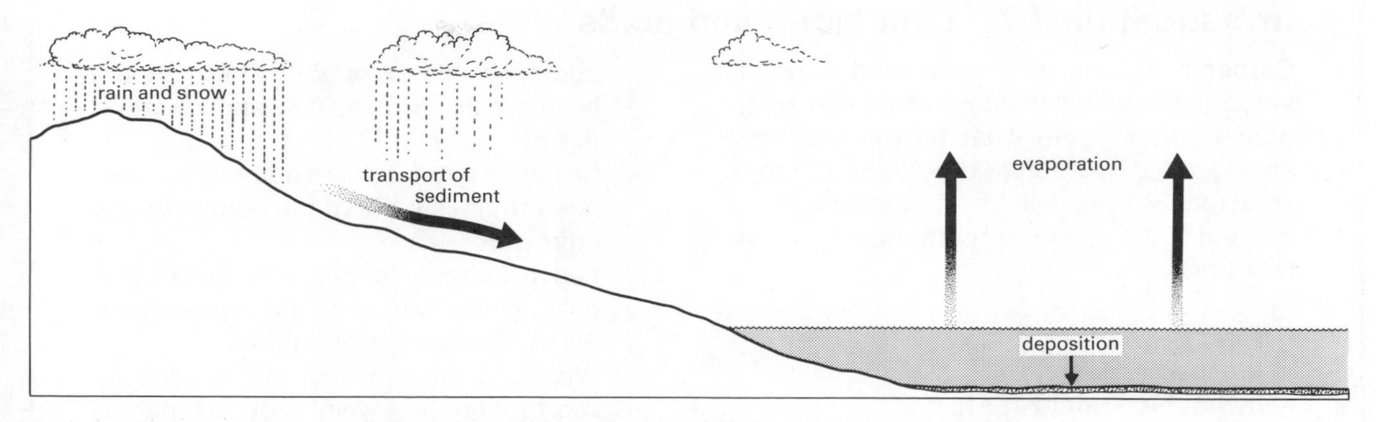

Fig. 1.51 The changing surface of the earth - how sedimentary rocks are formed

by the effects of the weather. This is called WEATHERING. The broken up bits are then carried away by rivers, glaciers and the wind. The process of breaking up and taking away is called EROSION. The bits are then deposited in layers to form sedimentary rocks.

This is a very simplified version. There are many different types of sedimentary rocks. The differences depend upon the bits from which they are made and, very importantly, the way in which the bits are deposited.

It is important for us to understand these differences. It is also important to understand weathering and erosion if we are to understand why the surface of the earth looks the way that it does.

WEATHERING

All rocks are affected by wind, rain, temperature and plants and animals. Some rocks take many thousands of years to break down. With others, changes can be seen in a few months. There are three main ways that rocks can be broken down. These are by physical, chemical and biological means.

Physical weathering

Physical weathering is caused by the action of water and wind and by the temperature changes that cause the rocks to expand and contract.

Look at Fig. 1.52. It shows steep slopes of rock fragments called SCREE. These are formed by a process called 'frost shattering'. Rock expands during the day when it gets hot. It then cools and contracts at night. This causes the rock to crack. Water then enters the cracks and freezes. As water freezes it expands. The cracks get bigger. Eventually chunks of rock are broken from the rock face.

Chemical weathering

Chemical weathering is caused by a chemical reaction between rainwater and rock. As the rain falls through the atmosphere it dissolves small amounts of carbon dioxide gas to form a weak acid called carbonic acid. Limestone rocks contain calcium carbonate and react with the carbonic acid. So each time the rain falls, a little of the limestone dissolves away.

Fig. 1.52 Physical weathering - scree slopes in the Scottish Highlands

Investigation 1.7 Limestone and acids

Carbonic acid is very weak and it reacts very slowly with limestone. In this investigation you will see what happens when a strong acid, hydrochloric acid, is added to limestone. You will also calculate the amount of calcium carbonate in the limestone.

Collect

crushed limestone
hydrochloric acid (1 M)
filter paper
filter funnel
250 cm^3 beaker
top-pan balance
safety spectacles

What to do

1 Copy out this table.

Mass of filter paper	=
Mass of filter paper + crushed limestone	=
Mass of crushed limestone	=
Mass of filter paper + residue	=
Mass of residue	=
Loss of mass of limestone	=

2 Measure the mass of the filter paper. Write down the mass in the table.
3 Put some crushed limestone on the filter paper. Measure the mass again and write the result in the table.
4 Work out the mass of limestone and write this in the table.
5 Put the crushed limestone in a 250 cm^3 beaker. Add about 50 cm^3 of dilute hydrochloric acid. Leave to stand until no more gas is given off. Check by adding a little extra acid. There should be a small solid residue and no more fizzing.
6 Carefully filter this residue. You may need to rinse it out of the beaker with a little water.
7 Leave the filter paper with the residue on it to dry. Measure the mass again. Fill in the result in the table.
8 Work out the mass of the residue by subtracting the weight of the paper. Write this in your table.
9 Now work out the loss in mass of the limestone during the experiment.

loss of mass
= mass of crushed limestone
− mass of residue

Write this in your table.
10 This loss in mass is due to the calcium carbonate being dissolved in the acid. You can now work out the percentage of limestone that is calcium carbonate by

% of limestone that is calcium carbonate

$$= \frac{\text{loss in mass of limestone}}{\text{original mass of limestone}}$$

Questions

1 What happened when the acid was added to the limestone?
2 What gas came off?
3 Describe how you could test for this gas.

Limestone is a very common rock in the earth's crust and the effects of weathering can be seen in many different places. You have probably seen limestone caves. Other features, include SINK HOLES. These are hollows formed when limestone dissolves in acid rainwater and then collapses to form odd looking pits.

Acid rain

In industrial areas the air may be polluted with sulphur dioxide. This dissolves in rain to form sulphuric(IV) acid. This is a much stronger acid than carbonic acid, and the 'acid rain' that falls on cities may seriously weather buildings that are made of rocks containing calcium carbonate. Many of the fine sculptures in Greece are made of marble which is a form of calcium carbonate. Figure 1.53 shows the serious problems that the acid rain is causing.

Fig. 1.53 The effects of acid rain - an eroded marble statue of a lion at Delos in Greece

Biological weathering

Biological weathering is caused by plants and animals. The roots of plants can work their way into small cracks and wedge the rocks apart. Burrowing animals can also be a factor. Their digging activities can expose new rock surfaces to be weathered.

Plants can also contribute to chemical weathering. Lichens produce acid substances that can lead to weathering. Other plants produce weak acids around their roots which can also dissolve away rocks.

EROSION

Erosion is the wearing away of rock *and* its transportation elsewhere. There are four main ways by which erosion takes place (Fig. 1.54).

It is erosion that shapes our landscape. Over very long periods of time erosion reduces the land to low-lying plains. Some rocks are harder and more resistant than others.

Erosion is rapid in steep areas with heavy rainfall. The fast moving water is able to carry rock fragments away in large quantities. Vegetation protects against erosion. In certain areas of the world, like the Himalayas, forests are being chopped down. This greatly increases erosion. Erosion is slow in deserts and cold low-lands.

The erosion rate for the land area as a whole has been estimated as 8.6 cm of depth per 1000 years.

Fig. 1.54 (a) (above) A rock stack eroded by waves, (b) (below) Wind-eroded rock in the Sahara Desert.

Fig. 1.54 (c) (left) A deep gorge eroded by a river (d) (right) A glacier eroding a mountain

Assignment – Using a stream table to investigate erosion and deposition

We can study the way in which moving water erodes and then deposits fragments in the laboratory. The diagram below shows a stream table.

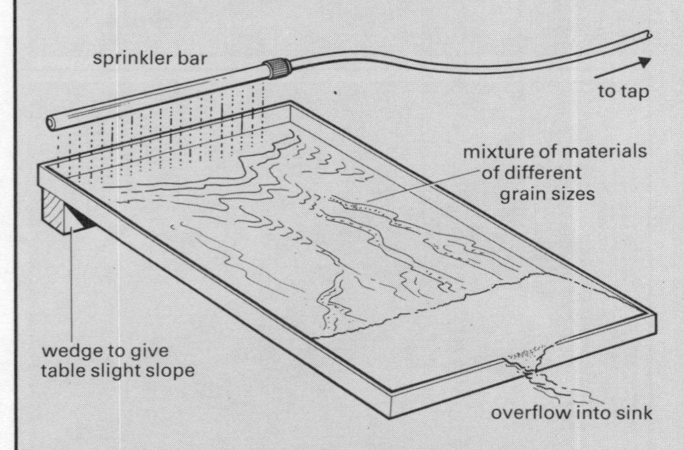

Set up the stream table as shown in the diagram. Make sure that the tray overflows to somewhere safe. Place a mixture of material of different grain sizes on the tray. You should use small pebbles, coarse sand, fine sand and chalk mixed well together. Allow the water to drop from the sprinkler bar onto the mixture.

You can use your stream table to investigate the following. Use diagrams where possible in your write-up.

1 *Erosion*

(a) When water channels form, are they straight or do they meander?

(b) Once a channel forms, does it stay the same or does it change? How does it change?

(c) How does the steepness of the table and the rate of water flow coming out of the sprinkler bar affect the erosion?

2 *Deposition*

(a) Where does the flowing water deposit fragments?

(b) Where on the stream table is the water flowing quickly and where is it flowing slowly?

(c) When the water has been flowing for a considerable time, look carefully at the material deposited at different places. Can you see any pattern? Is the speed of water flow related to the size of grains deposited?

SEDIMENTATION — PUTTING DOWN THE LAYERS

The assignment on page 44 shows a very important aspect of rock formation:

fast flowing water deposits large grains;
slow flowing water deposits small grains.

These grains or fragments are then compressed together to form sedimentary rocks.

Figure 1.55 shows how flowing water deposits different grain sizes in different places.

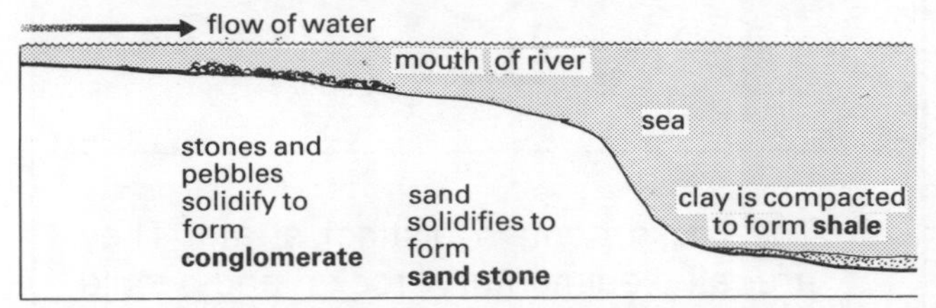

Fig. 1.55 The effect of water speed on grain size

The table below shows the grain sizes in these rocks.

conglomerate	sandstone	slate
larger than 2 mm	between 2 mm and 0.05 mm	less than 0.02 mm

These three types of sedimentary rocks are all formed from rock fragments compacted together. They are called MECHANICALLY FORMED SEDIMENTS.

Limestone is formed rather differently. Plants and animals on the sea bed take calcium carbonate from the water to form shells and skeletons. As the animals live and die, generation after generation is piled up on the sea bed. These deposits are compressed and eventually form limestone. Rocks formed in this way are said to be BIOLOGICALLY FORMED SEDIMENTS. Another example of this type of rock is coal, which was formed from plant remains in swampy forests millions of years ago.

There is a third type of rock which is formed in a different way. As well as carrying rock fragments, rivers also contain dissolved chemicals. When inland seas dry up these chemicals are deposited and can be compressed to form rocks. Rocks of this type are called CHEMICALLY FORMED SEDIMENTARY ROCKS.

FOLDING AND FAULTING

Sedimentary rocks are deposited in horizontal layers. This is shown in Fig. 1.56(a) Photographs. (b) and (c) show rock structures that are layered and are obviously sedimentary, but are not horizontal. They are examples of FOLDS. Folds

Fig. 1.56 Changing the horizontal layers by folding and faulting: (a) (above) horizontal strata, (b) (below) folded strata,

can be all sizes from huge arches many kilometres across to microscopic crinkles. They are formed over millions of years by great pressures in the earth's crust. Figure 1.56 (d) shows a FAULT. This is formed when the strata fracture. Again, faults vary in size from microscopic to many kilometres in length.

Fig. 1.56 (c) (above) nearly vertical strata, (d) (below) A fault in strata

Assignment – A geological history

The diagram shows rock strata in the Clydach Gorge in South Wales. (A sketch of this gorge appears on page 67.)

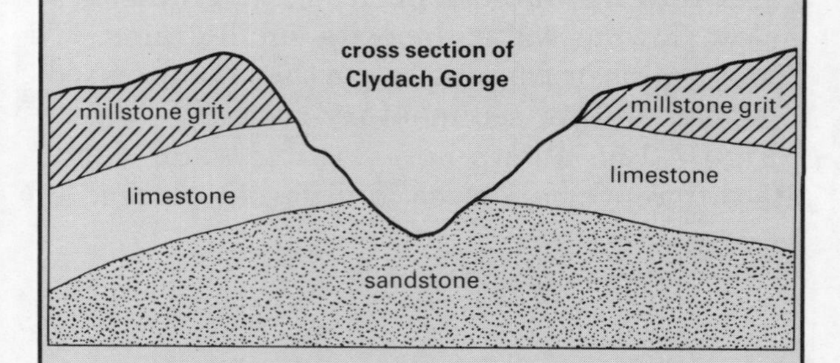

1 The rocks contain distinct strata. They are all sedimentary rocks and would have been deposited in horizontal layers. Which rock is the oldest? Which is the youngest?
2 The sandstone is fairly large grained and contains few fossils. What can you say about the way it was formed? Was the water still or moving?
3 The limestone has very fine grains and contains fossils of corals and shellfish called brachiopods. What can you say about how this was formed? Was the sea still or moving? Deep or shallow? Warm or cold?
4 The millstone grit is large grained with rounded pebbles of quartz in it. Describe how this could have formed. There are few fossils.
5 The rock strata were originally deposited horizontally. They are now sloping. How could this have happened?
6 How could the valley have been produced?

Now try and put these answers together to write a geological history of the area. Start with the oldest rock. Describe how it was formed. Be sure to back up your ideas with evidence.

Work your way up the layers. Say what must have happened to produce the landscape shown in the sketch.

ROCK CYCLE

We can now understand part of the pattern of changes on the earth's surface. The rocks in upland areas are weathered and eroded. The particles of rock are transported by wind and rivers and deposited to form sedimentary rocks. These are brought to the surface by folding and uplifting and the process starts again. This pattern of change is called the ROCK CYCLE. Figure 1.57 shows the part of the rock cycle that concerns sedimentary rocks. You will meet the full rock cycle later in this section.

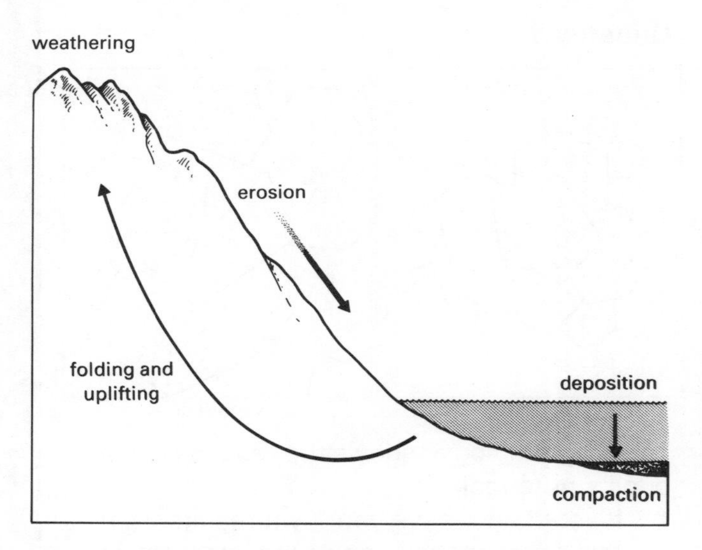

Fig. 1.57 Part of the rock cycle that concerns sedimentary rocks

HOT ROCKS

Figure 1.58 shows the volcano, Mount St. Helens. It erupted in 1980. Had the explosion happened in a heavily populated area, the loss of life would have been catastrophic. One and a half cubic miles of debris were thrown out of the volcano. Local forests and crops were destroyed. The dust settled on farms many miles away.

Figure 1.59 shows some features of a typical volcano.

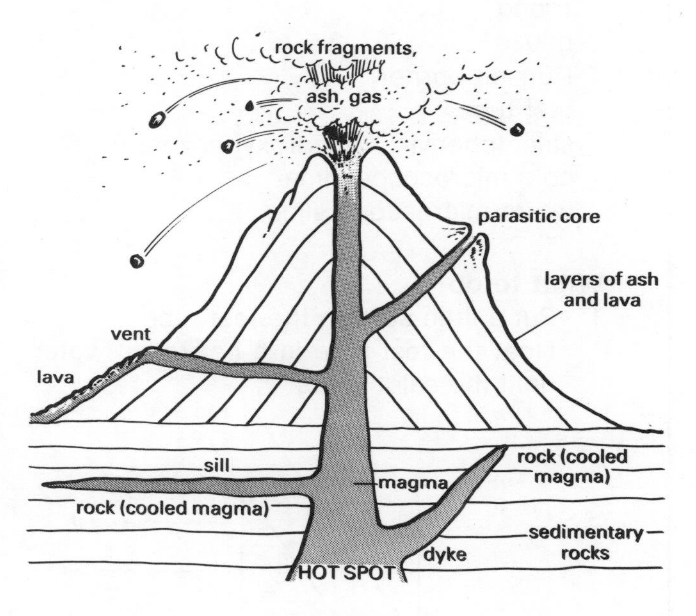

Fig. 1.59 Some features of a volcano

Fig. 1.58 The eruption of Mount St. Helens, in June 1980

Investigation 1.8
The effect of rate of cooling on crystal size

Igneous rocks are formed when molten magma cools and solidifies. Does the rate of cooling affect the crystals that form? In this experiment you will first of all melt a substance called salol and then see what happens when you cool it at different rates.

Collect

heatproof mat
tripod
gauze
Bunsen burner
test tube
salol (phenyl-1, 2-hydroxybenzoate)
cold microscope slide
warm microscope slide

What to do

1 Put a little salol in the test tube. Heat the test tube in a beaker of water until the salol melts.

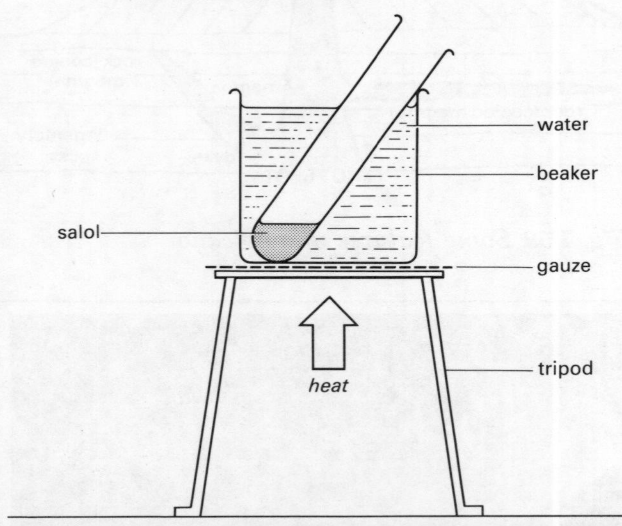

2 Put one drop of the melted salol on a cold slide and one drop on a warm slide. Watch carefully with a hand lens as crystals form.
3 Look carefully at the grain size of the crystals. Draw diagrams showing the size and shape of the crystals in each case.
4 Say what effect the rate of cooling has on grain size.

Questions

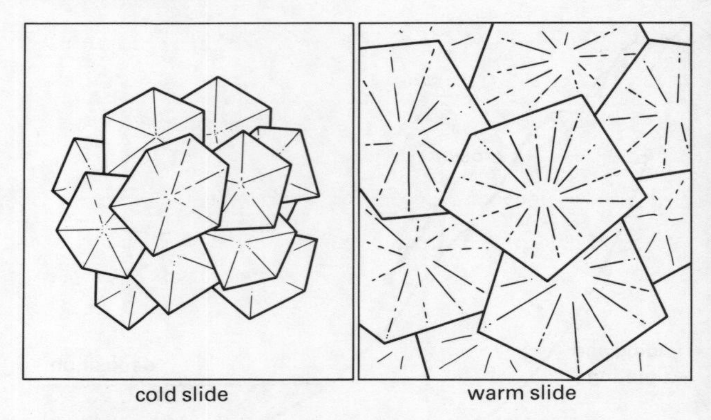

The diagram above shows the results from a pupil's notebook.

1 If you do not have your own results from this investigation, copy the diagram in your book.
2 How does the rate of cooling affect grain size?
3 Your answer to question 2 might be a guess. After all, you only have two results. Say how you could get other results with even more rapid cooling. Say what results you would expect if your answer to question 2 was correct.

Different types of igneous rocks

Figure 1.49 on page 40 shows basalt and granite. They are both igneous rocks but they look different.

Look carefully and you will see that they both contain interlocking crystals but the granite has much larger grain size. From Investigation 1.8 we can see that the granite must have cooled more slowly. It is formed from magma that does not reach the surface but cools underneath (plutonic igneous rock). Figure 1.60 shows how this might happen.

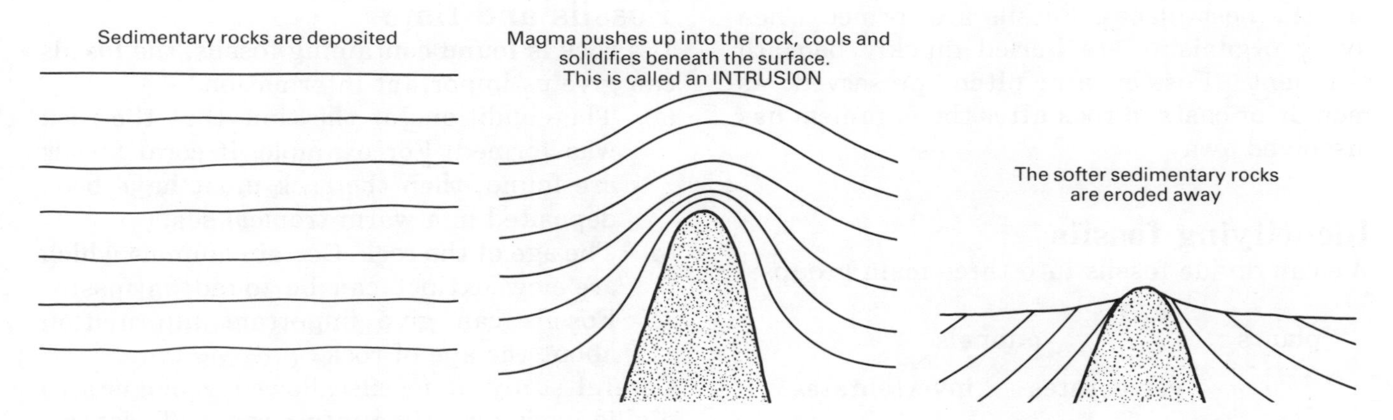

Fig. 1.60 The formation of granite

THE ROCK CYCLE

We are now able to complete the rock cycle.

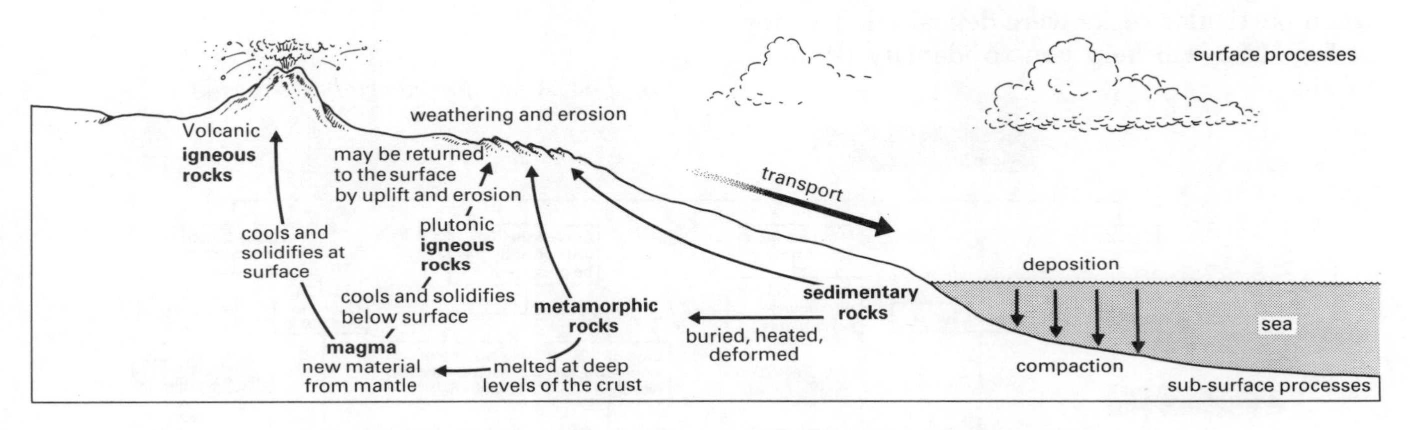

Fig. 1.61 The full rock cycle

FOSSILS — THE KEY TO THE PAST

You will be very lucky to find a fossil like the one in Fig. 1.62. More modest specimens are very easy to find if you know where to look. With a little bit of thought you can easily work

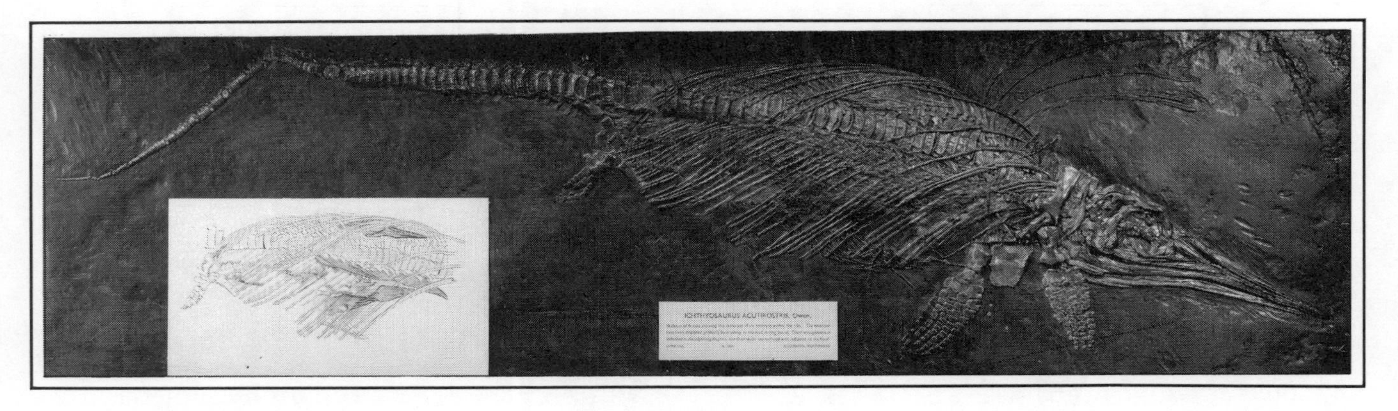

Fig. 1.62 A fossil fish

out the best places. Fossils are formed when living organisms are buried quickly beneath sediment. Fossils are often preserved as moulds or casts in rock after the organism has dissolved away.

Identifying fossils

We can divide fossils into three main groups:

plants — animals → vertebrates, invertebrates

Of the animal fossils, the invertebrates are far more common than the vertebrates.

Fossil identification is very important. Fossils give very useful information about when particular rocks were deposited. The key in Fig. 1.63 will help you to identify common fossils.

Fossils and time

If a rock is found containing fossils, the fossils can give us important information:

(a) The conditions at the time that the rock was formed. For example, if coral fossils are found, then the rock must have been deposited in a warm tropical sea.

(b) The age of the rock. Certain animals which are now extinct can be found as fossils. Fossils can give important information about the age of rock.

Careful study of fossils allowed geologists to divide geological time into periods. Today we can use radioactive dating to give actual dates in millions of years to these periods.

Fig. 1.63 A key for invertebrate fossils

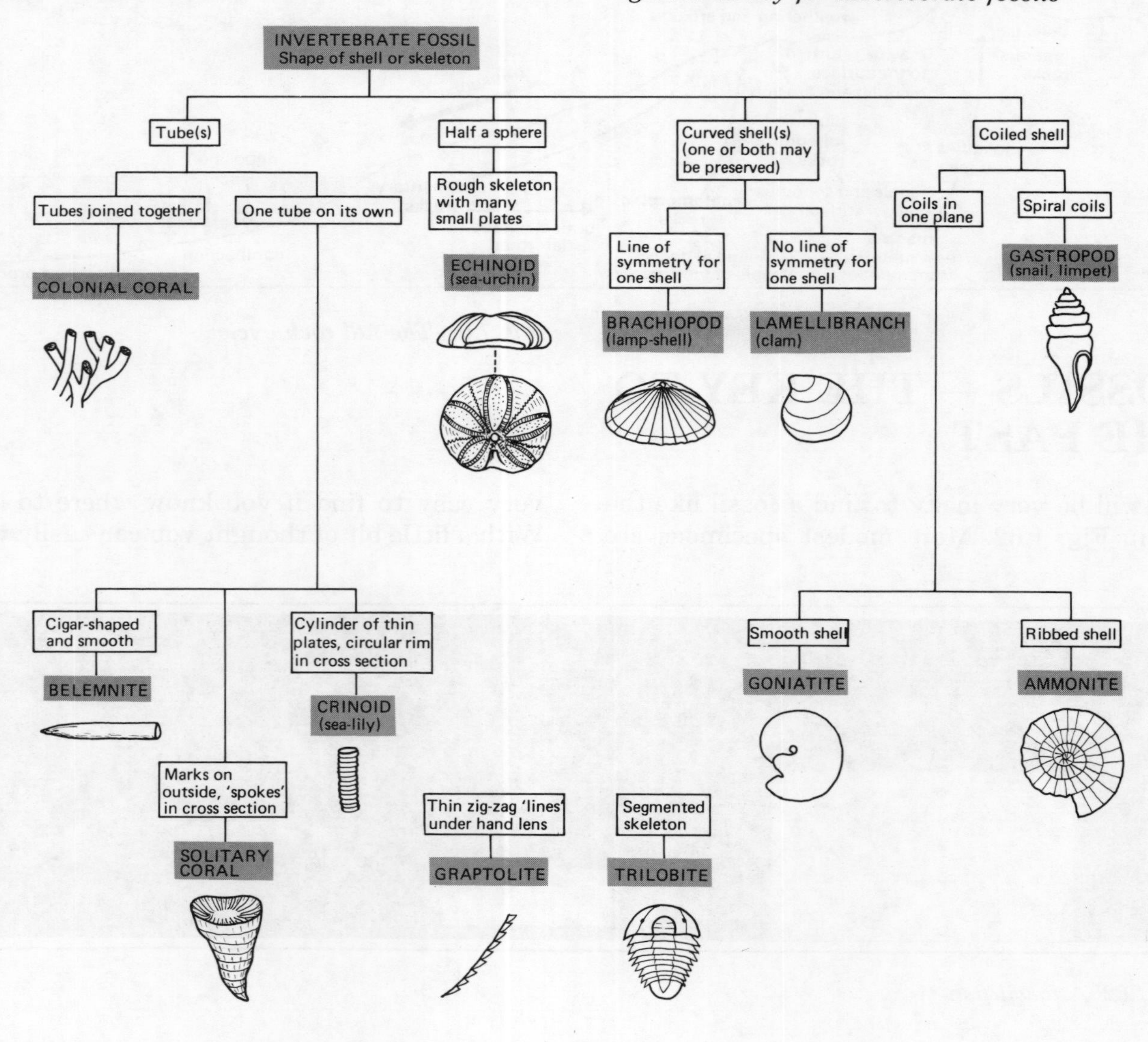

Assignment – Fossils and earth history

Look carefully at the diagram. It shows the time in earth history that various life forms existed. Invertebrate fossils are the most useful for rock dating. For each type of invertebrate the width of the column shows how common they were. Each type of invertebrate includes many different species, some of which only existed for a short time before becoming extinct.

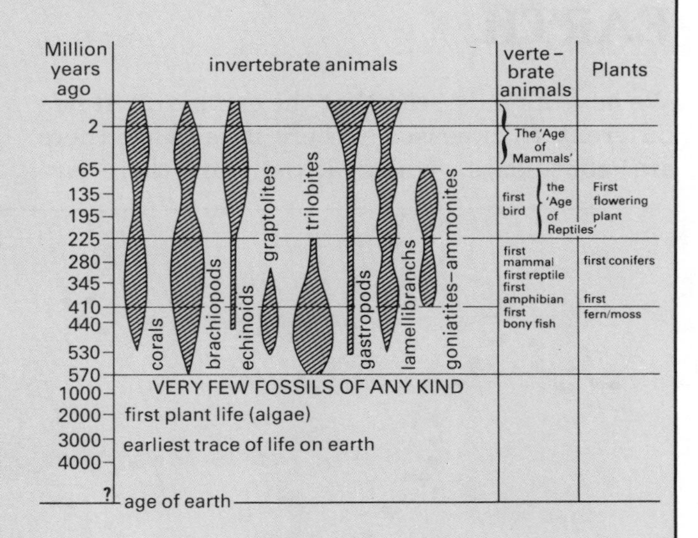

1 Which invertebrates are now extinct?
2 If you found a rock with graptolite fossils in it, what could you say about its age?
3 What are the oldest rocks that could contain coral fossils?
4 When did gastropods start to become increasingly common?
5 Certain fossils are found only in rock layers of one particular age. These are called index fossils. They can be used to identify specific rock layers. An example of this is the identification of oil-bearing rocks. Suggest some properties that an index fossil must have if it is to be used for this job.

Assignment – Fossils and past climates

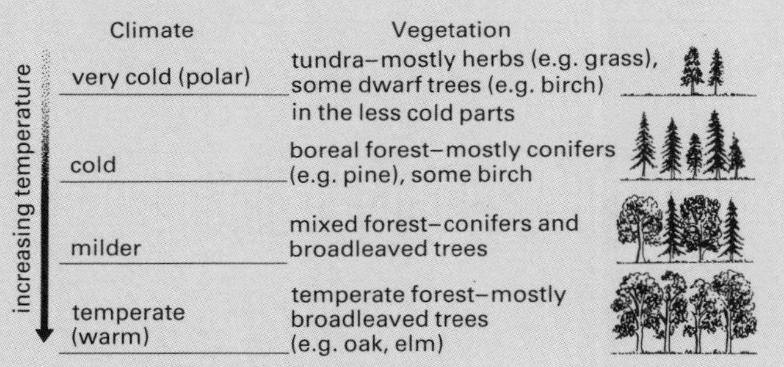

This diagram shows the different trees that grow in different climates. Various trees and plants have been preserved as fossils. The pollen grains produced by different plants have distinctive shapes. It is possible to tell from the shape of pollen grains what plants produced them. Pollen does not rot easily and the pollen from ancient forests is now fossilized and gives important information about past vegetation and climate. The diagram below shows the amount of pollen from different trees found in deposits in south-east England which are between 128 000 and 115 000 years old.

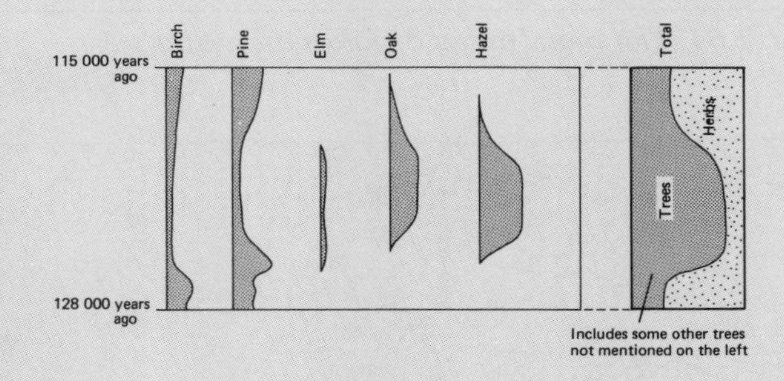

1 Use the information to describe changes in vegetation in south-east England over 13 000 years.
2 Describe how the climate must have changed.
3 Some gravels under Trafalgar Square about 120 000 years old contain the remains of elephants, hippos and rhinos. No such remains are found in older rocks. Explain this.
4 Would you expect to find the remains in rocks that are 115 000 years old?
5 Explain your answer.

THE PLATES OF THE EARTH

We have already seen that the earth's crust has different thicknesses. Where it is thick there are continents, where it is thin there are oceans. There is evidence to show that the crust is made up of different sections, and that these sections are on the move.

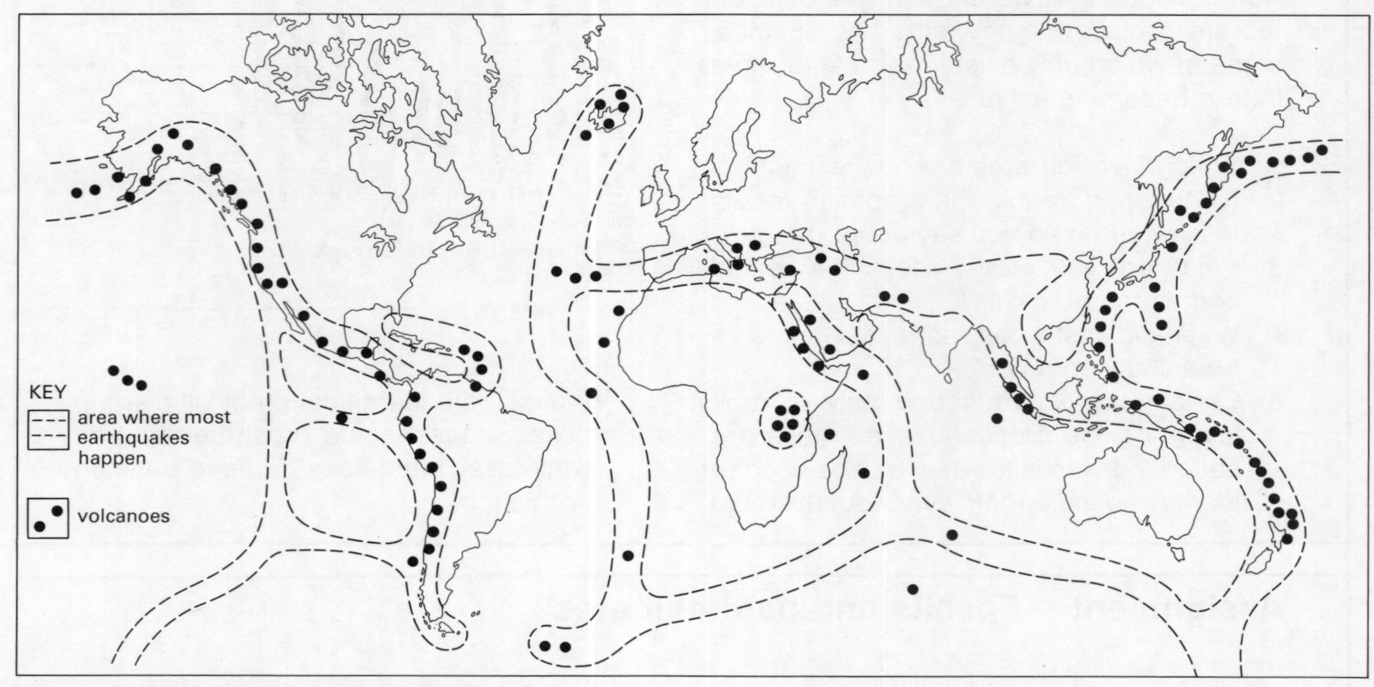

Fig. 1.64 The main areas of fold mountains

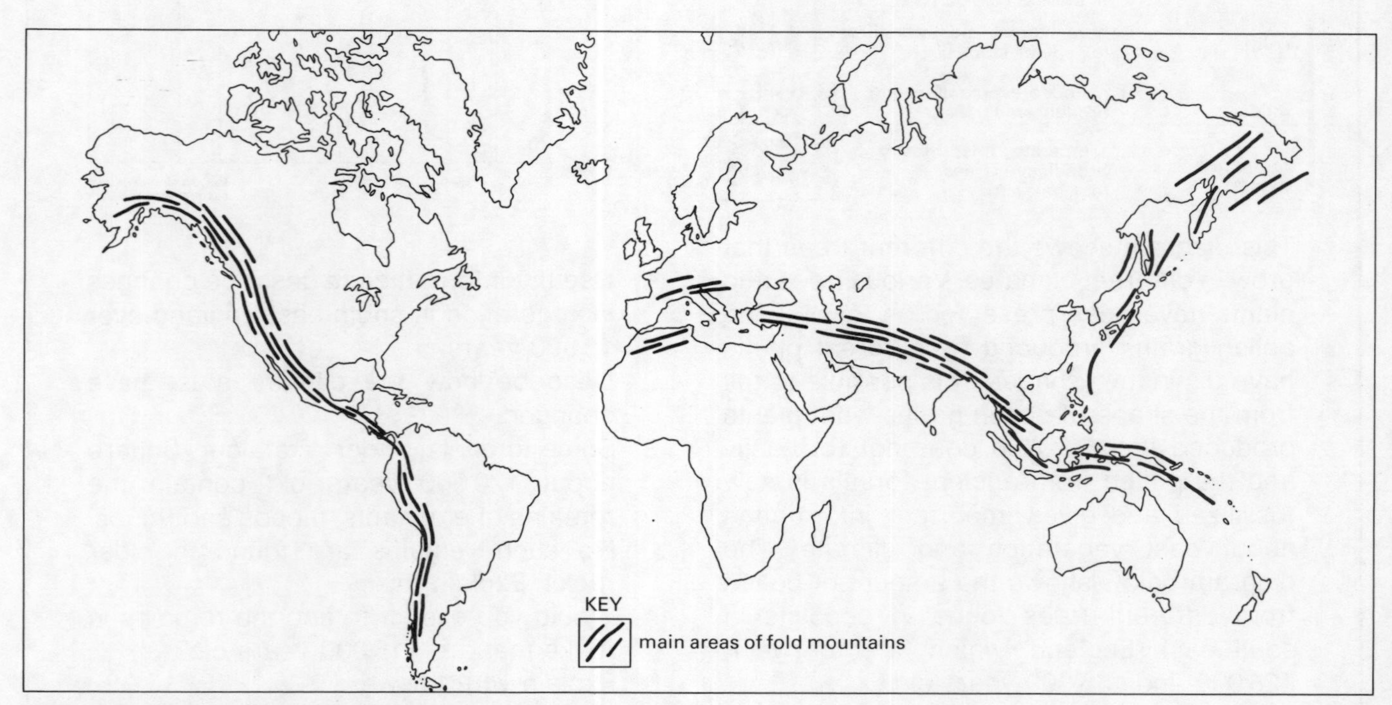

Fig. 1.65 The main areas of earthquakes and volcanoes

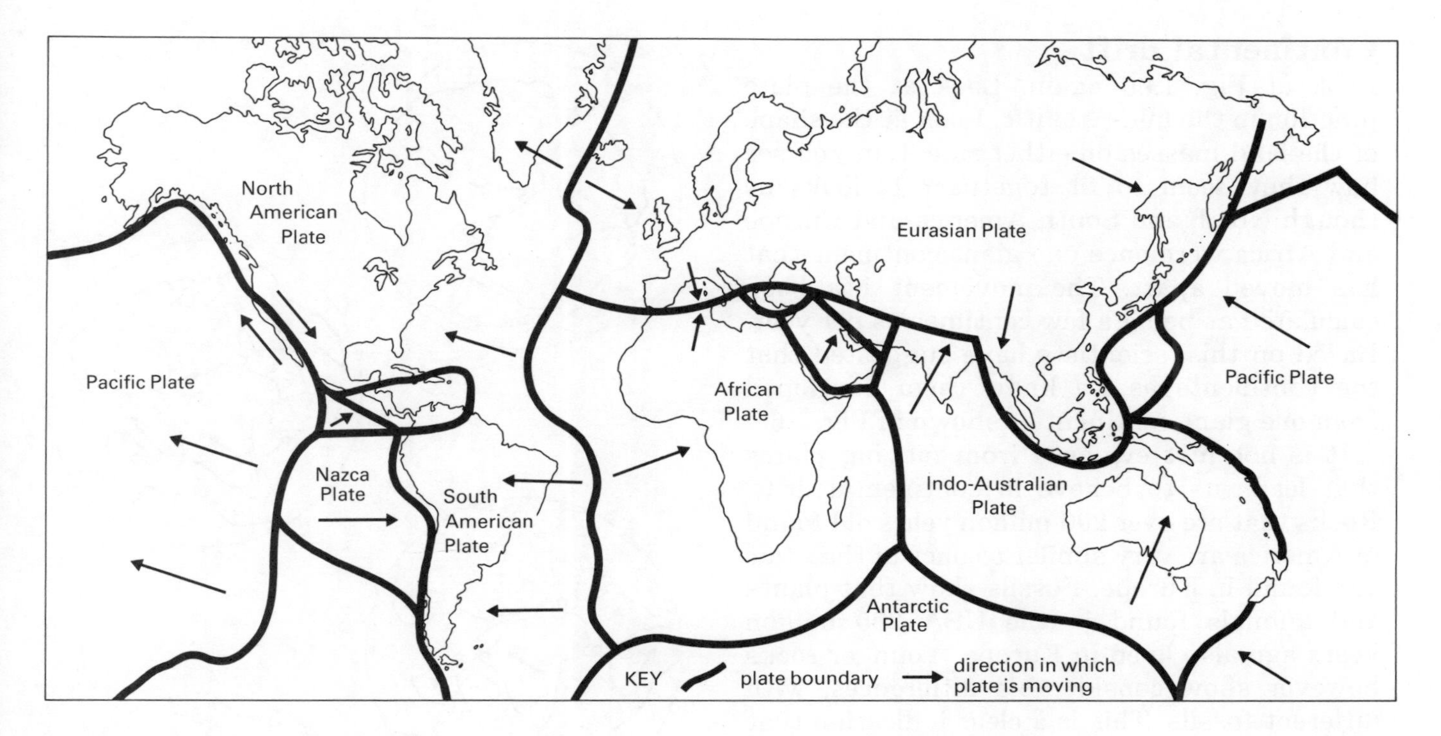

Fig. 1.66 The main plates of the earth's crust

Look at Fig. 1.64. This shows the main areas of fold mountains. These are mountains that have been formed by earth movements. They are rather like the ruck formed in a carpet when you push it with your foot.

Now look at Fig. 1.65. This shows the main areas of volcanic activity and the main areas where earthquakes occur. Can you see the pattern? How can this be explained?

Look again at the diagram of the earth's structure on page 39 (Fig. 1.48). You can see that the crust is a thin layer of solid on top of very hot liquid rock that forms the mantle. You can actually think of the crust as being rather like skin floating on hot custard. It is thought that the crust is not rigid but is made of a number of sections called PLATES that are moving around on top of the magma. By studying earth movements carefully the plates shown in Fig. 1.66 have been identified. The fold mountains, earthquakes and volcanoes all lie along plate junctions. This is where the crust is at its weakest and most unstable.

The arrows show the way the plates move. Look carefully and you will see that there are places where the plates move apart and places where they move together.

Figure 1.67 shows the three different ways that neighbouring plates can move.

Fig. 1.67 Moving plates

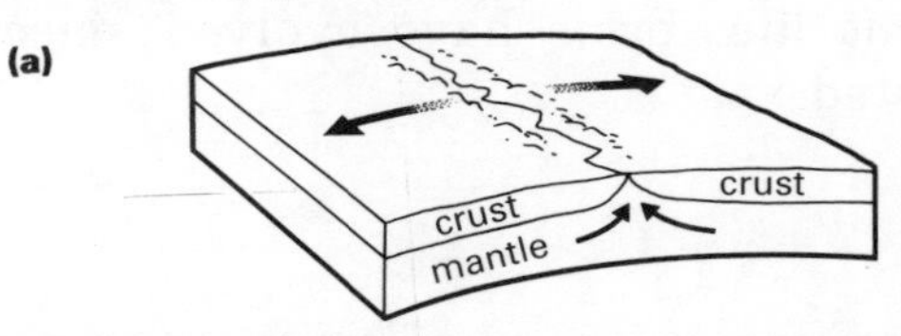

(a) Here the plates move apart. Magma flows up into the gap and new rocks are formed as a ridge on the ocean bed. This happens on the floor of the Atlantic Ocean.

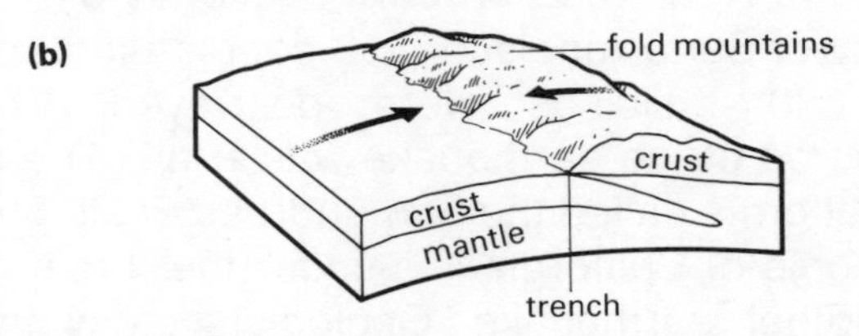

(b) Here the plates are pushing together. Fold mountains are formed and as the oceanic crust slides under the continental crust an ocean trench is formed. Magma also flows into the folded crust and volcanoes form.

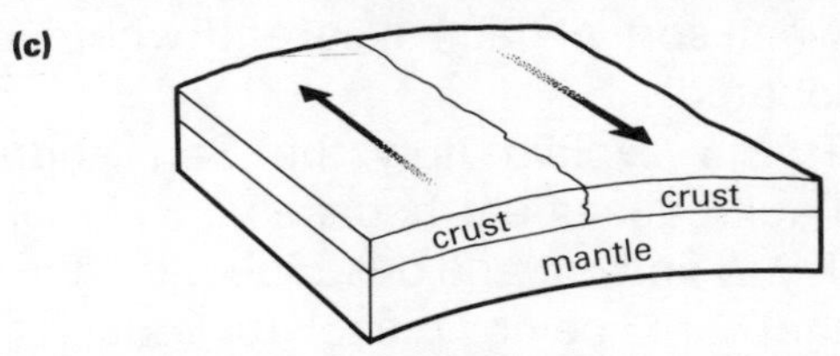

(c) Here the plates are sliding past each other. As in the other two cases, the weakness can lead to volcanoes and earthquakes.

Continental drift

Look at Fig. 1.66 again. Look at the plate junction in the mid-Atlantic. Look at the shape of the land masses on either side. Can you see how they seem to fit together? It looks as though North and South America and Europe and Africa were once one giant continent that has moved apart. The movement has been calculated as being a few centimetres per year. Based on this, scientists have suggested that the continents as we know them developed from one giant continent as shown in Fig. 1.68.

It is not just evidence from moving plates that leads us to believe in continental drift. Rocks that are over 200 million years old found in America are very similar to ones of the same age found in Europe. Fossils show that plants and animals found in the USA 200 million years ago also lived in Europe. Younger rocks however show considerable differences, with different fossils. This is a clear indication that once America and Europe were joined and that different life forms have evolved since they separated.

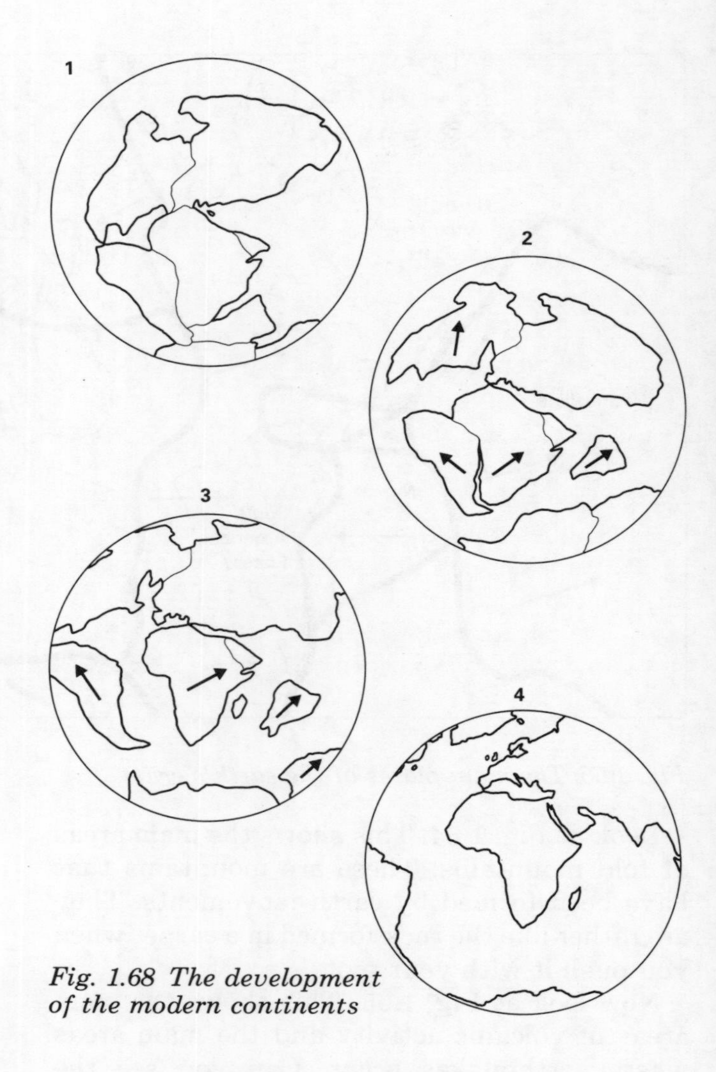

Fig. 1.68 The development of the modern continents

Assignment – The San Andreas Fault

On 18 April 1906 an earthquake shook the city of San Francisco. 452 people died and 28 000 buildings were destroyed. The source of the earthquake was a fault line in California called the San Andreas Fault. The people of California live under the threat of another earthquake. Geologists study the area carefully in order that predictions can be made about further earth movements.

1 Find out all you can about the San Andreas Fault. Draw a diagram to show what sort of fault it is and where it is located.
2 Try to explain how the San Andreas Fault causes earthquakes.
3 Try to find out and describe the attempts that are being made to reduce the sudden earth movements around the San Andreas Fault.

The surface and beneath

WHAT YOU SHOULD KNOW

1 Rocks are a mixture of chemical compounds called minerals.
2 Sedimentary rocks are formed when rock materials removed by weathering and erosion are deposited elsewhere. They often contain plant or animal fossils.
3 Metamorphic rocks are formed when rocks are changed by heat or pressure.
4 Igneous rocks are formed when a magma solidifies.
5 Plutonic igneous rocks solidify slowly underground and have large crystals.
6 Volcanic igneous rocks solidify at or near the surface and have small crystals.
7 Rocks are broken down by weathering.

8 There are three types of weathering - chemical, physical and biological.
9 The breaking down and removal of rock is called erosion.
10 When sediments are deposited, fast flowing water deposits large grains, slow flowing water deposits small grains.
11 Sedimentary rocks are deposited in horizontal layers, but folding may cause the layers to buckle and tilt.
12 Fossils give important information about the age of rocks and the way they are deposited.
13 The crust of the earth is made up of a number of plates that are moving.
14 Major earthquake and volcanic activity is found along plate junctions.
15 The continents were once joined. They are now drifting apart at a few centimetres per year.

QUESTIONS

1 (a) If dilute hydrochloric acid was added to limestone, what would happen?
  (b) Which other rocks would give similar results?

2 This diagram shows two igneous rocks.

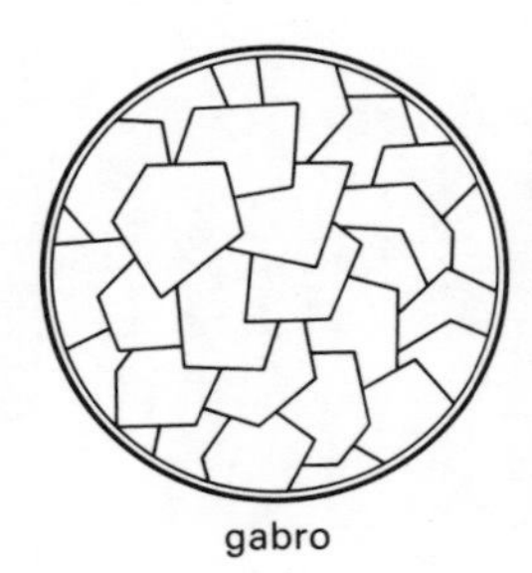

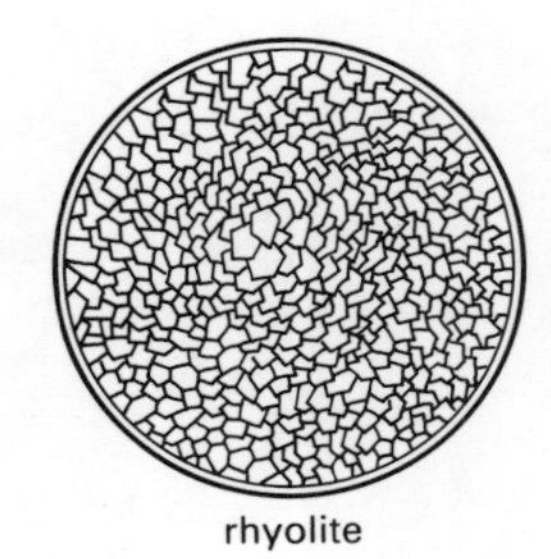

  (a) What is the name given to the molten rock from which these rocks were formed?
  (b) Which rock solidified more quickly? Explain your answer.
  (c) What type of igneous rock is gabbro?
  (d) What type of igneous rock is rhyolite?

3 The diagram shows three strata of sedimentary rock.

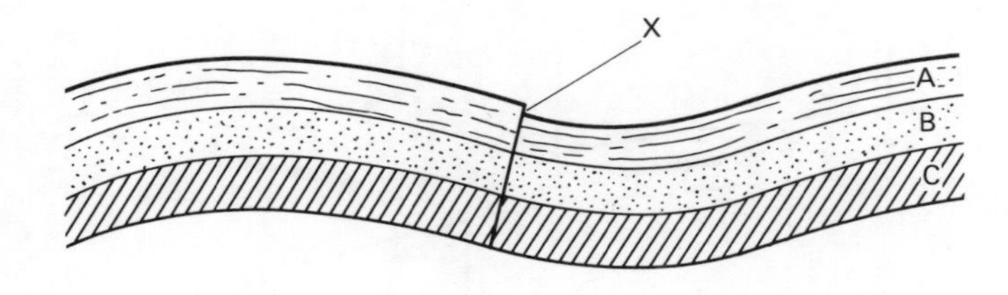

  (a) Which rock is the oldest?
  (b) Which rock is the youngest?
  (c) What is the name given to the feature at X?
  (d) What other change has taken place since the rocks were deposited?

4 The diagram shows a section through the rocks in County Antrim, Northern Ireland.

  (a) Which rock is the oldest?
  (b) How was chalk formed?
  (c) What was the basalt formed from?
  (d) Basalt has very small crystals. What does that tell you about the way that it formed?

5 The diagram shows a section through the rocks on Dartmoor.

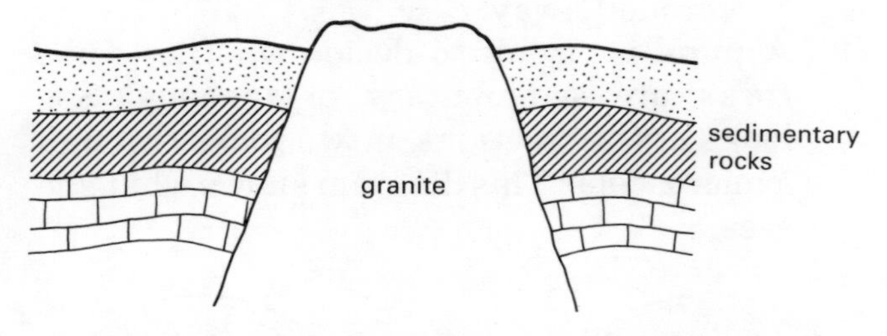

  (a) What type of rock is granite?
  (b) How was the granite formed?
  (c) Why are the layers of the sedimentary rock not straight?
  (d) Copy the diagram and mark on it a place where metamorphic rock is likely to be found.
  (e) The structure shown in the diagram is called an igneous intrusion. Draw diagrams to explain how this structure was formed.

6 The diagram shows two plates of the earth's crust moving together.

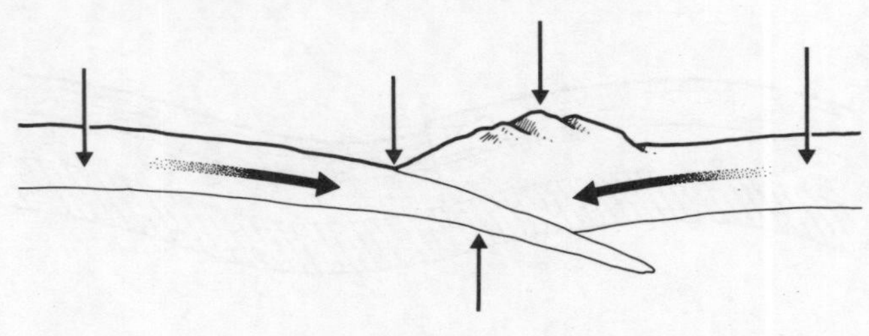

(a) Copy the diagram and complete it by adding the labels: continental crust, oceanic crust, mantle, ocean trench, fold mountain.

7

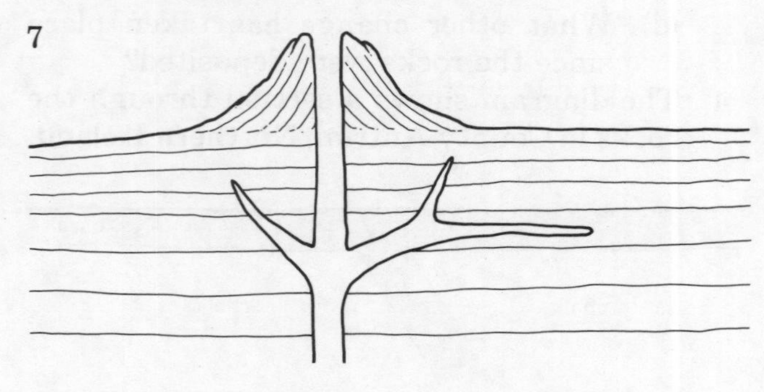

This diagram shows a volcano that is no longer active. Copy the diagram into your book and mark the following:

(a) the place where you are likely to find volcanic ash;

(b) the place where basalt is likely to be found;

(c) the material that is most likely to get eroded away.

8 A pupil is trying to decide whether some rocks are sedimentary or igneous. She looks at thin sections of two rocks through a microscope. This diagram shows what she sees.

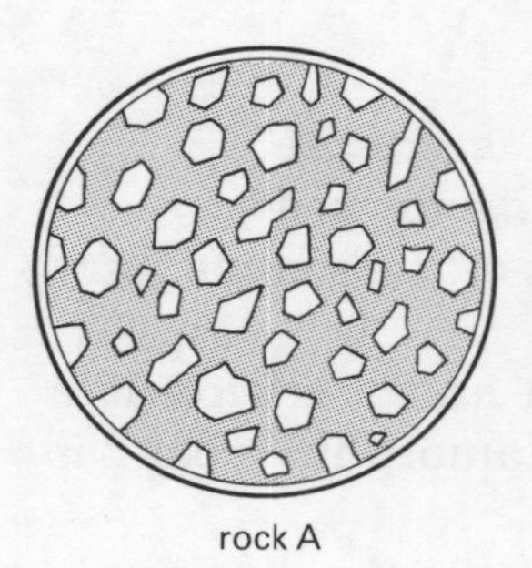

rock A

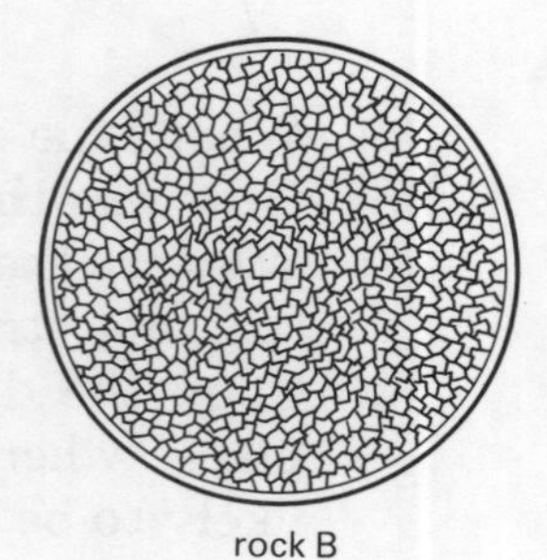

rock B

(a) Which rock would you say was igneous?

(b) Which rock would you say was sedimentary?

(c) Explain your answers.

9 What evidence do geologists have for believing that the continents are moving apart?

10 The diagram shows sets of footprints that have become fossilised in sedimentary rocks. From the evidence given, describe as carefully as you can the sequence of events that you think probably happened millions of years ago.

There are a number of possible answers to this problem. You could have a lively lesson discussing them.

THE EARTH'S RESOURCES

A resource is something that humans find useful in providing food, shelter, clothing and other items needed for life. Water, soil and metal deposits are all examples of resources. All of them are limited. Some resources, such as water, are RENEWABLE. This means that they can be recycled like water is through the water cycle. Others, like coal ,are NON-RENEWABLE. This means that once they have been used, they are gone and cannot be replaced. It is true that coal is being formed in swamp forests at the moment, but it is forming very slowly compared to the rate at which it is being used up.

The use of earth resources during the twentieth century has increased rapidly. Why do you think this is so? Some resources, particularly the energy sources such as oil and uranium, are nearing exhaustion. As the resources dwindle away they become more difficult, more expensive and require more energy to extract.

SOIL

Only about one-third of the earth is dry land. Much of this covered with a thin layer of soil. Human existence and the existence of many other organisms depends on this soil layer. Can you think why?

What is soil?

Figure 1.69 shows what happens if soil is shaken up with water in a gas jar and then allowed to settle. As the soil settles out, particles of different sizes form layers. The heaviest and largest particles sink to the bottom. The very small particles take a long time to settle and some remain suspended in the water. It is useful to classify particles according to size. The table on the next page shows how this can be done.

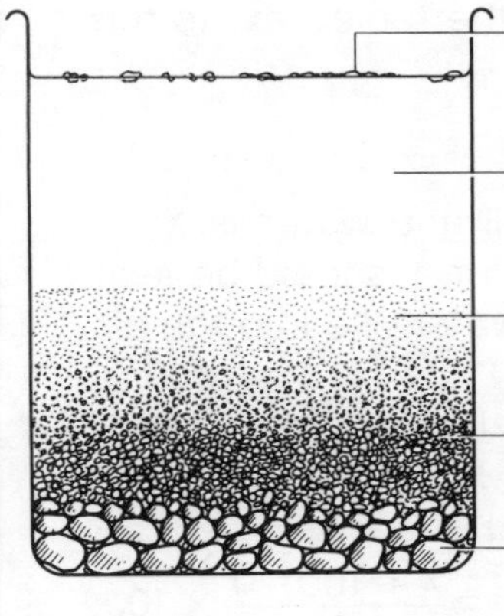

Fig. 1.69 Soil shaken with water and allowed to settle

Particle	Diameter/mm
stones	above 2
coarse sand	2 - 0.2
fine sand	0.2 - 0.02
silt	0.02 - 0.002
clay	below 0.002

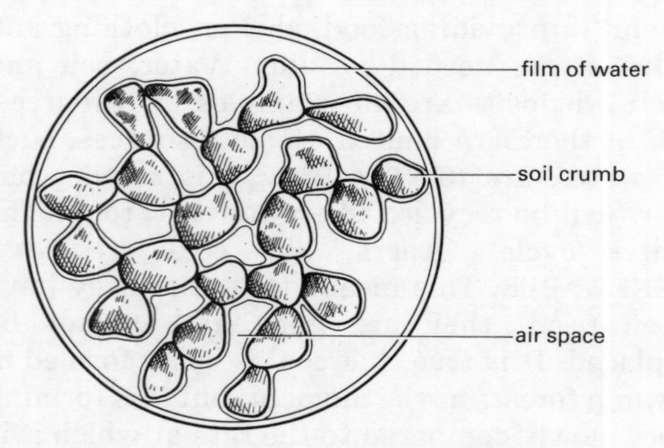

Fig. 1.70 Soil viewed through a microscope

These soil particles can be seen through the microscope (Fig. 1.70) They are usually joined together to form crumbs.

The microscope shows up two other very important things in the soil. *Air* is trapped in the spaces between particles and between crumbs. *Water* forms a film around the crumbs.

If you look again at Fig. 1.71 another important ingredient of soil can be seen. Floating on the water are decaying plants, called HUMUS. If you could test the water that you shook the soil in very carefully you would find that certain chemicals had dissolved in it. We can therefore identify another ingredient in soil - soluble minerals. These minerals can produce soils that are acid and alkaline.

We can see then that soil is made up of particles of different size, water, air, humus and soluble minerals. A good balance of these ingredients is essential to healthy plant growth. It is important to be able to measure the amount of each in the soil.

Investigation 1.9
Measuring the amount of air, water and humus in soil

A. Air

In this investigation you will measure the percentage of a soil volume that is air, by mixing soil and water together. The water fills the air spaces in the soil and some volume seems to disappear. The lost volume is equal to the volume of air in the soil.

Collect

small tin
100 cm³ measuring cylinder
250 cm³ plastic beaker, graduated
soil bucket to get rid of waste

WARNING - do not wash soil down the sink

What to do

1 Copy out the table below.

Original volume of soil	X	cm³
Volume of water		100 cm³
Total volume of soil and water after mixing	Y	cm³

2 Measure the volume of the small tin by filling it with water and tipping the water into a measuring cylinder. Write your result in the table next to the letter X.

3 Push the tin carefully into the soil to obtain a tin full. Do not squash the soil down too hard.

4 Tip soil into a 250 cm³ plastic beaker that has volume marks on it.

5 Add 100 cm³ of water. Mix thoroughly. Measure the volume of the mixture. Write the result in the table next to the letter Y.

Calculating the result

The total volume of soil and water is X + 100 cm³. Your answer for Y should be less than this because the water filled up the air spaces.

Volume of air in soil = lost volume
$= X + 100 - Y.$

Percentage of air in soil =

$$\frac{\text{volume of air}}{\text{total volume of soil}} \times 100 = \frac{X + 100 - Y}{X} \times 100.$$

B. Water

In this investigation you will find out the percentage of a soil mass that is water. You will do this by measuring the change in mass of soil when it is dried out.

Collect

metal tray
soil
access to top-pan balance

What to do

1 Copy the table below.

Mass of tray	P	g
Mass of (tray + soil)	Q	g
Mass of (tray + dried soil)	R	g

2 Weigh the metal tray. Write your result in the table next to the letter P.
3 Place enough soil in the tray to cover the bottom. Re-weigh. Write your result in the table next to the letter Q.
4 Carefully put the tray in a warm dry place for at least 24 hours.
5 Re-weigh the tray of dried soil. Write your result in the table next to the letter R.
6 Keep the tray of dried soil for the next experiment.

Calculating the result

The mass of the soil
= mass of (tray + soil)
− mass of tray
$= Q - P$.

Loss of mass of soil
= mass of (tray + soil)
− mass of (tray + dried soil)
$= Q - R$.

Percentage of water in soil

$$= \frac{\text{loss of mass of soil}}{\text{original mass of soil}} \times 100$$

$$= \frac{Q - R}{Q - P} \times 100$$

C. Humus

In this investigation you will find out the percentage of soil mass that is humus. You will do this by burning away the humus and measuring the change in mass.

Collect

tray of dried soil from previous experiment
Bunsen burner
tripod
heatproof mat
safety spectacles

What to do

1 Copy out the table below.

Mass of (tray + dried soil)	R	g
Mass of (tray + burned soil)	S	g

2 Heat the soil strongly with the Bunsen burner. Start by heating from below. Then carefully pick up the Bunsen burner by the base and heat from above. The soil must glow red.
3 Allow the tray of soil to cool and re-weigh it. Write the result down in the table. This is called S.

Calculating the result

Loss of mass on burning
= mass of (tray + dried soil)
− mass of (tray + burned soil)
$= R - S$.

Percentage of humus in soil =

$$\frac{\text{loss of mass on burning}}{\text{original mass of soil}} = \frac{R - S}{Q - R} \times 100.$$

Questions

1 A pupil carries out part (B) of this investigation but is rather careless and spills some soil after weighing the tray and soil. How would this affect the result? Would it make it too high or too low?
2 What have you assumed in part (C) of the investigation?
3 A pupil carries out part (C) and gets a result that is much lower than the results obtained by everybody else in the group. How could that be explained?

What is soil made from?

Now that we know what soil contains we can explain how it is formed. One more piece of evidence is needed. Figure 1.71 shows a SOIL PROFILE. This is a cross-section cut down through the soil. You can make one yourself by digging a trench.

Fig. 1.71 A soil profile - notice the change in size of rock fragments

The bottom of the spade marks the level of the underlying rock — called bedrock. Look carefully at the photograph. You can see that above the bedrock there are rock fragments of different sizes. The higher up above the bedrock, the smaller these fragments become.

The rock has been weathered physically, biologically and chemically. Plants have then grown in the young soil and added organic matter. Plants encourage the growth of animals. These animals break up the soil and add air to it. As plants and animals die they form humus. This sort of soil that forms above bedrock is called RESIDUAL SOIL. A different type of soil is found when the weathered rock is eroded away and taken somewhere else. This is called TRANSPORTED SOIL. The soil in meadows or river banks is of this type.

Assignment – Soil formation

This diagram shows how soil is formed.

rain
soil A
erosion
bedrock
cracks and hollows in bedrock
transportation

1. Look carefully at the soil A above the bedrock. This has been formed by the weathering of the bedrock. Describe how this weathering has taken place.
2. The plants that first colonised the young soil would have been very different from those shown in the diagram. Say why this is so and try to describe one feature of the early plants.
3. How would the plants shown in the diagram increase the break-up of the bedrock?
4. In what other way do these plants contribute to the soil?
5. The cracks and hollows shown in the bedrock indicate that a particular type of erosion is taking place. What causes this?
6. The bedrock that is eroded forms another type of soil that is not shown on the diagram. Use the diagram to write a few sentences explaining how this soil is formed.

Soil types and plant growth

There are, broadly speaking, three different soil types in Britain with very different types of vegetation growing on them.

Loam soils

These have about 50% sand, 30% clay and 20% humus. They are the most fertile soils and well suited to agricultural crops. They have good mineral content, are well drained, yet retain enough water and dissolved minerals for healthy plant growth.

Sandy soils

These contain over 70% sand with less than 20% clay. They are well drained and are easy to break down into small lumps for easy seed planting. The problem with sandy soil is that it retains little water, so plants may dry out during drought. Also, as water moves so quickly down through the soil it dissolves and carries away important minerals. This is called LEACHING and reduces the fertility of the soil.

Clay soils

These soils contain roughly equal percentages of clay, silt and sand. They retain water very well and have plenty of minerals. Clay soil is, however, very difficult to cultivate. It tends to get waterlogged and gets very sticky and heavy. Lack of air spaces means that it gets very cold in winter, and in summer it dries and forms rock-hard lumps.

Improving soils

Soils are ploughed by farmers in autumn so that the winter frost can break down the lumps. Sandy and clay soils can be improved by adding manure. Manure however is bulky, expensive and difficult to transport and apply.

In recent years farmers have used increasing amounts of factory produced or inorganic fertilisers. These dissolve quickly and the minerals they contain are readily available to plants. Fertilisers are also used to change the pH of the soil. They must be used carefully however. If over-used they can lead to the breakdown of soil crumbs, leaving the soil dry and dusty.

Soil erosion

It was failure to pay careful attention to soil texture that led to the great dust bowl disaster of America. When the North American prairie was first cultivated the soil was rich, dark and full of humus. After many years cropping, minerals were removed and humus was not added and the soil became thin and dry. In the 1930s, drought hit the mid-West. Fierce winds whipped up the dry unprotected soil in great dark clouds of dust. Houses were buried, people and cattle suffocated. The whole area became huge dust bowls. The precious soil that had taken thousands of years to form was blown away while the farmers looked on helplessly.

John Steinbeck described the events in the famous novel *The Grapes of Wrath:*

> 'Now the wind grew strong and hard and it worked at the rain crust in the cornfields. Little by little the sky was darkened by the missing dust and the wind fell over the earth loosening the dust and carried it away. The wind grew stronger. The rain crust broke and dust lifted up out of the fields and drove grey plumes into the air like sluggish smoke. The corn thrashed the wind and made a dry rushing sound. The finest dust did not settle back to earth now, but disappeared into the darkening sky . . .'

Fig. 1.72 A dust storm - the result of soil erosion

1
The soil has developed over thousands of years in the forest. Then the farmer cuts the trees down.

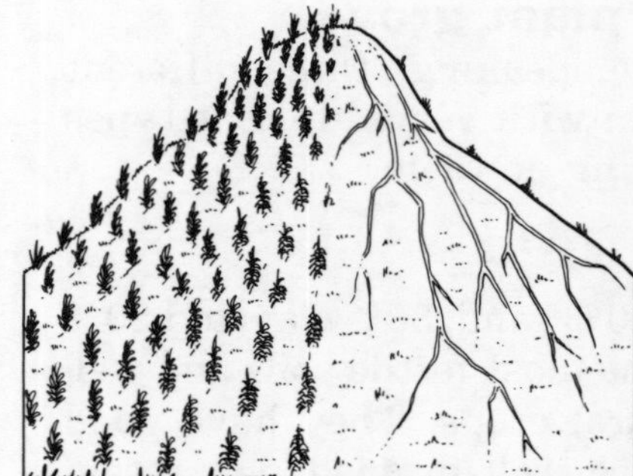

2
The farmer plants crops on the hillside. Ploughing breaks the soil down and nutrients leave the soil each year in the harvest. The soil becomes less porous and cannot soak up so much rainwater. Some rainwater therefore runs down the slope taking some of the topsoil with it.

3
The soil is not now good enough for crops so is given over to pasture. If the soil is grazed heavily bare patches appear. Rain runs down the animal tracks taking soil with it. When the pasture is no longer rich enough for cattle it is used by sheep or goats.

4
The sheep have stripped the remaining pasture. Rainwater has carved the animal tracks into gullies which become deeper as rain washes more soil down them. In the end all the top soil is washed away, leaving the rock bare. The once fertile hillside has turned into barren desert.

Fig. 1.73 Poor farming methods lead to soil erosion

This was a tragic example of soil erosion caused by the wind. Poor farming methods can also lead to soil erosion by water. As Fig. 1.73 shows, a fertile hillside can be turned into a barren desert. At the moment forests in the Himalayas are being chopped down at a tremendous rate. Soil erosion of this sort is a real danger.

Assignment - Designing an experiment

This question is a very important one. It is all about making sure that scientific tests are fair. The problem is this: farmers think that cabbages grow better in soils that are alkaline, that is, soils with a pH of around 8.5. It is very important to test if this is true. At the research station scientists divide up a patch of land into two sections. They now add lime to one half to give a pH of 8.5. They test the pH of the other half and find it is 7.

They now decide to plant cabbages in each half to see how well they grow.

Describe carefully what you would now do if *you* were in charge of the experiment. Point out any precautions that you would take to ensure that the test was fair.

Table 1.3 UK mineral production and UK imports of some mineral-based products, 1974

Mineral product	UK imports/ thousand tonnes	UK primary production/ thousand tonnes
Crude fertilizers	2078.7	
Stone, sand, gravel	590	162017
Sulphur minerals	1414.2	
Abrasives	119.5	
Asphalt	35.2	
Clay and other refractories	173.8	2277
Graphite	16.0	
Salt	23.7	3148
Asbestos	153.0	
Feldspar/fluorspar	197.1	5569 (fluorspar)
Iron ore	20291.6	3602
Iron/steel scrap	139.5	
Non-ferrous ores	1678.8	24.1
Non-ferrous scrap	138.1	
Uranium and thorium concentrates	0.8	
Coal, coke, etc.	3845.6	110200
Petroleum	113551.3	88
Petroleum products	15716.8	
Natural gas	724.5	159
Chemicals	2760.4	
Manufactured fertilizers	1560.4	
Lime, cement, building materials	555.5	
Limestone		100915
Chalk		20415
Igneous rock		41717
Gypsum		3567
Potash		250
China clay and potters clay		4298
Other minerals, celestite, barytes		327

MINERAL RESOURCES

Table 1.3 shows the main mineral resources taken from the earth in the UK, and also those imported in 1974. You can see that although large amounts of sand, stone, limestone and coal are produced, like many other countries Britain relies heavily on imports.

Minerals cannot simply be dug up and used immediately. Several processes have to be carried out before they can be used.

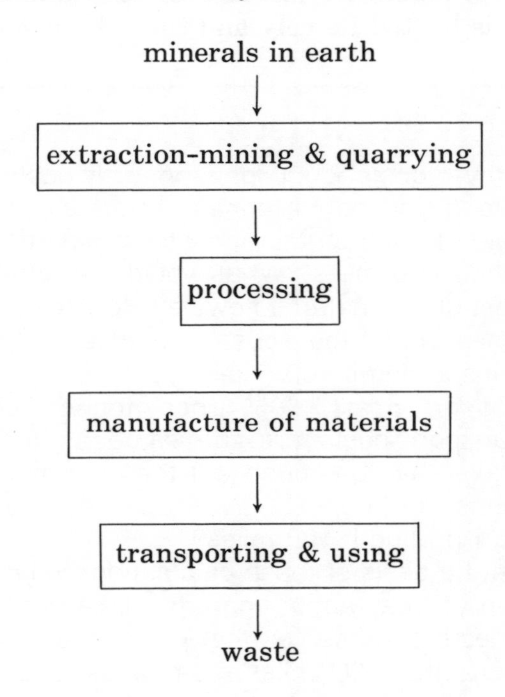

USING THE ROCKS

As soon as humans wanted to choose where to live instead of living where caves happened to be, they had to build a shelter or home. If these were to last a lifetime then stone was used. There is plenty of stone around, it is strong, it takes a long time to wear away, it does not rot and with care it could be split and shaped into convenient pieces.

Some of the most attractive buildings in the country are made from the natural stone of the area (Fig. 1.74)

If you look around the area that you live in you will probably find some buildings made of natural stone. If you visit a local building site, however, you would be most unlikely to see any

Fig. 1.74 Buildings made from natural stone: (a) (top) Cotswold limestone, (b) (centre) sandstone, (c) (bottom) granite

sandstone, slate or Cotswold stone. Most buildings these days are made of bricks or concrete. Why do you think this is? There is probably little natural stone in the area, crafts people are needed to shape it and it could be very expensive to make houses or flats out of it. You could ask your local builder for his opinion!

Bricks

These are made from an earth product, but it is processed to get it into its most useful form. The main constituent is *clay*. Nowadays bricks are made by machine. The soft wet clay is squeezed out in a rectangular column - rather like toothpaste coming out of a tube with a rectangular nozzle. This is then cut into correctly shaped bricks by wire stretched on a frame. These wet clay bricks are dried and then fired in a kiln. The size that is chosen is that which a builder can comfortably hold in one hand - at present, $8\frac{3}{4}$ in $\times$ $4\frac{3}{16}$ in $\times$ $2\frac{5}{8}$ in.

Cement

This is the builder's glue. It is also made from earth products. A mixture of clay and limestone is heated fiercely and then the resulting

Assignment - Aggregate

Concrete is the main building material of the modern age. It is made from cement and aggregate. This aggregate is a granular material that provides bulk and strength. The concrete glues it all together. Most aggregates used in Britain are obtained from sand, gravel, limestone, sandstone and igneous or metamorphic rocks. These rocks are all very plentiful and very easy to extract. You might think that aggregate should be very cheap. Compared to other materials it is. The price will depend on

(a) extraction costs in mining and quarrying;
(b) processing costs - crushing, grinding, grading and sizing;
(c) transport costs;
(d) the profits of the companies concerned.

Look at how the cost of limestone changes:

Value before mining	*Cost after crushing*
3p/tonne	£2.50/tonne

Cost at works of concrete blocks
£19/tonne

Transport costs form a very large amount of the final cost of the materials. Try to find out the cost of transporting gravel. You should find that it is about £1 to £3 per tonne for every 10 miles. So we can take a value of about 20p per tonne per mile.

1 Gravel costs £1.50 per tonne. It costs 20p to transport 1 tonne a distance of 1 mile. How much will it cost to transport 1 tonne 100 miles? What would the total cost of the material now be? How many times has the cost increased by transporting it 100 miles?
2 Copper costs £850 per tonne. If transport costs are the same as for gravel, by how much will the cost of 1 tonne of copper be increased by transporting it 100 miles?
3 On the basis of these calculations what can you say about aggregate quarries? Does it surprise you that you are rarely more than 20 miles from a sand or gravel quarry in the UK?
4 The annual demand for aggregates is about 200 million tonnes. What do you think most of this is used for?
5 There is plenty of rock that can be used for aggregate left in the earth. Do you think that any problems arise from this demand?
6 Two suggestions have been made for meeting mineral demands in the future.
   (a) Before any site is developed for building, the resources below it should be extracted.
   (b) Mining waste from spoil heaps and slag heaps should be used for aggregate.

   Say what the main advantages and disadvantages of these schemes are.

Fig. 1.75 Bricklaying - using cement to glue the bricks together

solid is finely ground into cement. Cement is mixed with sand and water to produce mortar. This is used for glueing bricks together. Figure 1.75 shows a builder spreading mortar on bricks.

Concrete

Cement is also used to produce the most important modern building material - concrete. To make the strong concrete used in building, cement is mixed with gravel or broken stones and pebbles (aggregate) together with sand and water.

METALS

Only a few metals such as gold and silver are found naturally occurring as elements. The others are found as compounds (mostly combined with oxygen in the form of OXIDES). These metallic minerals are called METAL ORES. Getting metals from ores involves two main problems. Firstly the ore must be located and mined. It must then be processed. Processing the ore to get the metal from the oxide is expensive. It requires expensive equipment, raw materials and also energy to make the chemical reaction work.

Both mining and processing are made more difficult by the fact that the precious metals are often mixed with a good deal of unwanted material. The GRADE of an ore tells us the percentage of valuable mineral in the deposit. The table below gives examples of grades.

	Grade %
Open pit copper mine, USA	0.63
Zambian copper mine	2.5
European lead mine	4.0
South African gold mine	0.0006
Cornish tin mine	0.8
Iron ore deposits	30–65
Aluminium (bauxite) deposit	20.0

Consider the copper mine in the USA. For each 0.63 tons of copper produced, 100 tons of ore have to be dug from the ground. It is not just the expense that causes concern. The impact on the environment can be very serious. The landscape can be scarred with open pit mines and huge heaps of unwanted material called SPOIL.

Sites like this look unattractive but that is not the only problem. Rain can wash toxic materials from the spoil heaps into local rivers and water supplies, which can prove a serious problem for wild life and humans. The problem has been particularly serious with uranium mines in Canada.

Fig. 1.76 An open pit mine - notice the huge spoil heaps

Processing ores

With the scientific knowledge that we have now we can design refineries that can extract metals from ores very efficiently. Thousands of years ago humans managed to extract metals without any idea of the chemical reactions involved. The metals would have been extracted first by accident. Probably some rocks were thrown on a fire and the following day the new metal was seen. From this, early humans worked out that two things were needed - heat and charcoal. They then designed bellows to make hotter fires and the metal ages began - first bronze, because tin and copper are easy to extract (for reasons that you will find out later), then iron. Perhaps you can think why we have only recently moved into the new metal age of aluminium.

We now know that when we heat a metal oxide with charcoal, the carbon in the charcoal takes the oxygen to form carbon dioxide.

lead oxide + carbon → lead + carbon dioxide

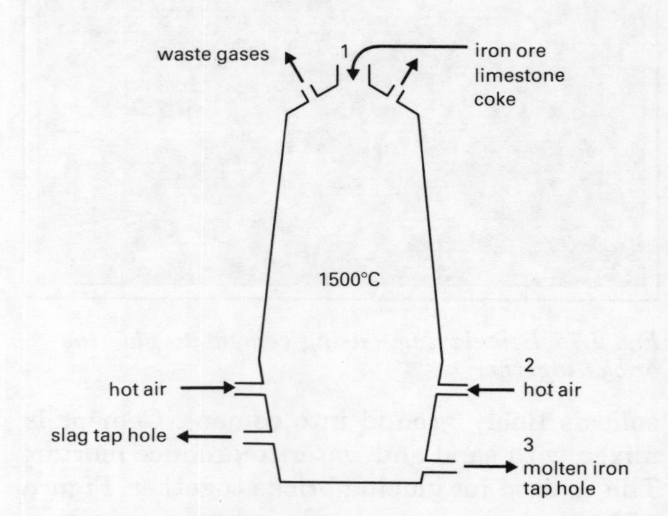

Fig. 1.77 The blast furnace

Investigation 1.10 Lead from lead ore

In this investigation you will extract lead from lead ore by mixing it with carbon and heating strongly.

Collect

- Bunsen burner
- heatproof mat
- tripod
- crucible
- pipeclay triangle
- spatula
- lead oxide
- carbon powder
- charcoal lumps
- tongs
- safety spectacles

WARNING – Make sure your laboratory is well ventilated

What to do

1. Put 3 spatulas of lead oxide onto some scrap paper with 1 spatula of carbon. Mix well.
2. Tip the mixture into a crucible and cover with charcoal lumps. Put the crucible on the pipeclay triangle. Heat strongly. Leave to cool.

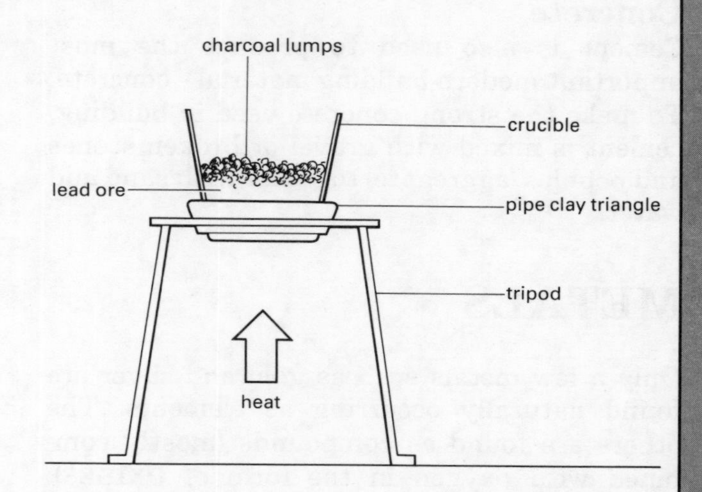

3. When the crucible is cool, carefully tip away the charcoal lumps. Look at the bottom of the crucible. Heat the crucible and carefully pour the lead onto a heatproof mat.

Questions

1. Write down and complete the equation lead oxide + carbon ⟶ . . . + ...
2. The charcoal lumps were used to prevent air coming into contact with the metal. Why was this done? (Think about your answer to question 1).

Iron from iron ore

It is more difficult to get iron from iron ore. The principle is the same: iron oxide + carbon → iron + carbon dioxide. However, very high temperatures are needed. This is done on an industrial scale using a blast furnace (Fig. 1.77).

Limestone, iron ore and coke are tipped into the top of the blast furnace (1). The coke takes the oxygen from the iron oxide, leaving iron behind. The limestone combines with the sandy impurities in the iron ore to form SLAG. Hot air is blasted in through the hole (2). The molten iron is tapped off from (3) after the slag has been removed from (4).

The iron that is taken from the furnace is cast into shapes called 'pigs' and so is called pig iron. This is a very brittle material and would shatter if dropped on a hard floor. Most of the iron produced is converted into steel, which is much tougher and a much more useful building material.

Other metals

Copper can also be extracted by heating its oxide with carbon. Some metals however cannot be extracted by this method. Aluminium is extracted by passing electricity through the molten ore. This requires a great deal of energy. Aluminium refineries are always found near cheap sources of electricity. The reason why certain metals are much harder to extract will be looked at in Unit 9.

Assignment – The iron industry

This sketch shows part of the Clydach Gorge in South Wales. It was a very important iron producing area in the middle of the 18th century but it ceased operation before 1900.

1. What local minerals would have led to the iron industry developing?
2. Why should the blast furnace at the Clydach ironworks have been situated right down in the valley bottom, by the river?
3. What was the purpose of the inclines and tramroads still clearly seen in the gorge today?
4. The limestone quarries continued to produce limestone after iron production ceased. What would this limestone have been used for?
5. The lime kilns stopped 'burning' limestone in the early 1930s but they were used again during the Second World War. Why was the limestone burned and why was there a high demand during the Second World War?
6. Iron was produced in the Clydach Gorge before coked coal was used. What would have been used in its place? (You can see the answer in the diagram).

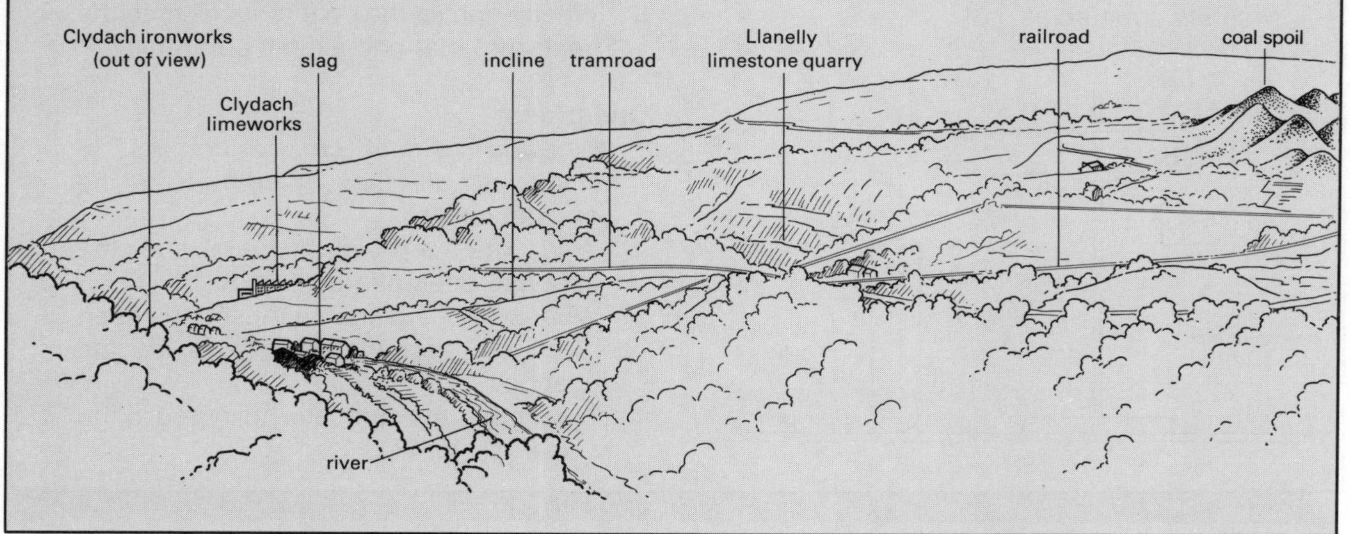

THE FOSSIL FUELS

Coal, oil and gas are earth products that are essential to modern industrial society. These fossil fuels provide us with energy and important chemicals.

Coal

Millions of years ago, during the Carboniferous age, much of the country was covered with forests of giant trees and tree-like ferns. Figure 1.78 shows a section through the earth beneath an ancient forest. As the vegetation died it decayed. Because the ground was very wet there was a limited amount of oxygen available for decomposition and so a layer of peat was formed.

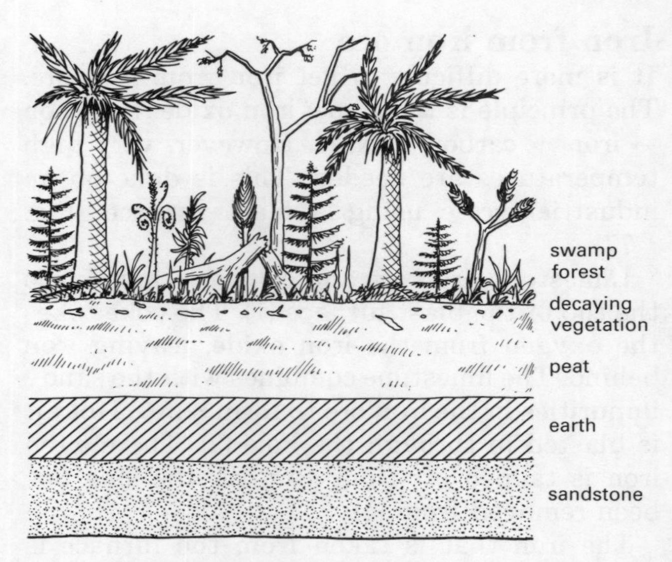

Fig. 1.78 Section beneath a prehistoric swamp forest

Investigation 1.11 Coal distillation

In this investigation you will heat coal in only a small amount of air.

Collect

clamp stand
heatproof mat
Bunsen burner
splint
pH paper
hard-glass test tube
test tube with side arm
delivery tube with 2 bungs
small pieces of coal
distilled water
safety spectacles

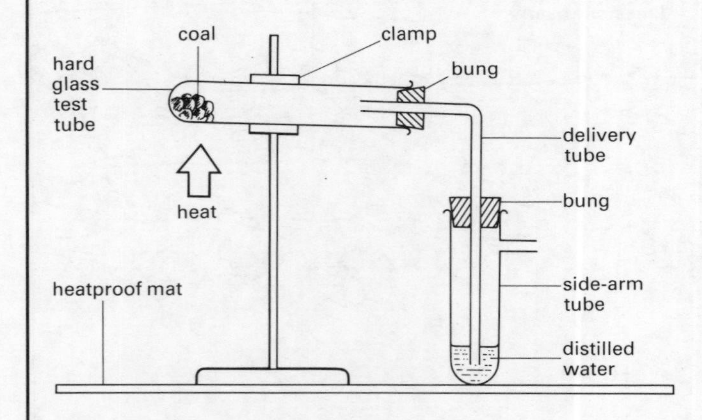

What to do

1 Set up the apparatus shown in the diagram.
2 Heat the coal gently at first, then more strongly.
3 Use the splint to light the gas coming out of the side arm.
4 When no more gas is produced, loosen one of the bungs, then stop heating. This prevents the water from being sucked back into the hot tube.
5 Look carefully at the water in the side-arm tube. Test the pH of the water using the pH paper.
6 When cool, tip the coal remains onto the heatproof mat and examine carefully.

Questions

1 What do the coal remains look like? In what way are they different from the original coal?
2 What was the pH of the water after the coal had been heated?
3 What sort of substance must have been given off from the coal to cause the change?
4 What other substance is collected in the side-arm tube?

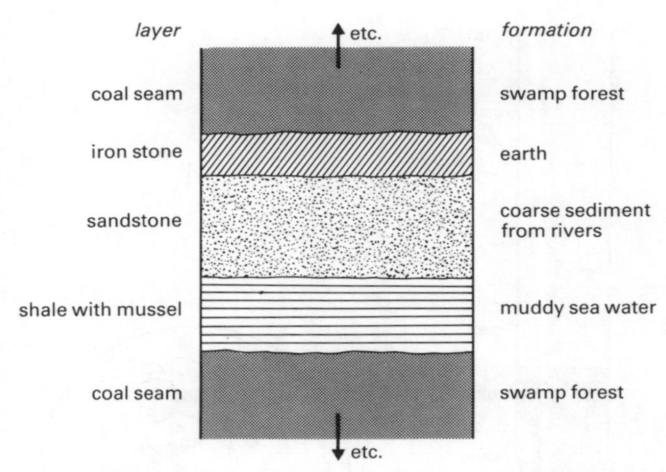

Fig. 1.79 Sequence of layers associated with coal seams

The land then subsided and the forest was covered by muddy sea water. This settled out to form shale with mussel bands running through. Rivers then have filled the area with coarse grained sediment which formed sandstone. This was then weathered and eroded to form soil on which a new swamp forest developed, and the sequence started again. The weight of sediment turned the peat into coal.

Figure 1.79 shows this sequence. The pattern, which is called a cyclothem, is repeated again and again. Coal is found in seams mostly less than 2 metres thick. Bands of ironstone are also found associated with coal in many areas; these were the main source of iron ore in Britain in the 18th century.

Coal can be burned as a fuel. It can also be used to produce important chemicals as Investigation 1.11 shows.

Coal products

Investigation 1.11 shows us the important products that can be obtained from the destructive distillation of coal.

Coke: This is the material left behind when coal is heated. A very important fuel and also used in iron extraction.

Gas: This is the substance that you ignited at the side arm. A fuel and also used in the production of paint and plastics.

Tar: This is the oily substance that collected in the side-arm tube. A fuel, and also used in the chemical industry.

Ammoniacal liquor: This is the substance that caused the distilled water to become alkaline. Used in fertilisers and household cleaners.

Oil

Like coal, oil has been formed from decayed organic matter. The difference is that oil has been formed from minute sea creatures which decayed over a long period in deep muddy seas. Further sediments were then deposited on top and the muds were compacted into shales. The oil was either trapped as shale oil or else squeezed out with any water into more porous rocks like sandstones and limestones. Porous rock is one that allows water to move through it. Oil is now found in porous rocks, often floating on top of water. Gas is often found above the oil. Figure 1.80 shows an example of an oil trap.

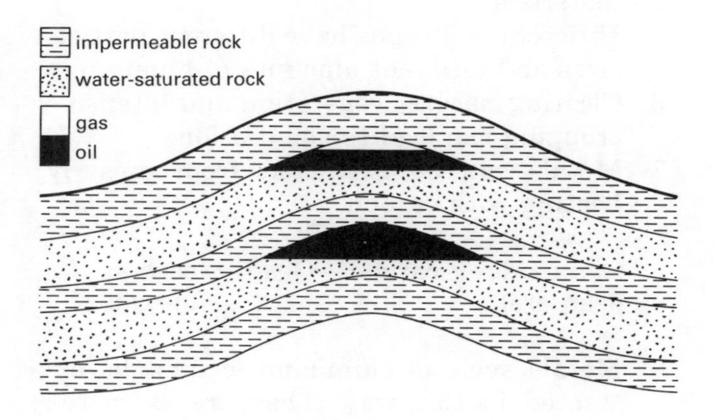

Fig. 1.80 An oil trap

The oil is trapped in the small gaps between the grains of the porous rock. It is stopped from moving down by the water in the porous rock. It cannot move sideways because of the layers of impermeable rock.

Assignment – Oil

Oil is a very precious resource and one which is rapidly running out. Use the reference books that you have available to find out more about oil. Write an account of how oil is located, how it is extracted and how it is transported. Use diagrams to help your explanation. Try to include any environmental problems that arise and describe how they can be avoided or overcome.

Refining of oil is covered in a later section of this book so you need not include that in your account.

WHAT YOU SHOULD KNOW

1 Some resources can be recycled; these are called renewable. Others cannot be recycled; these are called non-renewable.
2 Soil is formed from weathered rock. Rock can be weathered by chemical, physical or biological means.
3 Plants decay and provide humus for the soil.
4 Soil also contains water, air and dissolved materials.
5 Different soil types have different particle sizes and different amounts of humus.
6 Clearing natural vegetation and intensive cropping can lead to soil erosion.
7 Metals can be extracted from ores by heating with carbon.

lead oxide + carbon → lead + carbon dioxide
(lead ore)

8 Metals such as aluminium cannot be extracted in this way. They are extracted using electricity and a great deal of energy is required.
9 Mining can ruin the landscape with open pits and spoil heaps and toxic materials can be washed into water supplies.
10 Coal was formed millions of years ago from the vegetation of swamp forests.
11 Coal can be burned as a fuel or split up into important chemicals. The processing of coal is called destructive distillation.

QUESTIONS

1 A pupil shakes some soil with water and leaves it to settle. Two hours later the tube looks like this - the drawing is from her notebook.
(a) What type of soil sample did she have? Explain your answer.

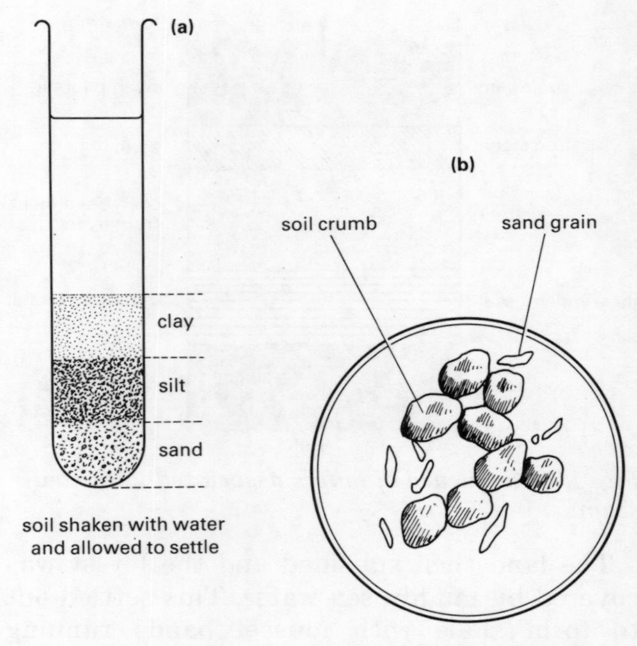

(b) On the diagram on the left she has missed out an important soil ingredient. Redraw the diagram showing this.
(c) Draw the microscopic view in your book and include on it a scale to show the size of the grains. Explain what a soil crumb is.

2 A pupil mixes 100 cm^3 of soil with 100 cm^3 of water and finds that the total volume is 150 cm^3.
(a) Explain carefully why some of the volume has disappeared.
(b) Use the information to calculate the percentage of air in the soil.

3 Copy and complete this passage:

Soil is formed from rock fragments. Rocks are broken down by ________, ________, and ________ means. This process is called ________. Soil also contains decayed plant and animal material called ________. Essential ________ are also found in soil. They may cause the soil to have a high pH and so be ________ or a low pH and so be ________.

4 These reactions show what happens when metal ores are heated with carbon. Copy and complete them.
(a) lead oxide + ______ ⟶ lead + carbon dioxide
(b) iron oxide + carbon → ____ + ____

5 Look carefully at Table 1.3 on page 62.
   (a) What does 'non-ferrous ore' mean? Give some examples.
   (b) The figures given for petroleum are very different from today's figures. Explain why this should be so.
   (c) Find out the uses of asphalt, fluorspar and gypsum.
   (d) Which minerals listed are most in danger of running out? Explain your answer.

6 The following graphs show limestone and clay production in the UK.

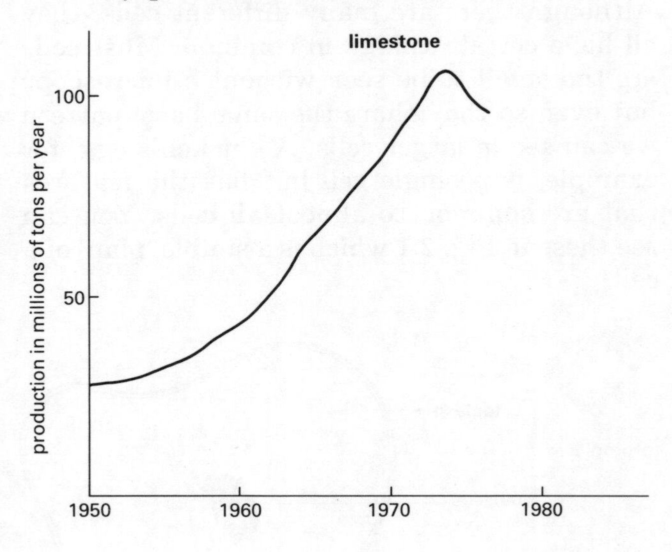

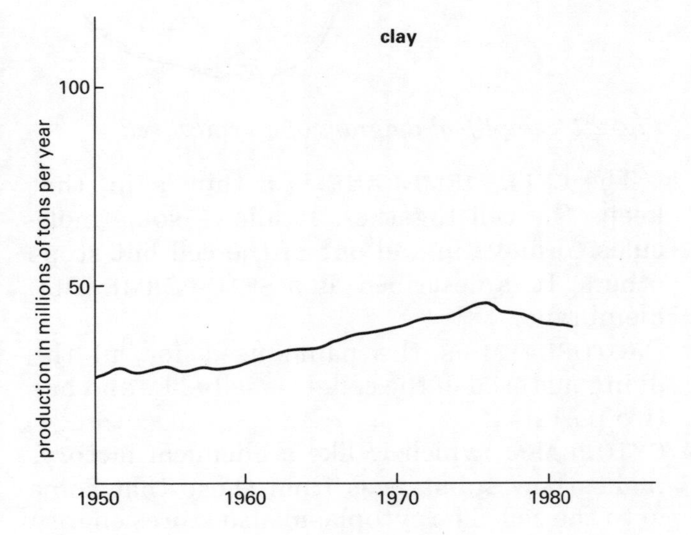

   (a) Describe as carefully as you can how the production of limestone has changed over the period shown in the graph.
   (b) What are the main uses of limestone?
   (c) Both limestone and clay are important materials and are used in building. Try to explain why limestone production has increased much more than clay.

7 With the aid of diagrams explain how iron is extracted from iron ore in the blast furnace.

8 Write a few sentences on each of the uses of limestone. You should be able to think of at least three different uses.

9 Explain why wherever you are in Britain you are rarely more than 20 miles from a gravel quarry.

10 In what ways is mining for metal ores harmful to the environment? What steps could be taken to reduce the problems?

11

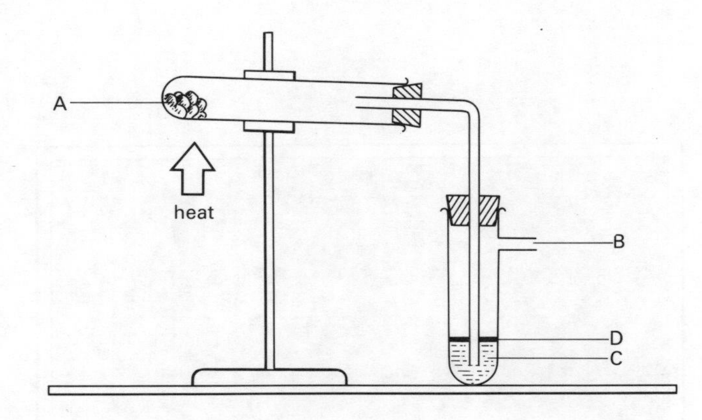

   (a) The diagram shows the apparatus used for destructive distillation of coal in the laboratory. Draw the diagram and name the substances collected at A, B, C and D during the experiment.
   (b) State two uses of each product.
   (c) Explain carefully why it is necessary to remove one of the bungs before you stop heating in this experiment.

2. Cells and Organisms

CELLS AND CELL STRUCTURE

All living organisms are made up of the building blocks we call CELLS. These cells can be arranged in many different ways to build up the large number of different animals and plants that exist. Plants and animals are usually made from millions of cells. The types of cell, and the way they are put together, decide what the organism will be like. Although there are many different cells, they all have certain things in common. Most cells are too small to be seen without a microscope but even so they share the same basic pattern we can see in larger cells. A chicken's egg, for example, is a single cell but has the features that are common to almost all cells. You can see these in Fig. 2.1 which is a simple 'plan' of a cell.

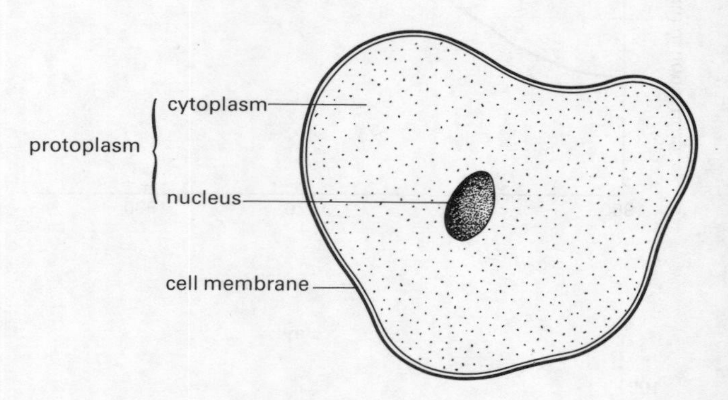

Fig. 2.1 Simplified diagram of a typical cell

The CELL MEMBRANE is a thin 'skin' that keeps the cell together. It allows some molecules to move in and out of the cell but stops others. It is described as a SEMI-PERMEABLE membrane.

PROTOPLASM is the name used for all the living material in the cell. It is jelly-like and has two parts:

CYTOPLASM, which is like a chemical factory, makes new substances from those that come in to the cell. The cytoplasm also stores energy and releases it;

NUCLEUS, which acts as the 'brain' of the cell and controls everything that happens in it. The nucleus also holds 'building instructions' stored on *chromosomes* (see Unit 8).

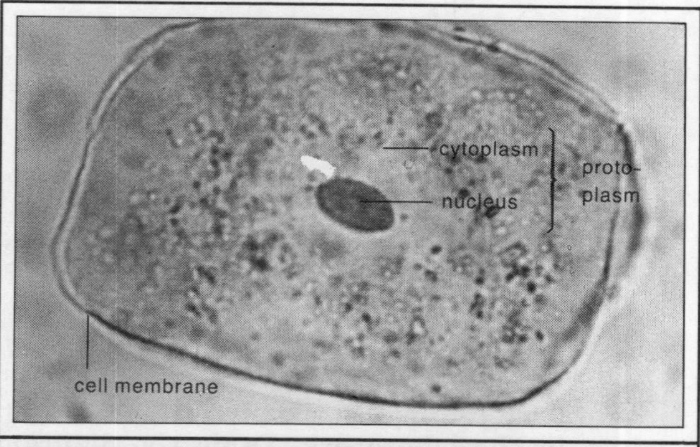

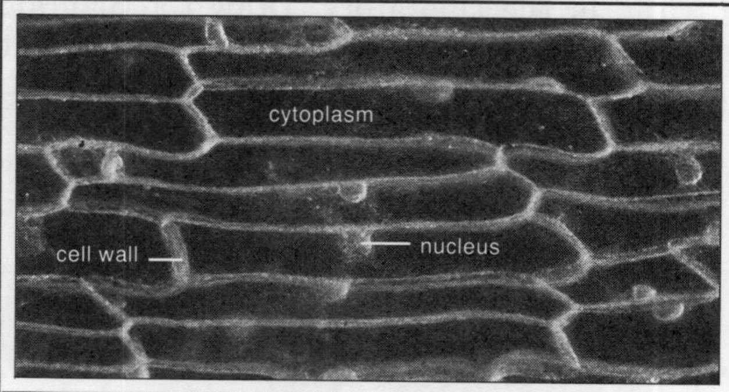

Fig. 2.2 Cells from (a) (above) human cheek (b) (below) onion

VACUOLES are tiny drops of liquid that are often in the cytoplasm of a cell.

You can look at some cells for yourself. Your teacher may show you how to scrape some cells from the inside of your cheek - with a clean wooden spatula. If you place the cells on a microscope slide and add a few drops of a stain like methylene blue, to make them easier to see, then look at them through a microscope, you should see something like Fig. 2.2(a).

Investigation 2.1 Making a slide of cells from an onion

The very thin membrane between two layers of an onion can make a very clear slide and will let you see a group of typical plant cells.

Collect

- part of a raw onion
- scalpel or knife
- forceps or tweezers
- microscope slide and cover slip
- iodine solution as a stain

What to do

1 Carefully cut a small 'window' in the onion.

WARNING – Scalpels are sharp – remember to cut away from you.

Lift out the square of onion from the first layer.

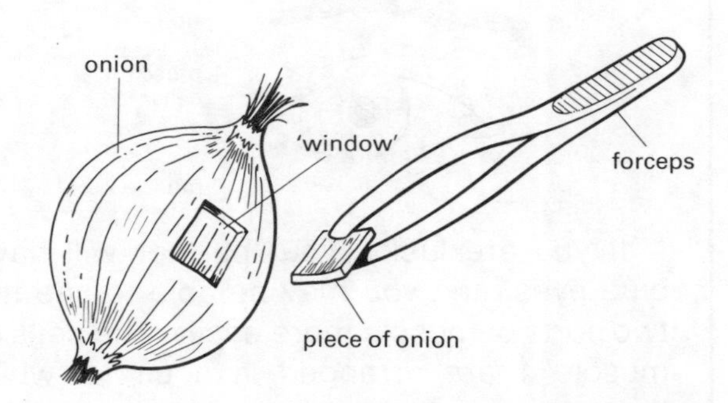

2 Examine the bottom of the square and the 'window'. On one of these you should be able to see the very thin membrane between the two layers. Use the forceps to lift the membrane away gently.

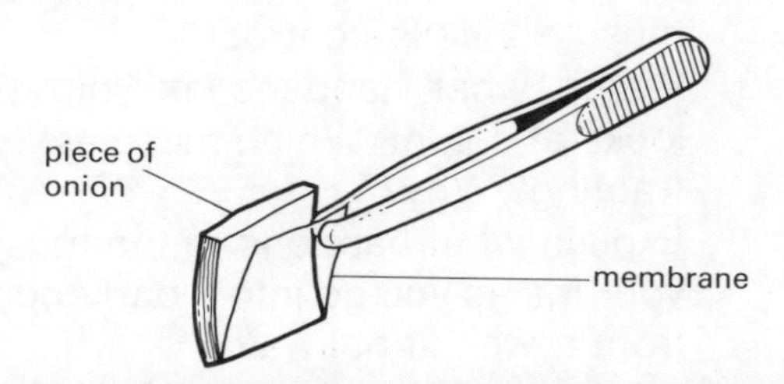

3 Lay the membrane flat on a clean microscope slide, and put on a few drops of iodine to stain the cells.

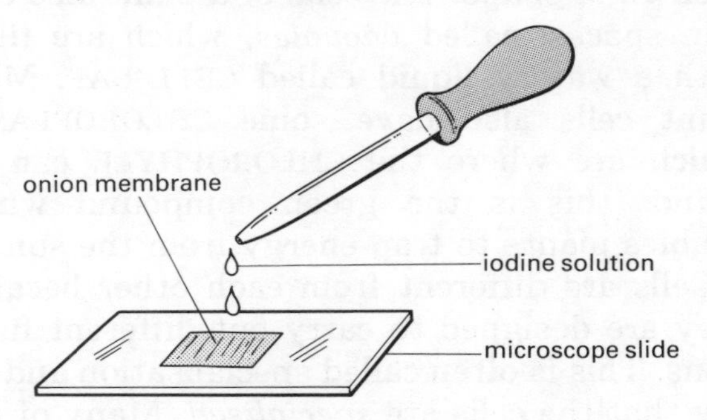

4 Lower a cover slip onto the onion membrane so there are no air bubbles trapped under it. If you are not sure how to do this your teacher will help you.

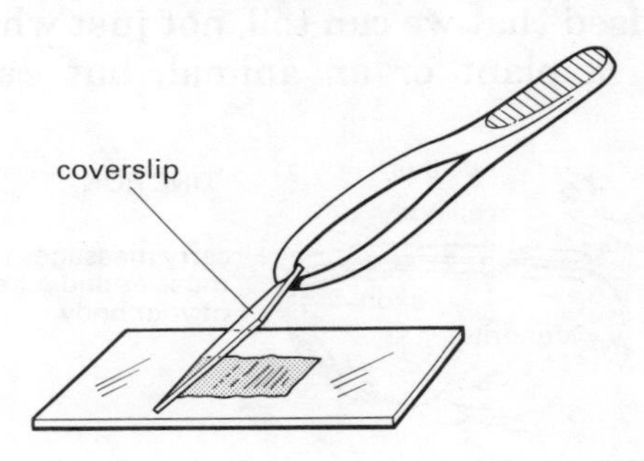

5 Look at your sample of onion cells under a microscope. Make a drawing of a group of the cells. Label the main parts of the cells: Fig. 2.2(b) should help you.

Questions

1 Compare the cells from an onion with those from inside a human's cheek. Make a list of the things they have in common.

2 What have onion cells got that are not visible in the cheek cells in Fig. 2.2(a)?

3 Suggest a reason for the difference in shapes of the two sorts of cell. It may help if you think of the jobs each cell has to do.

CELLS FOR SPECIAL JOBS

As you saw in Investigation 2.1, cells from a plant are different from animal cells. Most plant cells contain some things that are not in animal cells. They have a thick CELL WALL outside the cell membrane; it is made of a substance called *cellulose* and helps the cells keep their rigid shape. The cells of a plant also contain spaces, called *vacuoles*, which are filled with a watery liquid called CELL SAP. Most plant cells also have some CHLOROPLASTS which are where the CHLOROPHYLL can be found: this is the green compound which enables plants to trap energy from the sun.

Cells are different from each other because they are designed to carry out different functions. This is often called specialisation and we say that the cells are *specialised*. Many of the special features of plant cells, for example, are connected with the way that plants use energy from the sun to make food. (You will find out more about this in Unit 9.) Sometimes a cell is so specialised that we can tell, not just whether it is from a plant or an animal, but *exactly* where it belongs in the organism.

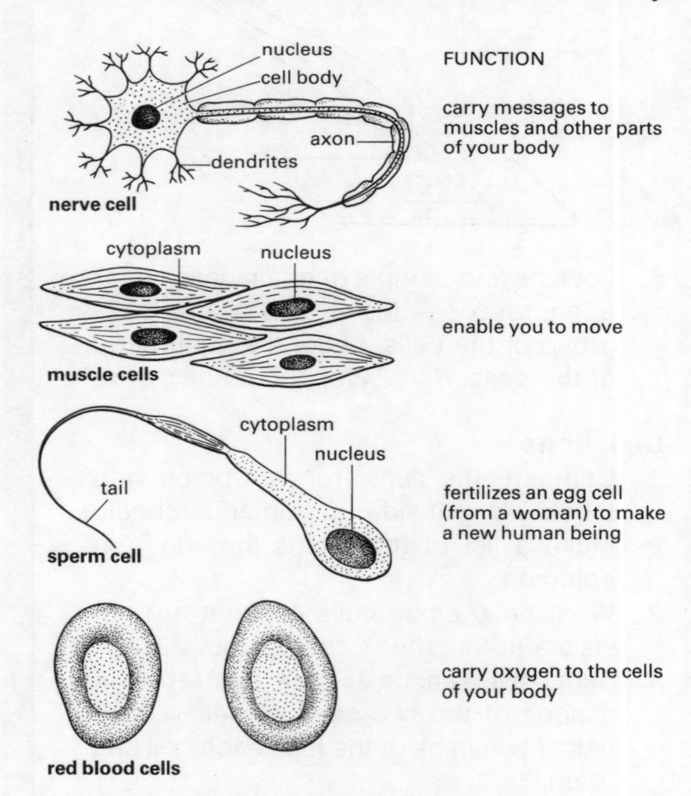

Fig. 2.3 Cells with different functions

The cheek cells in Fig. 2.2 are one example of specialised animal cells. Their function is to make a smooth lining for the inside of our cheeks. Fig 2.3 shows a few more specialised cells that we humans have.

All human cells are built from carbon, oxygen and hydrogen, together with smaller amounts of nitrogen, calcium, phosphorus, potassium, sulphur, sodium and chlorine. They also contain tiny amounts of iron, iodine, copper, zinc, cobalt, manganese and magnesium. Combining these elements in different ways makes all the different cells we need to live and grow.

Assignment – Looking into your eyes

Light gets into your eye through a hole called the pupil. Around the pupil are muscles in the part that we call the iris; these control the amount of light entering your eye.

Choose a partner, and look into each other's eyes: you should see something like this diagram.

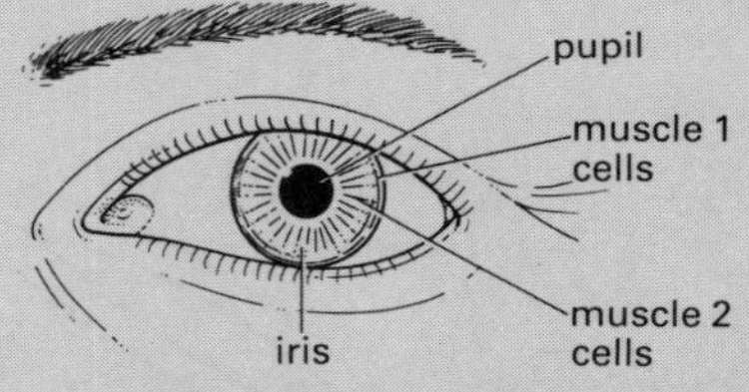

If you are lucky, your partner will have blue eyes, and you may be able to see the two sorts of muscle more easily. The cells of muscle 1 are arranged in a circle, while those of muscle 2 are like the spokes of a bicycle wheel.

1. What will happen to the size of the pupil when the cells of muscle 1 contract (get shorter)?
2. What will happen to the pupil size when muscle 2 cells contract?
3. Watch what happens as your partner looks at a light. Which muscles are contracting? Why?
4. Explain what happens to the muscles in your iris as you go into a darkened room from a well lit hallway.

CELLS ALONE

Most cells are part of a bigger living organism, but some cells can live alone. These single-celled organisms are microscopic: they live in water, damp soil and inside the bodies of animals. There are many different types of these uni-cellular organisms and they are usually labelled as members of a group called the *Protista*, as you will see in Unit 6.

Amoeba

This microscopic, uni-cellular organism lives in ponds and damp soil, where it feeds on microscopic plants. As you can see in Fig. 2.4, amoeba feeds by surrounding its food with cytoplasm. The food is then *inside* the cell in a space called a FOOD VACUOLE. It is held there until it is digested and small enough to be absorbed into the cytoplasm. The amoeba leaves behind any waste or indigestible material as it moves on.

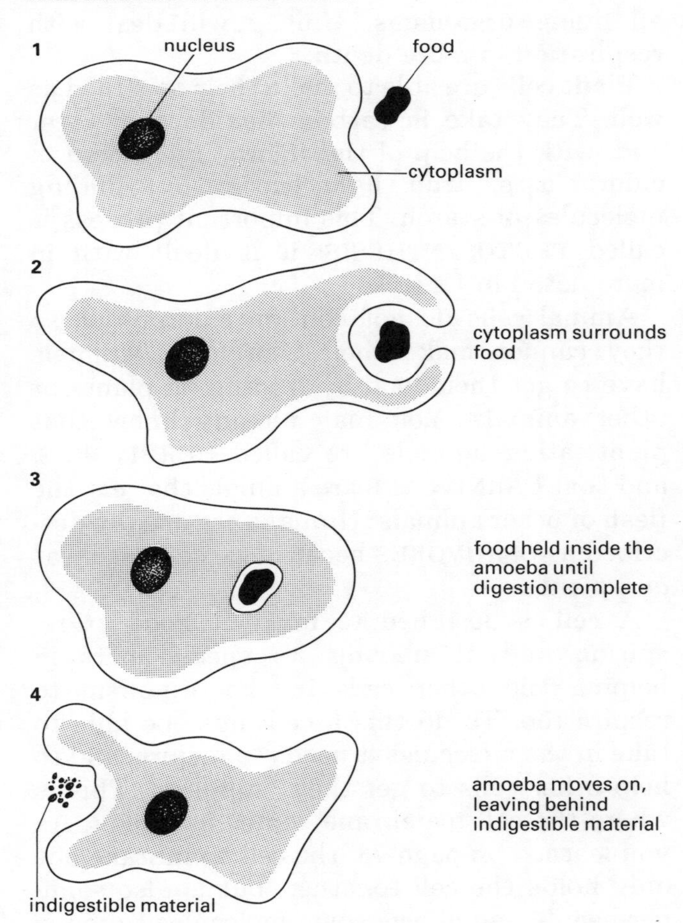

Fig. 2.4 An amoeba feeding

Paramecium

As you can see in Fig. 2.5, this uni-cellular organism is a little more complex than an amoeba. It can move itself along by 'swimming' with the hairs called CILIA, which also help to push food into the paramecium's gullet. Food collects in a vacuole at the bottom of the gullet. This food vacuole moves around in the cytoplasm until the food has been digested. Then a temporary gap in the cell membrane allows any waste material to leave the cell.

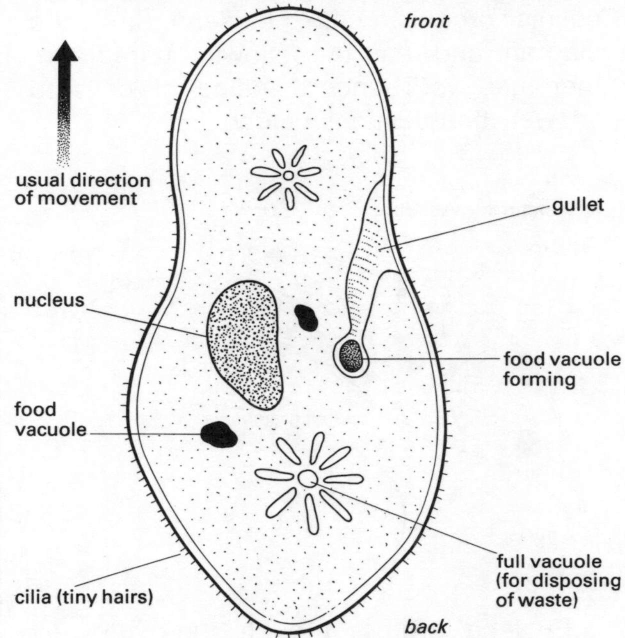

Fig. 2.5 Diagram of the unicellular organism called Paramecium

Yeast

Yeast is a uni-cellular organism that is famous for its ability to turn sugar into an alcohol. This process is called FERMENTATION and is the way that a yeast cell gets its energy. It breaks down sugars to release energy. You will learn more about this in Unit 8.

The yeast cells use the energy to grow and reproduce, making new yeast cells. We use the waste products, ethanol and carbon dioxide, in many different ways. One of these is in bread-making, where yeast is used as a 'raising agent': it is allowed to ferment with sugar until the carbon dioxide produced makes the bread dough much larger. The dough 'sets' as the bread is baked and the spaces filled with carbon dioxide make a fine network of holes throughout the bread. The more of these tiny holes there are in the bread, the lighter the texture of the loaf when it has been cooked.

Fig. 2.6 shows a close up view of some yeast cells.

Assignment – Euglena, an example of a uni-cellular organism

The diagram shows the structure of the uni-cellular organism called euglena. Study the diagram and read the following paragraph, then use your understanding of cells to answer the questions below.

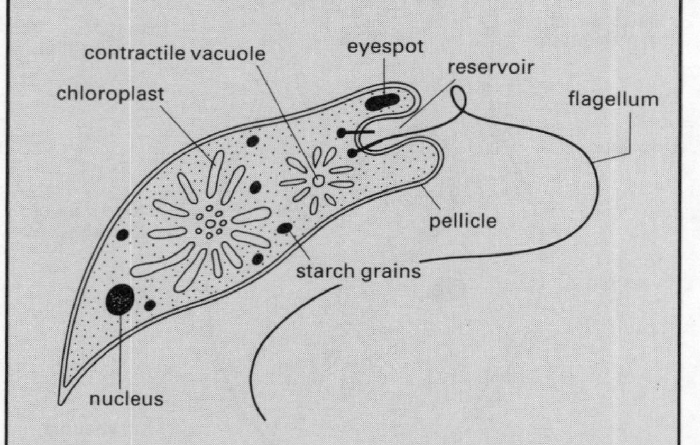

Euglena swims through the water by lashing its flagellum. This makes it rotate as it moves along. When it stops moving, euglena can change its shape. It is sensitive to light as it has an eyespot. It prefers to be in the light, because the chloroplasts enable the sun's energy to be trapped and used to make food, which is stored as grains of starch. The euglena's contractile vacuole controls the amount of liquid in the cell: it starts quite small and gradually fills with water. When it is full, the contractile vacuole empties the water into the reservoir.

1 What features does euglena have that are those of an animal cell?
2 Which of euglena's characteristics are those of a plant cell?
3 It has been discovered that euglena can live without light, provided there is a supply of food. Suggest how this is possible.

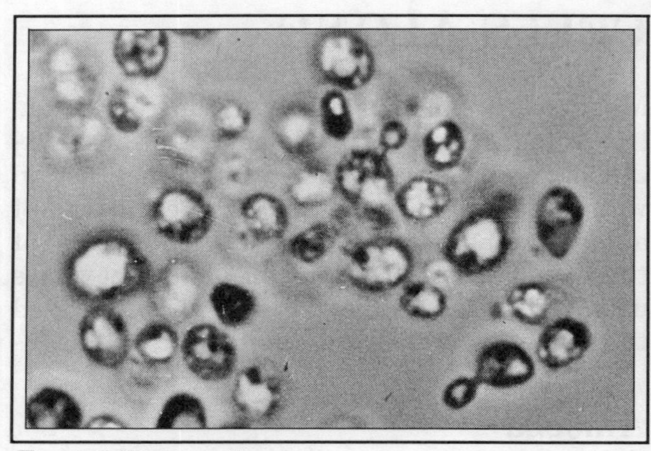

Fig. 2.6 Yeast cells

WHAT CELLS DO

We need to look at the processes that cells can carry out, because these are the basis of life. All cells get the energy they need to live by combining food molecules with oxygen. This process is called RESPIRATION and is a feature of all living organisms. Unit 9 will deal with respiration in more depth.

Plant cells are able to make their own food as well. They take in carbon dioxide and water and, with the help of the chlorophyll stored in chloroplasts, 'trap' light energy by building molecules of starch. This important process is called PHOTOSYNTHESIS: it is dealt with in more detail in Unit 9.

Animal cells do not contain chlorophyll, so they cannot make their own food. Animals have to get their food by feeding on plants or other animals. You may already know that plant-eating animals are called HERBIVORES, and that CARNIVORES are animals that eat the flesh of other animals. Humans are usually described as OMNIVORES because we can eat meat *and* plants.

A cell is designed to make it good at respiring and at playing its special part in helping the other cells in the organism to respire too. To do this a cell must be able to take in the molecules it needs to respire, and to help other cells to get their 'supplies'. This is where the cell membrane comes into play. As you learned on page 73, the cell membrane not only holds the cell together, but is also semi-permeable and allows some molecules to move in and out, but not others.

Investigation 2.2 Semi-permeable membranes at work

There are several ways to investigate the way a cell membrane works: this is one that you could try. You will use a tube made of semi-permeable material.

What you need to know

Starch molecules are large.

Iodine turns blue-black when it meets starch.

Collect

a piece of dialysis tubing about 15 cm long

1% starch solution

iodine solution

dropper

test tube

elastic band

What to do

1 Soak the dialysis tubing in water for a few minutes until it is soft.

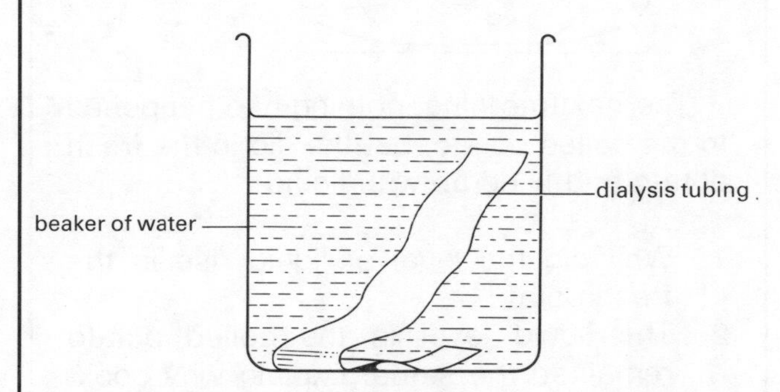

2 Tie a knot in the bottom of the tubing and use a dropper to half-fill it with starch solution.

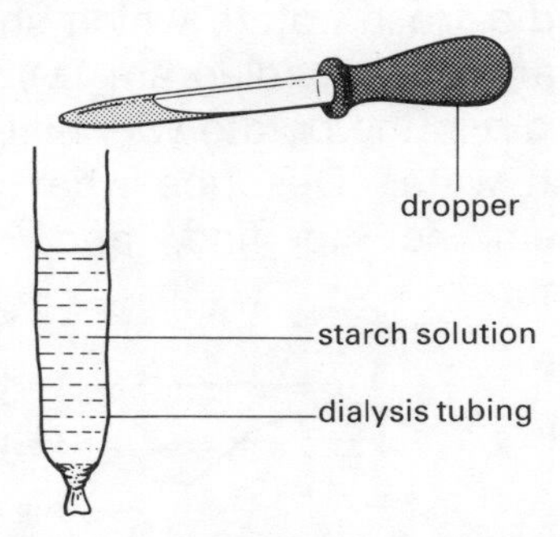

3 Put the dialysis tubing into the test tube, and hold down the top of the 'bag' with an elastic band.

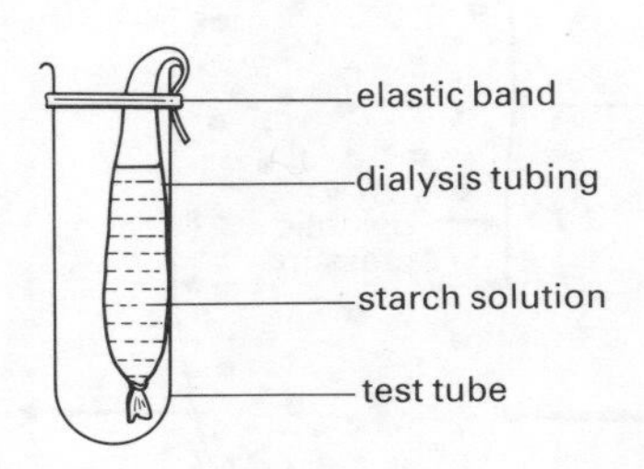

4 Carefully rinse the test tube and the outside of the 'bag' of starch with some clean water. This should get rid of any starch that you spilled outside the dialysis tubing.

5 Fill the test tube with water, to about two-thirds of the way up. Add a few drops of iodine solution.

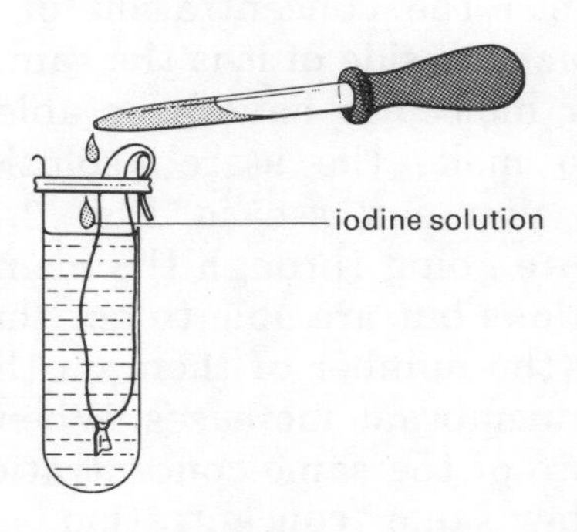

6 Leave the test tube in a test tube rack for about 15 minutes. Make a note of any changes that you see.

Questions

1 Dialysis tubing is semi-permeable, and will allow small molecules through it. Which molecules are the most likely to pass through the membrane, starch or water?

2 Why is it important to rinse the outside of the tubing, and the inside of the test tube at step 4?

3 After about 15 minutes the starch solution in the bag goes blue, while the iodine solution in the rest of the test tube stays yellow. What has happened?

4 Not only does the solution in the 'bag' turn blue but the level of the liquid in the dialysis tubing is higher. Explain what must have happened to cause this.

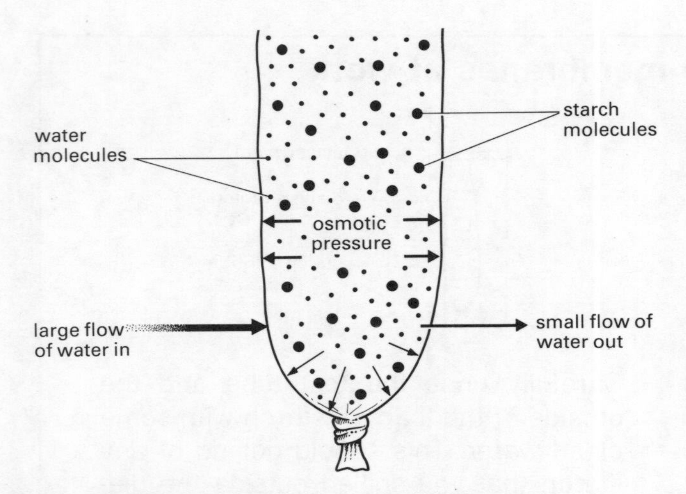

Fig. 2.7 Molecules moving during osmosis

A semi-permeable membrane like the dialysis tubing allows small water molecules to pass through until the concentration of the two solutions on each side of it is the same. In this case, water molecules have been able to pass through to make the starch solution more dilute. As you can see in Fig. 2.7, water molecules are going through the membrane in both directions but are able to get through so easily that the number of them on the starch side of the membrane increases. When the two solutions are of the same concentration, there will be the same concentration of water molecules on both sides of the membrane. Though they will still move back and forth, the overall number of molecules of water on each side will stay the same. The way in which two solutions change until they have the same concentration when water molecules pass through a semi-permeable membrane between them is called OSMOSIS. As you will see, osmosis is an important process for many cells, especially those involved in carrying water and minerals into plants.

Plants lose water from their leaves in a process called TRANSPIRATION. The water evaporates from the surface of leaves through gaps called stomata (see Fig. 2.19 on page 88), though some water evaporates from the surface cells of plant stems as well. The cells which have lost the water will contain cell sap that is more concentrated than the sap in cells below the surface. You can probably guess what happens: water molecules pass from cells with more dilute cell sap to replace some of the water that has been lost from the surface cells.

Assignment – An experiment using potatoes

A pupil had an experiment to do as homework. She cut a potato in half, and cut a dip in the unpeeled end of each half. She then trimmed some peel from the cut ends.

One potato half was boiled, then both halves were set in saucers. The dips in the potatoes were half filled with sugar solution and water was poured into the saucers (diagrams (a) and (b)).

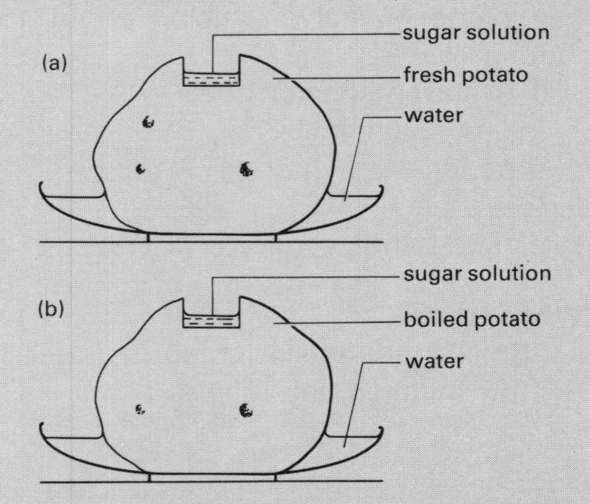

The next morning, nothing had happened to the boiled potato, but the dip in the fresh potato had filled up with liquid.

1 Why did the level of liquid rise in the fresh potato?
2 The liquid level in the boiled potato remained the same. Explain why cooking the potato prevented the change in level.
3 The pupil repeated the experiment, but used a fresh potato which she stood in sugar solution (diagram (c)). this time the dip in the potato was half filled with fresh water. Describe what you would expect her to find, and suggest a reason.

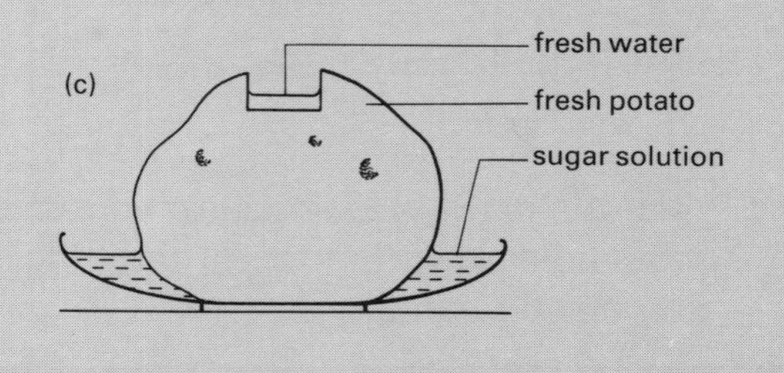

This is a 'knock on' effect and as each cell loses water to the next one along it replaces that water with some from the cell before it. In this way osmosis keeps the cells filled with cell sap, which makes the cells firm. Firm cells help to keep the plant upright. Without enough water cells become 'floppy' and the photograph in Fig. 2.8 shows one result - a wilting flower.

Transpiration and osmosis play a vital part in the way plants take in water and minerals through their roots and carry these to each plant cell. The rate at which a plant loses water by transpiration can be tremendous, especially if the air around it is warm and dry. An oak tree, for example, can lose as much as 680 dm^3 of water in one day's transpiration. The great

Fig. 2.8 A plant, (a) turgid, (b) wilting

Investigation 2.3 How plants take up water

A. Leaves

Collect

short leafy stem
test tube
a solution of dye e.g. the red stain eosin or red ink
hand lens

What to do

1. Choose a stem long enough to stand in the test tube.
2. Half fill the tube with dye solution.
3. Place the plant stem in the tube so that at least 5 cm of it are covered by the solution.
4. Put the tube in a place that is well lit and airy, and leave it for a few hours.
5. When you return, use a hand lens and a bright light to help you see and look at the plant leaves. You should be able to see where the dye has travelled.

Questions

1. Why did the plant need to be left in an airy place?
2. What effect would you expect there to have been on your experiment if you had placed your plant in a damp atmosphere?
3. You should find that the dye had travelled along the veins of the leaves. What does this tell you about the function of leaf-veins?

B. What about the stem?

Your teacher may let you do this or show you a slide that has been prepared already.

Collect

the plant stem from part (a)
hand lens
scalpel
microscope slide
microscope

What to do

1. Take the plant stem out of the test tube.
2. Cut off the piece that was standing in the dye.
3. Cut a thin slice across the stem, and another longways, through the centre of the stem.
4. Make a microscope slide of each of these slices and look at the cells that the dye solution has travelled through. These are called *xylem* cells and Fig. 2.24(a) on page 91 shows how they are arranged in the stem of maize.

Investigation 2.4 Measuring the transpiration rate of a plant

Collect

piece of plant that has a long cylindrical stem
about 5 cm of plastic tubing, just wide enough to fit on your plant stem
piece of capillary tubing about 40 cm long
water container (big enough to hold the plant stem)
knife
beaker of boiled, cooled water

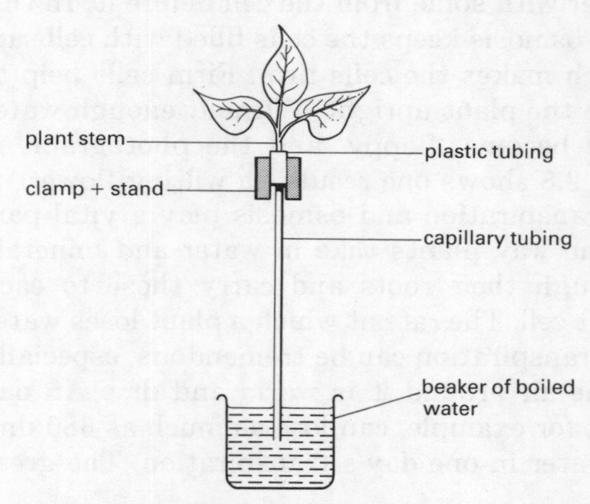

What to do

(Your teacher may have done steps 1 and 2 for you.)

1 Use the knife to cut the stem. Cut much more than you will need.
2 Plunge the stem in water *right away* and make a slanting cut about 15 cm from its bottom while it is still under water.
3 Put the plastic tube into the water with the stem. Make sure the tube is full of water, and that no water gets on the leaves. Carefully put the tube over the cut end of the stem. Leave them both in water.
4 Put one end of the capillary tube into the water, and fit the other end of the plastic tube over it. Leave them in the water.
5 Set up a clamp and stand ready to hold the tubes and stem. Put the beaker of boiled water ready.
6 When everything is ready, put one finger over the end of the capillary tube to stop any water getting out or any air getting in.
7 Keep your finger over the end of the tube while you get it arranged as the diagram shows.
8 Blot off any water that has got on the leaves.
9 Put two marks on the tube, about 20 cm apart.
10 Lower the beaker so the capillary tube is out of the water. As soon as a bubble of air gets into the tube, put the beaker back.
11 Measure the time it takes for the bubble to pass from the bottom mark to the top one.
12 When the bubble passes the top mark, pinch the plastic tube to squeeze the bubble out of the capillary tube.
13 Repeat steps 10-12 at least three times more. You can try the experiment with a cover over the apparatus, to prevent draughts, or with a hairdrier blowing to increase the draught around the plant.

Questions

1 It is important to prevent air getting into the stem as you prepare the apparatus.
 (a) Which steps did you take to stop air getting into the stem?
 (b) Why must air not get into the stem?
2 If you do repeat the experiment with a hairdrier blowing, you should find that the bubble moves faster. Explain why this happens.
3 If the stem that you used had many more leaves on it, what effect would this have on the rate of transpiration?
4 Plants that live in dry climates, have leaves designed to slow down transpiration. Look at a cactus.
 (a) Explain why plants which live in a dry climate need to slow down their rate of transpiration.
 (b) How is a cactus designed to make transpiration slower?

amount of water an oak tree can lose shows how important it is for a plant to have a good system for replacing that water. This is where osmosis comes into play again.

Figure 2.9 shows some of the cells in a plant root, the surrounding soil and the route water takes. The main job of a plant's root hairs is to hold the plant in the earth, which they do by getting between soil particles. The concentration of substances in the cell sap of these root cells is greater than in the soil water, so water is pulled into the root cells. This happens by osmosis. Osmosis is also the way by which water is carried up through the plant.

Investigation 2.4 was just one way of measuring the rate at which plants lose water by transpiration. For quicker, more accurate results an improved version of the apparatus, called a POTOMETER, is often used. You can see one in Fig. 2.10 and you should be able to see how it has been made from your simple version.

The combination of osmosis and transpiration is very important for plants. However, it is worth remembering that *all* cells can carry out osmosis. As you will see later, osmosis plays an important part in the life of animals as well. Two other things that cells do are dealt with in Unit 8, which shows you the ways cells divide for growth or for reproduction.

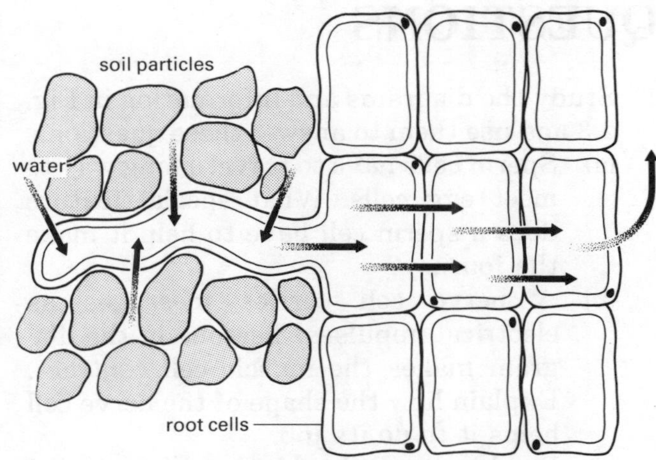

Fig. 2.9 Simplified diagram of a root. Arrows show the way water and minerals enter the root

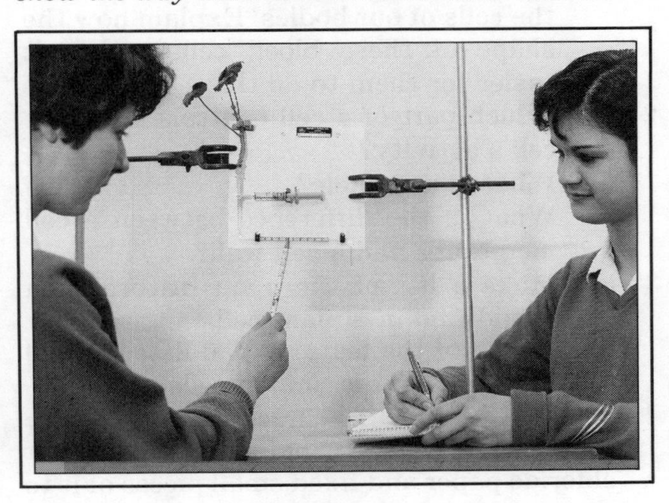

Fig. 2.10 Students using a potometer

Cells and cell structure

WHAT YOU SHOULD KNOW

1. All living organisms are made up of cells.
2. Cells are made from protoplasm, part of which is called the nucleus. The nucleus controls all that happens in the cell
3. A thin membrane holds a cell together. It is semi-permeable and lets some molecules through, but not others.
4. Plant cells are different from animal cells in the following ways:
   they have a thick cell wall made of cellulose;
   they have chloroplasts that contain chlorophyll;
   they contain vacuoles filled with cell sap.
5. Cells are specialised, or designed, to carry out particular functions.
6. Some organisms are uni-cellular and consist of a single cell. Other organisms are made from millions of cells.
7. All cells can carry out these functions:
   respiration - the use of oxygen to release the energy stored in food
   osmosis - the movement of molecules through the semi-permeable cell membrane.
8. Plant cells can also use energy from the sun to make food by a process called photosynthesis. Chlorophyll is necessary for this to happen.
9. Plant cells can lose water in a process called transpiration.
10. Transpiration and osmosis play an important part in the way water and minerals are carried through a plant.

QUESTIONS

1 Study the diagrams and information in Fig. 2.3 and use them to answer these questions.
(a) Sperm cells have to travel a long way to meet egg cells. What special feature does a sperm cell have to help it make the journey?
(b) A nerve cell carries messages as electrical impulses. The one in the diagram makes the muscle cell contract. Explain how the shape of the nerve cell helps it to do its job.
(c) Red blood cells have to be able to travel through narrow blood vessels to reach the cells of our bodies. Explain how the shape of these blood cells makes it easier for them to do this.

2 (a) Which part of a cell controls all of the cell's activity?
(b) What is a vacuole?
(c) What is the difference between a cell membrane and a cell wall?

3 (a) Make a list of the main features you would find in a plant cell.
(b) Which of the features you listed in 3(a) are *not* found in animal cells?

4 A pupil wanted to investigate the way plants lose water. She used dry cobalt chloride paper and fixed small pieces of it to the top and bottom of a leaf.
When she came back and removed the paper, the pupil found that the cobalt chloride paper had turned pink. This meant that there was water on the leaf.
(a) Where had the water come from?
(b) Which one of the pieces of cobalt chloride paper would you expect to have the most pink colour?
(c) Give a reason for your answer to (b).

5 A pupil tried a different experiment to investigate osmosis. Instead of using potato 'cups' as in the assignment on page 78, he cut four potato chips, making sure they were the same length. He stood each chip in a beaker. The pupil then poured a different liquid into each of the beakers and left them all for several hours. The table shows what he found when he returned and measured the chips. Use these results to answer the questions.

	Starting length/cm	Length after several hours/cm
Chip 1	5.0	5.0
Chip 2	5.0	5.2
Chip 3	5.0	5.4
Chip 4	5.0	6.0

(a) Which chip was probably standing in a beaker of water?
(b) Which beaker contained the most concentrated sugar solution?
(c) If the pupil had used a chip 5 cm long, but from a boiled potato in beaker 2, what result would you have expected when he came to measure it at the end of the experiment? Explain your answer.

6 Disease-causing bacteria can be killed by using strong salt solution as an antiseptic. Use your understanding of osmosis to explain how these uni-cellular organisms are affected by such a strong solution.

TISSUES AND ORGANS

As well as doing some things alone, groups of cells can work together to carry out a special function. Such a group of specialised cells is called a TISSUE. A group of muscle cells, for example, is called muscle tissue. They work together with bone tissue to make us move.

MUSCLES AND MOVEMENT

The place where two bones meet is called a JOINT, and they are joined by tissue called LIGAMENT. At some joints the bones are held together so tightly that they cannot move. This type of joint is described as FIXED, and our skull bones are joined in this way. In other joints the bones have spongy tissue called CARTILAGE between them. As you can see in Fig. 2.11, our backbone contains joints like this. Each of the bones called VERTEBRAE is separated from the next one by a disc of cartilage tissue. This type of joint allows a little movement, so that we can bend over, but also keeps our vertebrae bound together so that the backbone supports the body when we stand up.

Most joints can move more freely and there are two important joints of this kind, as you can see in Fig. 2.12.

Fig. 2.11 Human backbone (vertebral column)

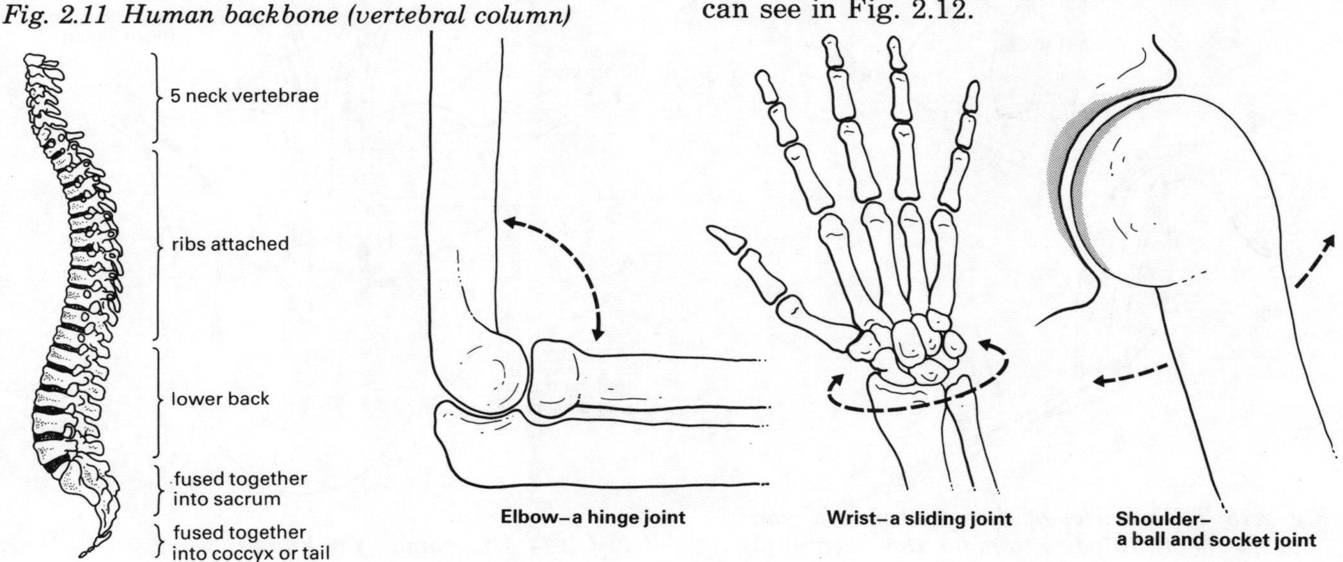

Hinge joints: the bones can move back and forward like a door swinging on its hinge. We have hinge joints in our knuckles, elbows and knees.

Ball and socket joints: the bones can move backwards and forwards *and* up and down. We have ball and socket joints in our shoulders and hips.

Our wrists contain a different type of joint. It is called a *sliding joint*, and the arrangement of bones allows us to make twisting movements with our hands. Just think of the way that your hand can turn a door knob, and think which other part of your body you can twist in the same sort of way.

How the joints move

Muscles are joined to bones by tendons. Tendon tissue does not stretch, so when a muscle contracts and pulls on a tendon the tendon pulls on the bone. Muscles contract, getting shorter and fatter; when they stop contracting, or *relax*, they need another muscle to pull them back to where they started. Our muscles are often arranged in pairs like this, with one pulling one way and one the other. Fig. 2.13 shows how such a pair of muscles moves our forearm.

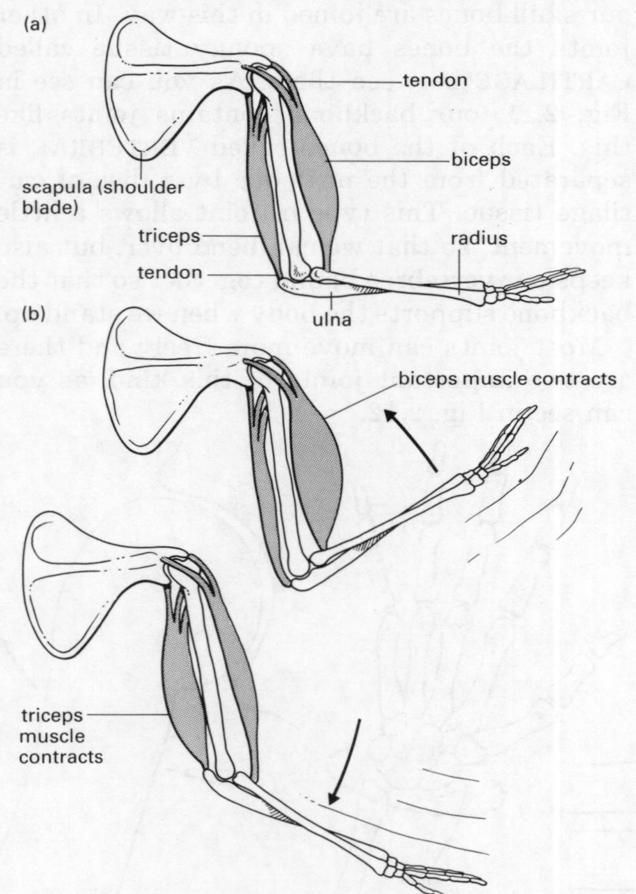

Fig. 2.13 The muscles of the upper arm (a) and how they work in pairs to move the forearm (b)

There are many other teams of tissue that work together like this. Both plants and animals have many different 'tissue teams', but we shall be most concerned with those in animals, and humans in particular. When many groups of tissue work together to carry out a special function, they are called an ORGAN. Our eye contains many sorts of tissue, each of them playing a part in helping us to see, so our eye is called an organ. You will learn more about this particular organ in Unit 3 (page 154). We have many other organs and if one stops working we can die. Because they are so important we shall look at each of the most vital organs in this section.

THE HEART

A human heart is really two pumps which work together, pumping blood. You have about 5.5 dm^3 of blood, carrying food, oxygen and water to every cell in your body.

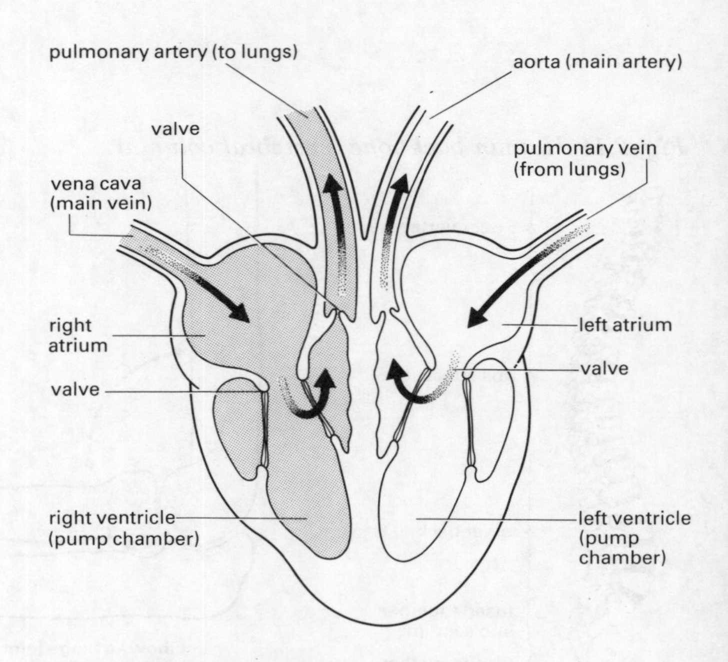

Fig. 2.14 Diagram of the human heart

The left hand side of your heart pumps this blood right around your body. Blood supplies the cells with the food, oxygen and water that they need. The cells return waste such as carbon dioxide to the blood. This food is returned to the right hand side of your heart, which then pumps it to your lungs. When the blood reaches your lungs, it loses carbon dioxide and picks up more oxygen. (How this happens is dealt with on page 86.) Oxygenated blood then makes its way back to the left hand heart pump, and is soon on its journey to the cells once more.

Assignment – Check up on your heart

Your heart pumps, or *beats*, about 60-150 times a minute. Where an artery passes over a bone you can feel the blood surge through on its way from your heart. This surge is called a PULSE, and you can feel a pulse beat each time your heart beats. If you count the number of beats you can get an idea of how hard your heart is working.

First of all, find a pulse point by putting your fingers on your wrist or on your collar bone. When you are on the right spot you should feel a little flutter. If you use your thumb to search for a pulse you will find them everywhere, because there is a pulse in your thumb!

1 Once you have found your pulse, count the beats for one minute.
2 Walk around and count again.
3 Try a more energetic activity like walking upstairs or jumping up and down. Count your pulse rate again.
4 Sit down and find out how long it takes for your pulse rate to get back to normal.

Question

1 Explain why your pulse rate is faster when you are active.
2 Athletes who train a great deal develop strong heart muscles and their pulse rate is often very slow.
Explain why their pulse rate can be much slower than an ordinary person's, even after they have both been active.

THE 'CLEANING' ORGANS

Cells make substances that your body does not need as well as those that it needs. Most of these unwanted substances end up in your blood, which then has to be 'cleaned'.

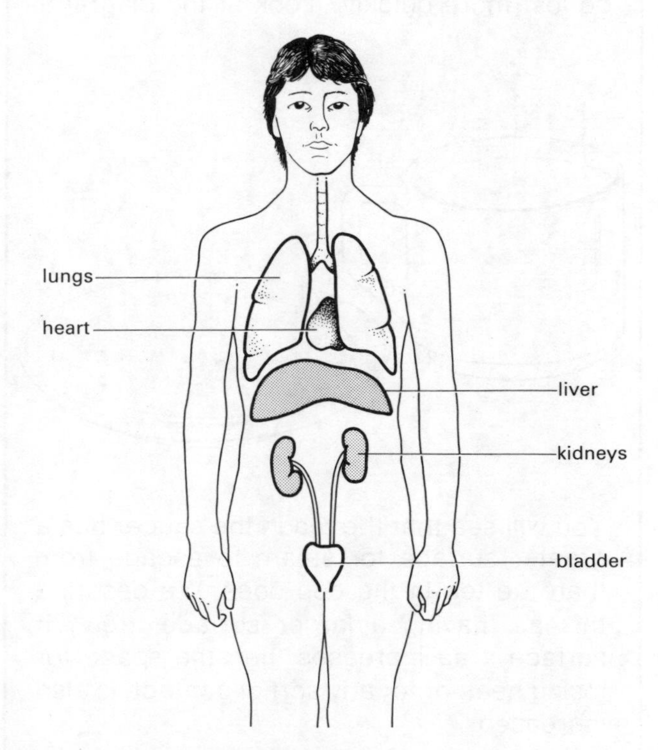

Fig. 2.15 The 'cleaning' organs

In Fig. 2.15 you can see where the main cleaning organs are, they are the *lungs* and *kidneys*, and most important of all, the *liver*. Lungs take out carbon dioxide from deoxygenated blood and replace it with oxygen. Kidneys take out extra salts, water and compounds rich in nitrogen (urea in particular). Your liver controls the level of many chemicals in your blood, amongst them glucose from digested food.

Tissues in organs such as these are arranged to give the greatest possible contact between blood and the tissue. Having a large tissue surface in contact with blood vessels means that all the waste can be taken out. It also means that as much oxygen as possible gets into the blood through lung tissue. Surface area is important for plants too! Can you think why?

Assignment – What is surface area?

Have you ever had a cup of tea that is too hot? Have you ever poured some of the hot tea into a saucer to help it cool quickly? Did you realise that you were giving your tea a larger surface area to allow heat energy to be lost more quickly? Look at the diagram.

You will see that the tea in the saucer has a greater surface for steam to escape from than the tea in the cup does. We describe this as 'having a larger surface area'. If surface area increases then the space for losing heat, or for any sort of contact, is also increased.

Questions

1 Frank had to cook an evening meal for the family. He arrived home late, and had to hurry. His mother had asked him to cook some steak, with boiled potatoes as one of the vegetables. To save time, Frank chopped the potatoes into small pieces and beat the steak to make it flatter and wider.
   (a) Explain why small chunks of potato should cook more quickly.
   (b) How would beating the steak help Frank's meal to cook more quickly?

2 (a) If you have a sweet tooth, which would you use to make your tea sweet more quickly, cube or granulated sugar?
   (b) What else could you do to help the sugar dissolve?

Lungs

If your lung tissue were to be opened out and spread flat, it would cover a tennis court! The tissue is very thin and 'folded' to make millions of tiny air spaces called ALVEOLI. These give your lungs a spongy appearance but, more important, make a very large surface for contact with the blood vessels. As you can see in Fig. 2.16, every air sac, or alveolus, is covered with the narrow blood vessels called capillaries.

Blood capillaries are so narrow that their walls are just one cell thick. This means that it is fairly easy for some substances to diffuse in and out of capillaries. When blood enters your lungs from your heart it is rich in carbon dioxide. As it reaches the capillary network

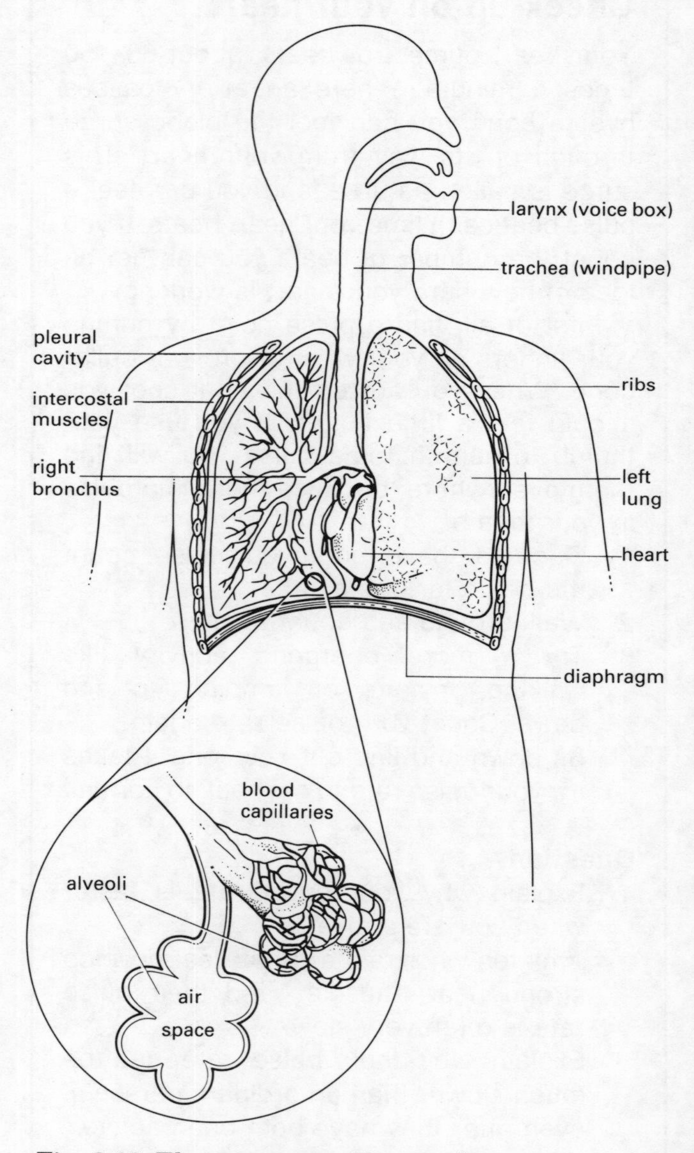

Fig. 2.16 The structure of human lungs

around an air sac (as in Fig. 2.16) the blood has a greater concentration of carbon dioxide than the air in the air sac. Inside the air sac there is a greater concentration of oxygen than in your blood. So two things happen:

(a) carbon dioxide diffuses from the capillary into the air sac;

(b) oxygen diffuses from the air sac into the capillary.

The blood in the capillary is also a little richer in water than the air in the air sac, so that water passes into your lungs and is breathed out with the carbon dioxide. The blood in the capillary is now rich in oxygen and eventually meets with the oxygenated blood from other lung capillaries as these tiny blood vessels join up on the way to your heart. From there oxygenated blood is pumped to the cells around your body to be used for respiration (Unit 9).

Kidneys

Humans have two kidneys, and most of the tissue in them is involved in filtering the blood. A pair of healthy kidneys can filter about 60 dm^3 of blood in an hour, which means that it takes about five minutes to filter all the blood in your body!

Your kidneys have to purify the blood that flows through them, by taking out harmful substances and removing them from your body. This is called EXCRETION and one of the main substances removed this way is UREA, which is mixed with the excess water from your blood to make a solution called URINE. The other job that your kidneys have to do is to keep the level of water in your blood constant.

In order to do their job, your kidneys need a good blood supply. It has been worked out that they have about 160 kilometres of blood vessels between them. As blood flows through kidney tissues, many of the substances in it are forced into the kidney. Some, such as water and glucose, are absorbed back into the blood a little later. Once these reach the correct concentration in the blood, no more will be reabsorbed from kidney tissues. The excess water and glucose, and harmful substances such as urea, travel along to a tube called the URETER, which takes the mixture to your BLADDER. The mixture is called urine and is disposed of when you next visit the lavatory.

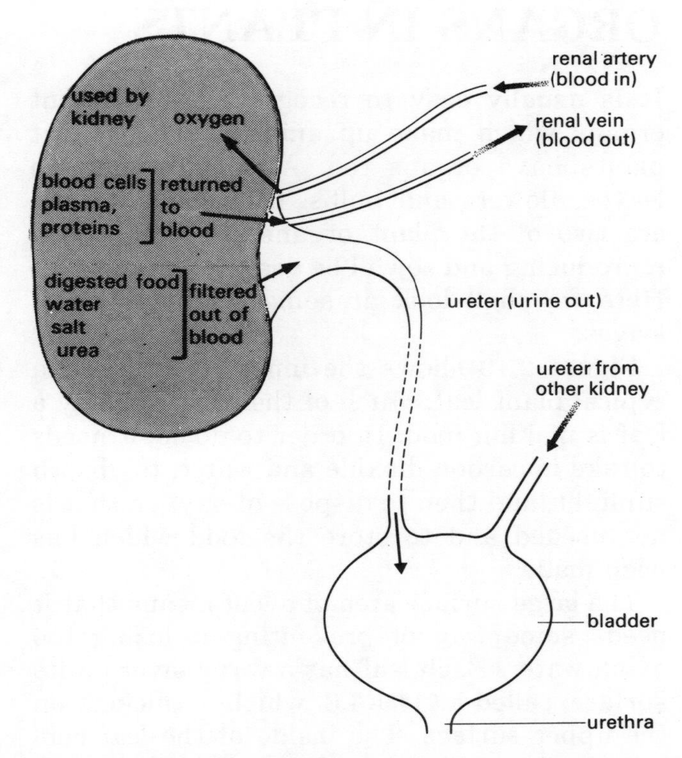

Fig. 2.17 Simplified diagram of human kidney

Liver

Your liver is like a chemical processing factory, and Fig. 2.18 shows some of the important tasks it carries out. These are just a few of the functions of the liver; there are many, many more. Some you will meet later (Unit 8), but many are too complex for us to consider at this level.

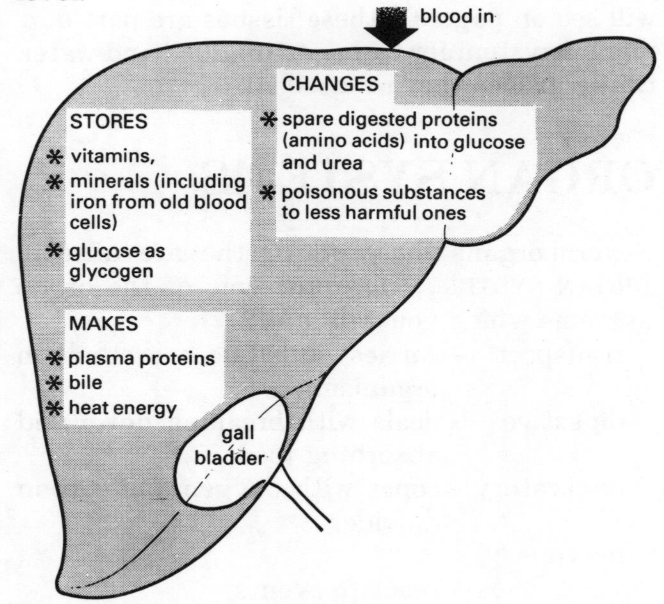

Fig. 2.18 What your liver does

ORGANS IN PLANTS

It is usually easy to recognise the different organs which make up animals like us. But plants have organs too. Amongst them are leaves, flowers and bulbs. Bulbs and flowers are two of the plant organs concerned with reproducing and so will be dealt with in Unit 8. Here we shall look at some of the tasks of leaves.

Figure 2.19 shows the inner structure of a typical plant leaf. Much of the work done by a leaf is making food. In order to do so, it needs to take in carbon dioxide and water, to absorb sunlight, and then to dispose of oxygen that is not needed and to store the food which has been made.

The large surface area of a leaf means that it needs some way of preventing it losing too much water. Each leaf has a waxy layer on its surface, called a CUTICLE, which is thickest on the upper surface. The inside of the leaf contains cells with chloroplasts and air spaces between them. These air spaces allow gases to move between the cells. In the lower surface of the leaf there are many pores, called STOMATA, which can open and close to control the amounts of gases and water passing in and out of the leaf. The veins in each leaf are made from two sorts of tissue: PHLOEM and XYLEM. Xylem tissue brings water and minerals to the leaf from the roots; phloem tissue carries away the food that has been made by the leaf. As you will see on page 91, these tissues are part of a plant's system for transporting food and water to the places they are needed.

ORGAN SYSTEMS

Several organs that work together are called an ORGAN SYSTEM. Here are some of the organ systems which you will meet:

- transport – carries substances around an organism;
- digestive – deals with breaking down and absorbing food;
- respiratory – copes with oxygen and carbon dioxide;
- nervous, hormone – react to events

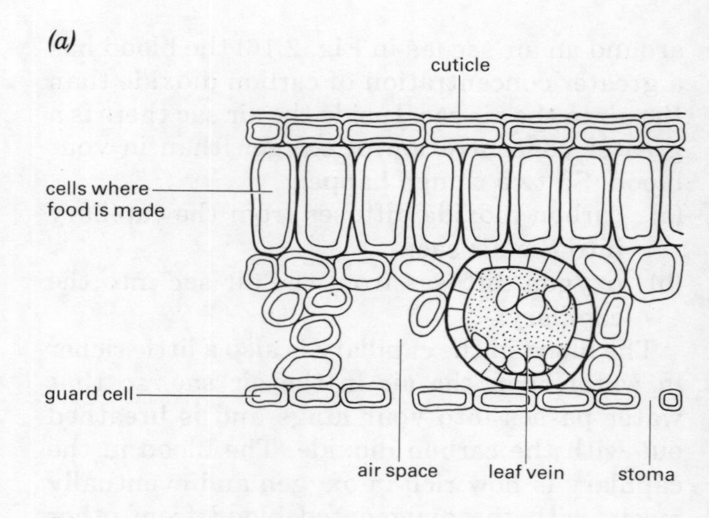

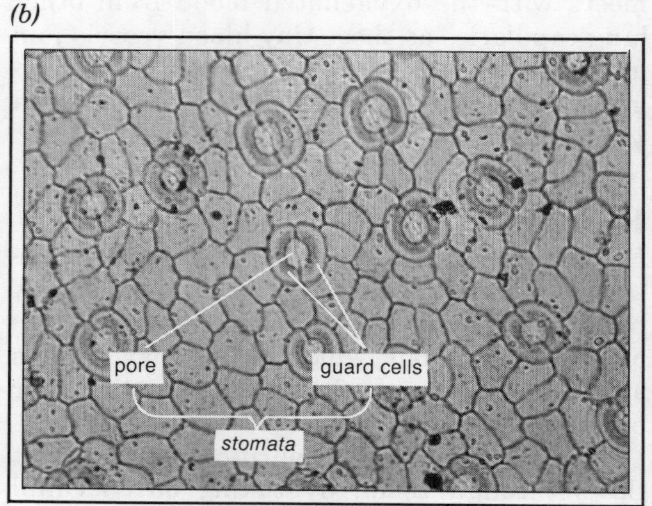

Fig. 2.19 (a) Cross section of a leaf. (b) Underside of a leaf

In this unit we shall consider the first three organ systems in the list. The others will be dealt with in Units 7 to 9.

The human digestive system

The organs that you can see in Fig. 2.20 work together to break down food into molecules simple enough to get into your bloodstream. If you look carefully at the diagram, you can see that your digestive system is like one long tube. It starts at the opening we call a mouth and makes a passageway for food down through your stomach and intestines. By the time food reaches your intestines it is ready to be absorbed into your bloodstream and only solid waste leaves your digestive system through the opening called the anus. Another name for your digestive system is the ALIMENTARY CANAL.

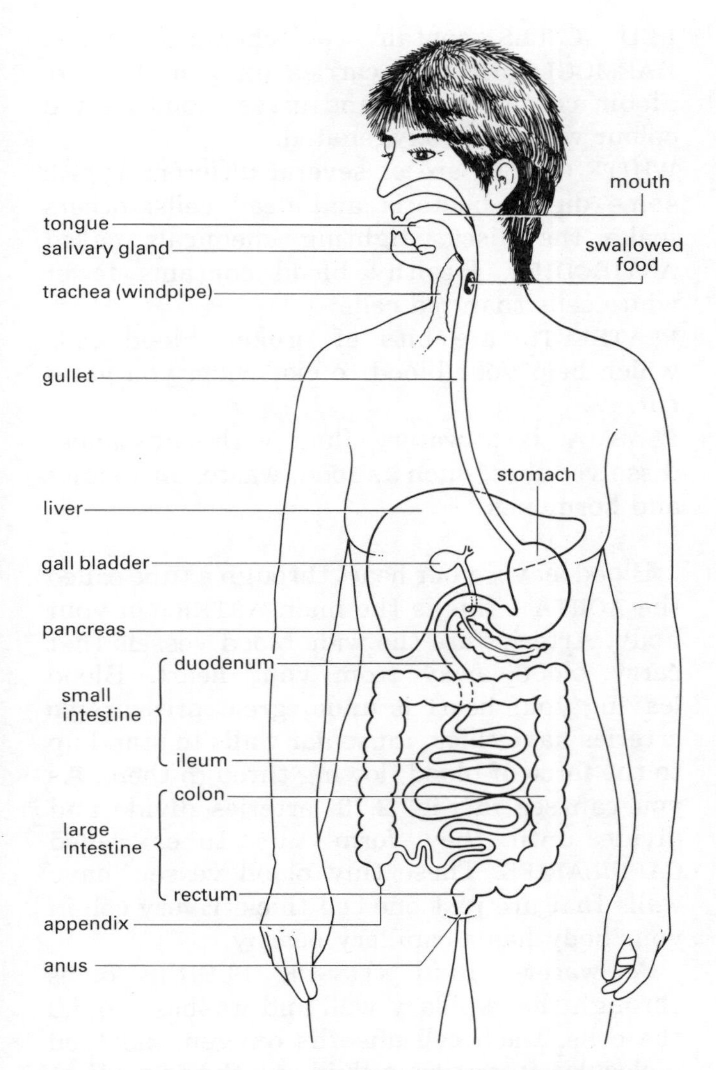

Fig. 2.20 Diagram of the human digestive system

How food is digested

Your teeth begin the task of digesting your food: chewing breaks up food into smaller chunks. At the same time, you have probably noticed that your mouth 'waters'. A solution called SALIVA mixes with food in your mouth and the *enzyme* in saliva starts breaking down the starch molecules in your mouthful of food. You will learn a little more about the way enzymes work in Unit 9: they are chemicals that break down food molecules until they are small enough to be absorbed into your blood system. Each enzyme attacks a different food and works under different conditions, so your digestive system is doing a slightly different job at each place along the way. Table 2.1 shows the most important stages in the way your food is digested.

Table 2.1 Stages in human digestion

Part of digestive system	Major events
Stomach	Gastric juices from cells in the lining; food mixed with gastric juices; enzymes activated by hydrochloric acid (both in gastric juices); enzymes act on protein in food
Small Intestine	1 Bile (from the gall bladder) - neutralises stomach acid, breaks fat into small droplets 2 Pancreatic juice (from pancreas) - enzymes to digest starch, fat & protein 3 Digested food absorbed into bloodstream, undigested fat into lymph system
Large Intestine	Water absorbed, undigestible substances e.g. cellulose expelled through the anus

As you can see, your stomach and small intestine are where most of your food is broken down. By the time food has reached your small intestine it is getting nearer to being absorbed into your bloodstream. Most of what you ate is absorbed through the walls of your small intestine. The wall has a large surface area, thanks to millions of tiny 'fingers' called VILLI that contain blood capillaries and lymph vessels (Fig. 2.21).

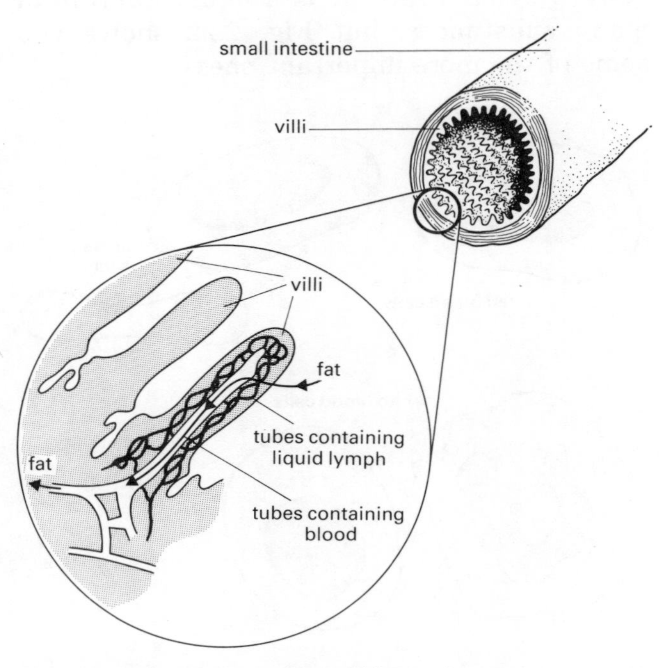

Fig. 2.21 A section of small intestine

Digested food molecules are simple enough to pass through the intestine wall and into the blood capillary. What makes food molecules move into the solution called blood? Look back to Investigation 2.2 and see if you can explain the way food is absorbed. Your small intestine also contains some droplets of fat that are not digested: these are absorbed into a liquid called LYMPH. The lymph vessels from your intestine wall carry away the fat, and empty it into the bloodstream at your neck.

Once digested food molecules have been absorbed into your bloodstream, they are carried around your body to be used or to be stored until they are needed. We have already dealt with the way your liver organises the storing of food (page 87). Now we shall look at the way food is transported to the liver, or to other cells to be used in respiration (see also Unit 9).

Transport systems

Animals and plants have organ systems that carry (transport) oxygen and food to cells and waste materials away from them. In humans the transport system is made of blood, blood vessels and a pump called a heart (page 84).

Transport in humans

Blood is a tissue that is pumped around your body by your heart. It is a liquid mixture of many substances, but Fig. 2.22 shows you some of the more important ones.

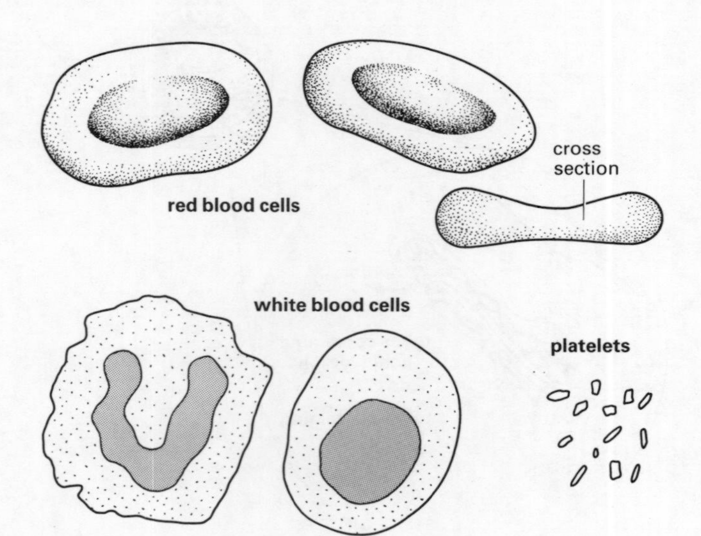

Fig. 2.22 What blood contains

RED CELLS contain a chemical called HAEMOGLOBIN that carries oxygen. Haemoglobin contains iron, and gives blood its red colour when it is oxygenated.

WHITE CELLS are of several different types: some digest bacteria and dead cells; others make the disease-fighting chemicals called ANTIBODIES. Healthy blood contains fewer white cells than red cells.

PLATELETS are bits of broken blood cells which help your blood to clot when you get a cut.

PLASMA is a watery fluid with substances dissolved in it, such as food, waste, antibodies and hormones.

Blood leaves your heart through a tube called the AORTA - this is the main ARTERY in your body. Arteries are the wide blood vessels that carry blood away from your heart. Blood leaving your heart is under great pressure so arteries have thick, muscular walls to stand up to the force of blood flowing through them. As you can see in Fig. 2.23, arteries divide and divide until they form tiny tubes called CAPILLARIES. These tiny blood vessels have walls that are just one cell thick. Every cell in your body has a capillary nearby.

A watery fluid (TISSUE FLUID) comes through the capillary wall and washes around the cells. Each cell absorbs oxygen and food molecules from tissue fluid. At the same time cell waste passes into the tissue fluid and from there into the capillary. The blood inside the capillary is now 'deoxygenated' and is carried to veins as the capillaries join up. As veins are the blood vessels that take blood back to your heart, they do not have to stand great pressure (why?) and have fairly thin walls. Some veins have one-way valves to stop blood flowing backwards. Figure 2.23 shows the design of one of these valves, which you could find in one of the veins in your leg. Think why veins in your legs need these valves.

Transport in plants

Plants take in water and minerals through their roots. The water is used to make the sugars that the plant cells need for the energy-releasing process called respiration. The minerals are used in the leaves to make the proteins that the plant needs to make new

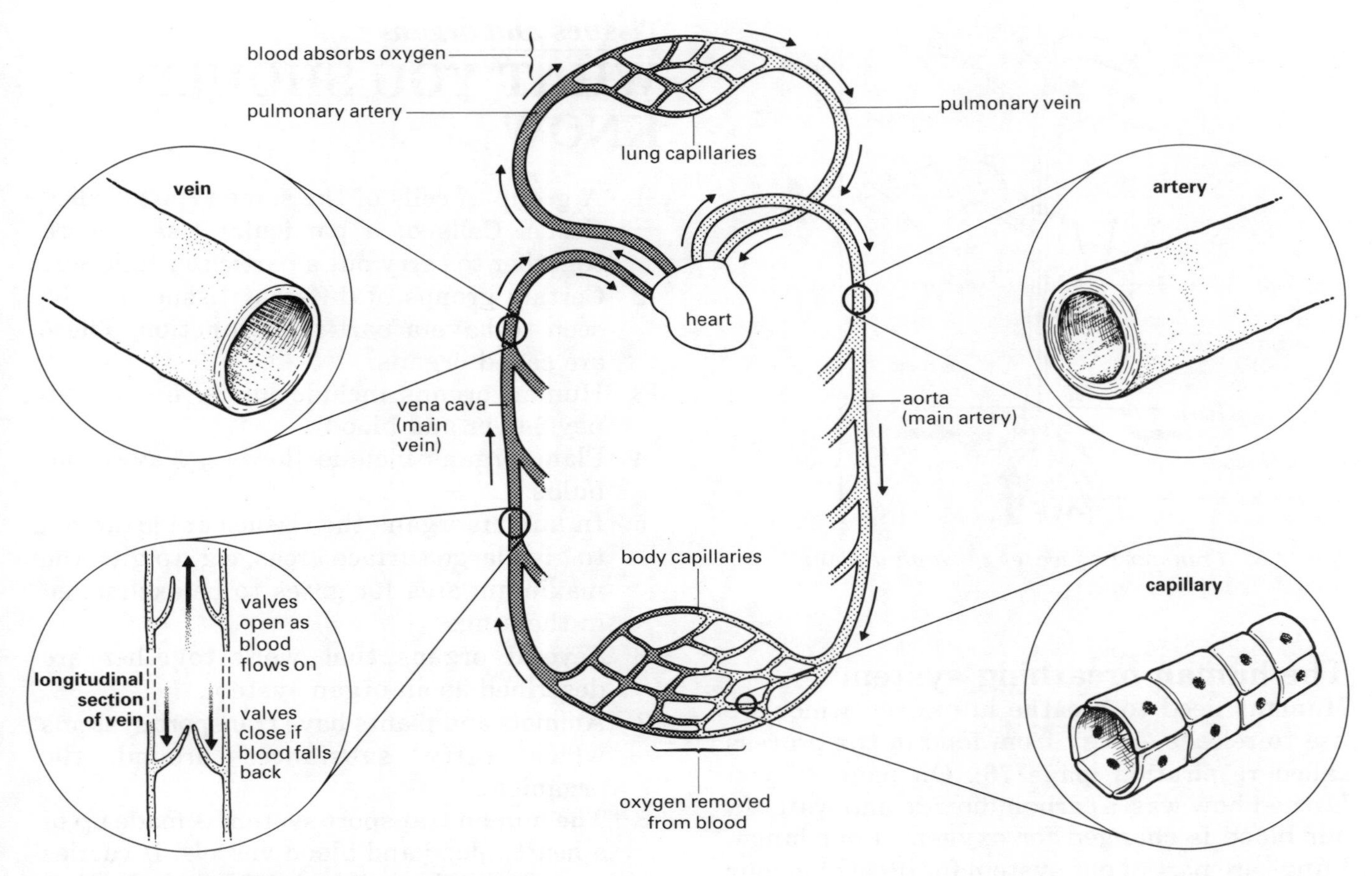

Fig. 2.23 Circulation of blood in humans

tissue. Earlier, we dealt with the way these substances are moved around by processes such as transpiration and osmosis. These are important in a plant's transport system, which is made of two sets of tissue. One, phloem tissue, carries food made by the plant; the other, xylem tissue, carries water and minerals.

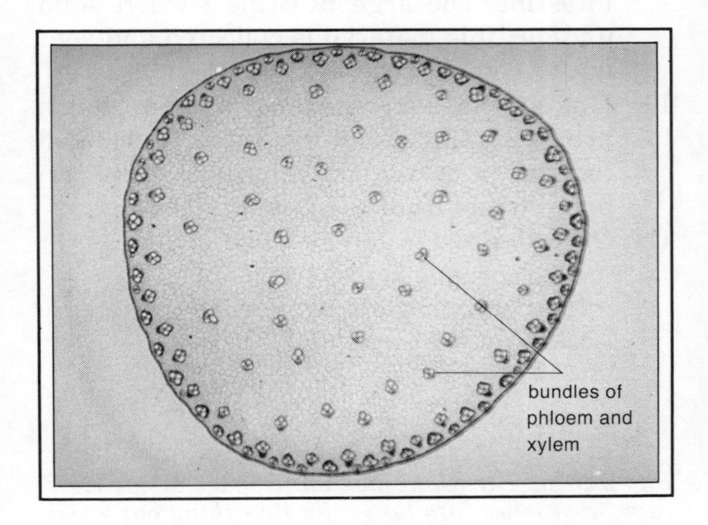

Fig. 2.24 Magnified cross sections of (a) (left) a plant stem (maize) and (b) (right) a tree trunk

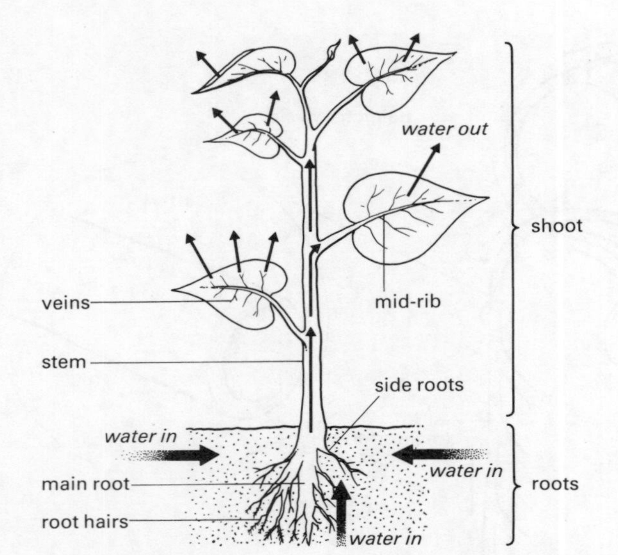

Fig. 2.25 Transport of water through a plant

The human breathing system

Humans need to breathe in oxygen which we use to release energy from food in the process called respiration (page 76). On page 86 you learned how waste carbon dioxide and water in our blood is changed for oxygen in our lungs. Lungs are part of our system for breathing (our RESPIRATORY SYSTEM). The rest of the system works to get air in and out of our lungs. If you look back to Fig. 2.16, you can see that lungs are inside an airtight 'cage' called the THORAX. The bars of the cage are your ribs, which are held together by INTERCOSTAL MUSCLES. The bottom of the cage is made of tough tissue called the DIAPHRAGM. These all work together as Fig. 2.26 shows.

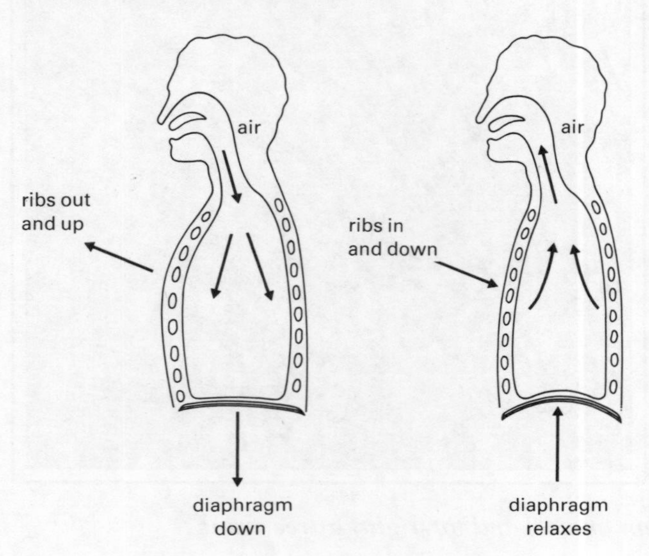

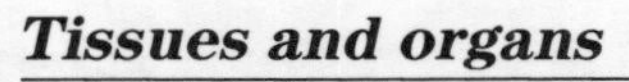

Tissues and organs

WHAT YOU SHOULD KNOW

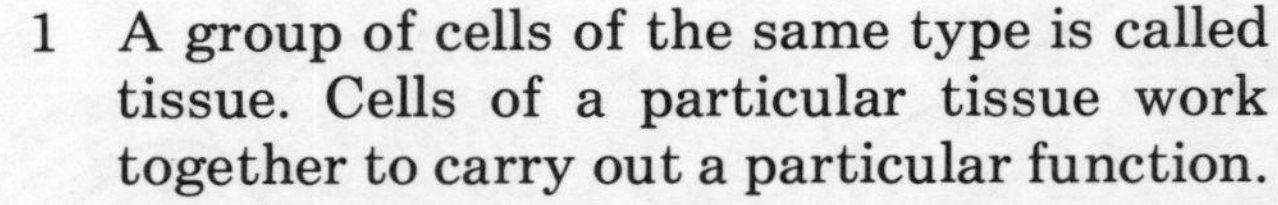

1 A group of cells of the same type is called tissue. Cells of a particular tissue work together to carry out a particular function.
2 Certain groups of different tissues can be seen to have a particular function. These are called organs.
3 Human organs include lungs, heart, kidneys, liver and blood.
4 Plant organs include flowers, leaves and bulbs.
5 In human organs the tissues are arranged to give large surface areas, e.g. to give the maximum area for gases to be exchanged in the lungs.
6 Several organs that work together are described as an organ system.
7 Animals and plants have transport systems which carry substances around the organism.
8 The human transport system is made up of a heart, blood and blood vessels. It carries oxygen, carbon dioxide, food and other useful substances around the body.
9 The plant transport system is made from tissue that carries food (phloem) and tissue that carries water and minerals (xylem).
10 Humans have a digestive system where food is broken down and absorbed into the blood stream. This happens mainly in organs called the stomach and the small intestine. The large intestine is where solid undigestible material is collected and sent out of the body.
11 Humans have a respiratory system, which brings air into the lungs to be exchanged for carbon dioxide. It then expels the carbon dioxide from the lungs.
12 Other human organ systems include
  the reproductive system;
  the nervous system;
  the excretory system.

Fig. 2.26 (a) Breathing in - more space inside rib cage, air rushes into lungs, (b) Breathing out - less space inside rib cage, air forced out of lungs

QUESTIONS

1 (a) What are ligaments?
(b) What are tendons?
(c) Explain how each of them is designed to help your joints to move.

2 (a) Name two places where you would find a ball and socket joint.
(b) Name two places where you would find a hinge joint.
(c) (i) What sort of joint links the bones in your back?
(ii) Explain how these joints let you bend over without collapsing.

3 (a) What do we mean when we say that food is *digested*?
(b) What is the name of the chemicals in our body which help to digest food?
(c) How is our small intestine designed to absorb food?

4 (a) What is the difference between a vein and an artery?
(b) Often people suffer from varicose veins as they grow older. This complaint happens when leg muscles are unable to support veins: a small portion of vein may 'sag', trapping blood, which then prevents blood flowing through easily.
(i) Why do leg veins need the help of muscles for support?
(ii) Draw a diagram showing the one-way valve that you could find in a vein and use it to explain how blood can get 'trapped' when the vein sags.
(iii) How would the situation you described in (ii), prevent blood flowing easily?
(iv) Suggest two means by which people could avoid developing varicose veins.

5 South American Indians, who spend their lives high in the mountains, have more red blood cells than people living at low altitudes.
(a) What is the job of red blood cells?
(b) What does the information in the question tell you about the air high above sea level?
(c) The Indians' bodies have adapted to life at great heights. When the Olympic Games were held in Mexico, high above sea level, many athletes went to Mexico months before the Games to train.
(i) Explain why this was necessary
(ii) What might have happened if they had not taken this precaution?
(iii) What extra safety precautions should be taken at sporting events so high above sea level?

6 The information in the table below shows the results of an investigation into a pupil's breathing pattern. Use the information to answer the questions.

	Composition of air/%		
	Breathed in	Breathed out during sleep	Breathed out while running
Nitrogen	78	78	78
Oxygen	21	17	12
Carbon dioxide	a little	4	9
Water vapour	a little	saturated	saturated

(a) Explain why the pupil breathes out less oxygen than he breathes in.
(b) Why does he breathe out more carbon dioxide when he is running than when he is asleep?
(c) Explain how there is extra water vapour in the air he breathes out.

Cells and Organisms

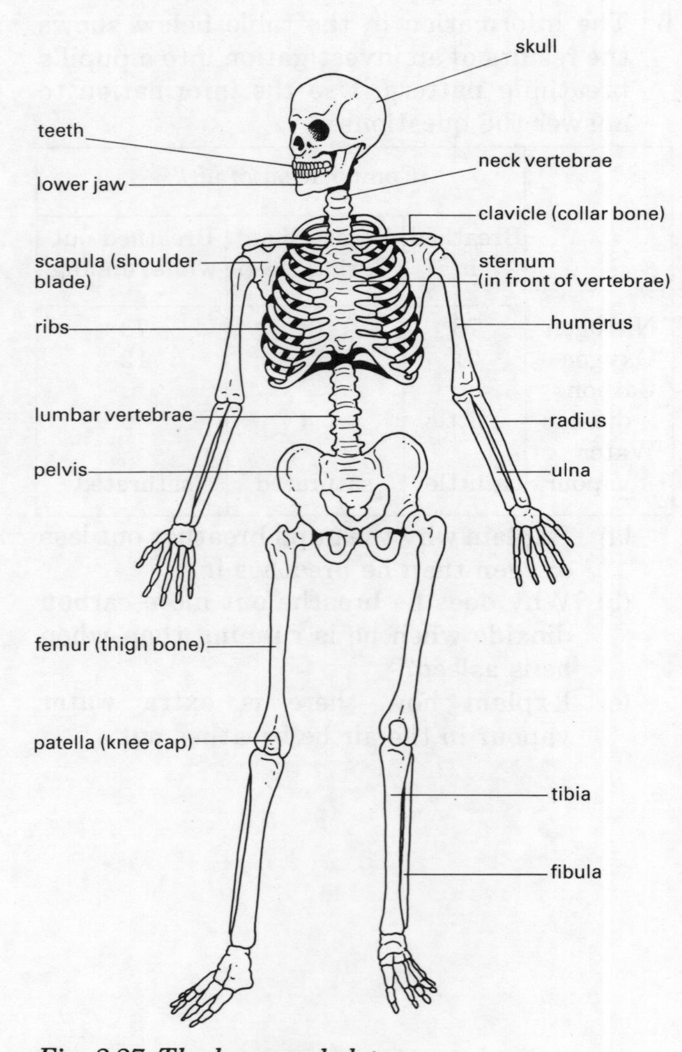

Fig. 2.27 The human skeleton

COMPLETE ORGANISMS

Each organ and organ system that makes up a plant or an animal is designed to do a special task that helps the organism to live. In the earlier parts of this unit you have seen how the organs and the cells in them are designed and organised to do their job. However, once we look at a complete organism there are some more design problems that need solving. Because organisms are alive, they need to move and grow without collapsing. In this section we shall look at some of the ways animals and plants are built.

SUPPORT IN PLANTS

Osmosis and transpiration work together to keep plant cells firm. As long as the cells in the stem of a flower are full of cell sap, the stem will keep the flower upright. As you saw in Fig. 2.10, a shortage of water means the cells get floppy and the flower wilts. As plants become bigger, their stems get bigger to support them. Plants such as trees which do not die at the end of the growing season are called PERENNIAL. Perennial plants usually have 'woody' stems which are strong and help to support the plant. The wood of a tree is mostly a substance called LIGNIN, which thickens many of the cell walls in the stem of a perennial plant as you can see in Fig. 2.24(b) on page 91.

SUPPORT IN ANIMALS

Animal cells do not have a rigid cellulose cell wall, so the cells themselves are not as 'firm' as those in a plant. Animals made from more than one cell usually have a SKELETON to give them support. Animals like humans have a skeleton made from hard bone tissue inside their bodies. Humans are VERTEBRATES, which means that they have a backbone as the main part of the skeleton inside their bodies. Some animals, described as INVERTEBRATES, have an external skeleton to support their bodies. In

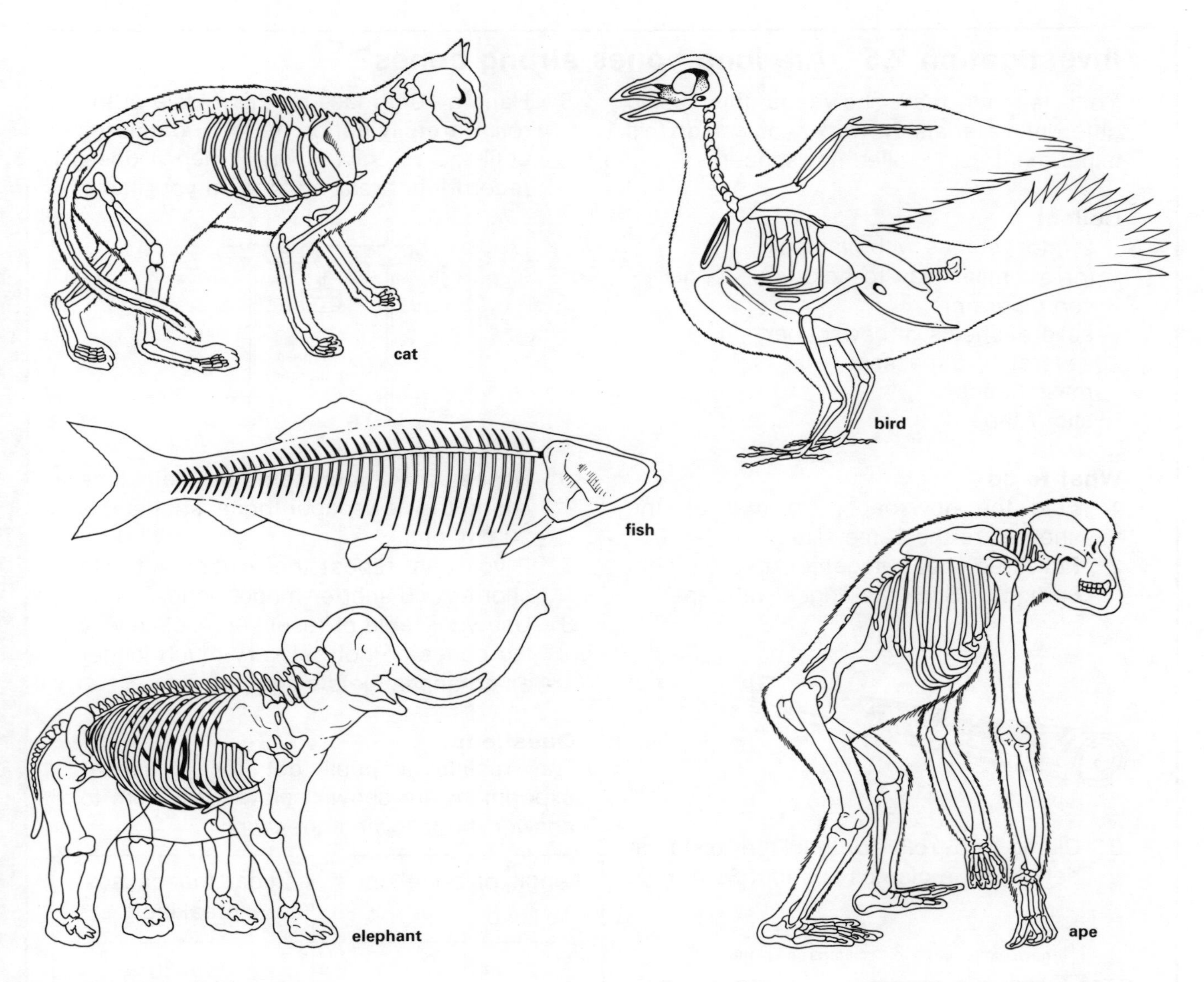

Fig. 2.28 Support in animals

Unit 6 you will meet some of the invertebrates.

In animals with a backbone, the shape of the backbone decides whether they can stand upright. As you can see in Figs. 2.27 and 2.28, the design of a human skeleton means that we can stand upright and move around on two legs rather than four.

What a skeleton does

Supports	—	so organs are not squashed,
	—	so animal can stand.
Protects	—	e.g. skull protects our brain, ribs protect our lungs.
Movement	—	muscles, bones and the joints between them make movement possible.

THE SPECIAL PROBLEMS OF LARGE ORGANISMS

Look at the animals in Fig. 2.28. They have been drawn the same size, to make it easier to compare how they are built. You should be able to see that large animals, like the elephant, have thick legs compared to smaller animals. Inside their legs, their leg bones are also thicker to help carry the extra weight. In order to understand *exactly* what happens as organisms get bigger, we need to look at the way mass and size are related.

Investigation 2.5 Are long bones strong bones?

Your teacher may show you this, using different materials. However you can do this experiment using rolled newspaper.

Collect

2 retort stands with clamps
(or a similar way to hold your 'bones' - see diagram)
several sheets of newspaper
several 10 g masses
mass hanger
sticky tape

What to do

1 Fold the newspaper so that all the sheets are the same size.
2 Put three sheets together and roll them up tightly. Stick the edges with tape.

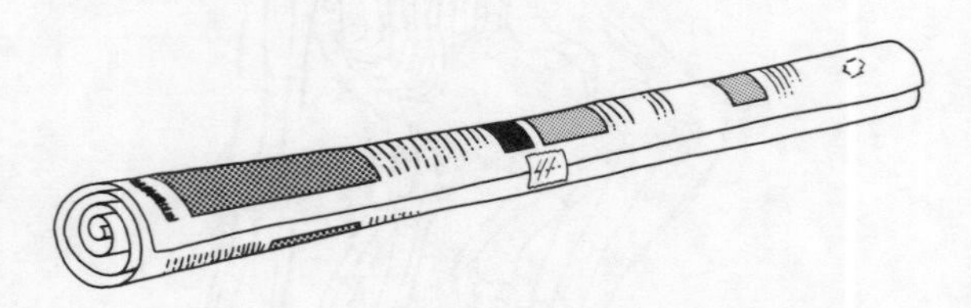

3 Clamp the roll so that the distance between the clamps is about 30 cm.

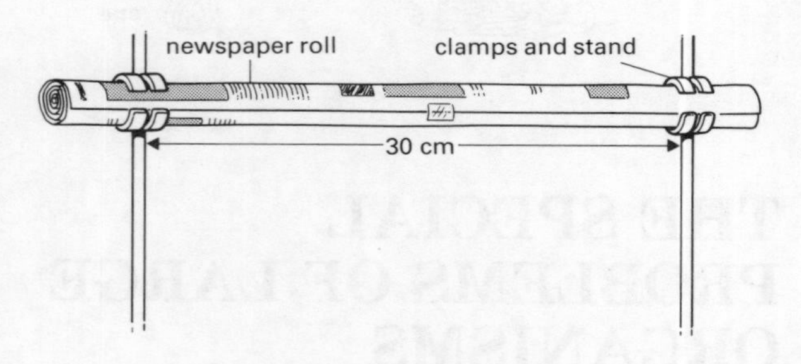

4 Copy this table into your book.

Length of roll/cm	Load at which it gives/g
30	
20	

5 Hang a 10 g mass on the middle of the roll. Carefully put on more 10 g masses until the roll gives way. When it does, record the 'breaking load' in your table

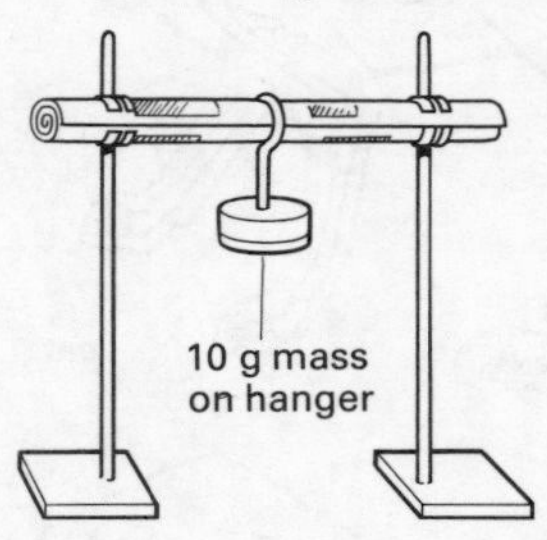

6 Repeat the experiment, but this time clamp the newspaper 'bone' so that it is 20 cm long.
7 If you can, repeat the experiment with shorter and shorter model bones.
8 Use your table of results to explain why leg bones do not get very much longer as an animal gets bigger.

Questions

The results a pupil got for a similar experiment are shown below. Use them to answer the following questions.

Length of 'bone'/cm	Load that causes breaking/g
40	30
30	40
20	60
10	80

1 What pattern does the strength of a 'bone' show as it gets longer?
2 Explain why animals that grow tall often have very long legs when they are young.
3 Which type of leg bone is more likely to break in an accident, a long one or a short one?

Putting on weight

If we look at the way animals are built, and how they cope with their size, we can also understand some of the ways other organisms support extra mass.

Look at the zebras in Fig. 2.29. The adult is much bigger and heavier but her legs are not much longer than her foal's, though they are a bit thicker. Leg bones are the main supports in an animal's legs, and the bones would need to get longer if the animal's legs were to grow.

Fig. 2.29 The baby zebra has longer legs in proportion to its size than its mother does

In Investigation 2.5 you saw that it takes a greater force to break a short bone than to break a long one. However, hanging masses on the 'bone' was putting force on it in a different direction than normal. In an animal's leg the force acting on it comes downwards, so the bone must stand up to that force. The next investigation should show you how the long bones in limbs are built to take the weight of an animal.

You may have seen long bones on sale at the butcher's shop, in which case you will know that they are hollow. In case you have never seen one, Fig. 2.30 shows you what a femur (thigh or hindleg bone) looks like. If these bones were the same length but solid, would they be heavier or lighter than the hollow version? If they were the same length *and* had the same mass, would they be thicker or thinner?

Investigation 2.6
Bone strength

You will need the same apparatus as for Investigation 2.5.

What to do

1. Set up the clamps in the same way as in Investigation 2.5.
2. Use the newspaper to make a hollow roll.
3. Clamp the hollow roll carefully, so there is 30 cm between the clamps.
4. Put on 10 g masses as before until the roll gives way.
5. If you can manage it, make rolls of different widths by rolling them more loosely, and repeat the experiment.

Questions

1. Compare your results from this investigation with the ones you got for Investigation 2.5. Which sort of 'bone' can bear a greater load?
2. The pupil carried on the experiment she began in Investigation 2.5 and found that 'hollow' bones carried a bigger load before they broke. This is what she wrote in her book: 'Long bones cannot carry as much weight as short bones. Baby horses have legs that are almost as long as their parents' legs. All the horses have leg bones that are hollow inside. This is because their legs do not grow much and solid bones would be too heavy for baby horses to move.'
   (a) Is she right when she says 'long bones cannot carry as much weight as short bones'?
   (b) Is the explanation for hollow bones correct?
   (c) Re-write the paragraph correctly.

In the last two investigations you should be able to see that a hollow bone is much stronger than a solid bone of the same length. This means that this sort of bone can support a heavier body, without having the extra load a solid bone would mean. Also, heavy animals get more support from short thick legs. In

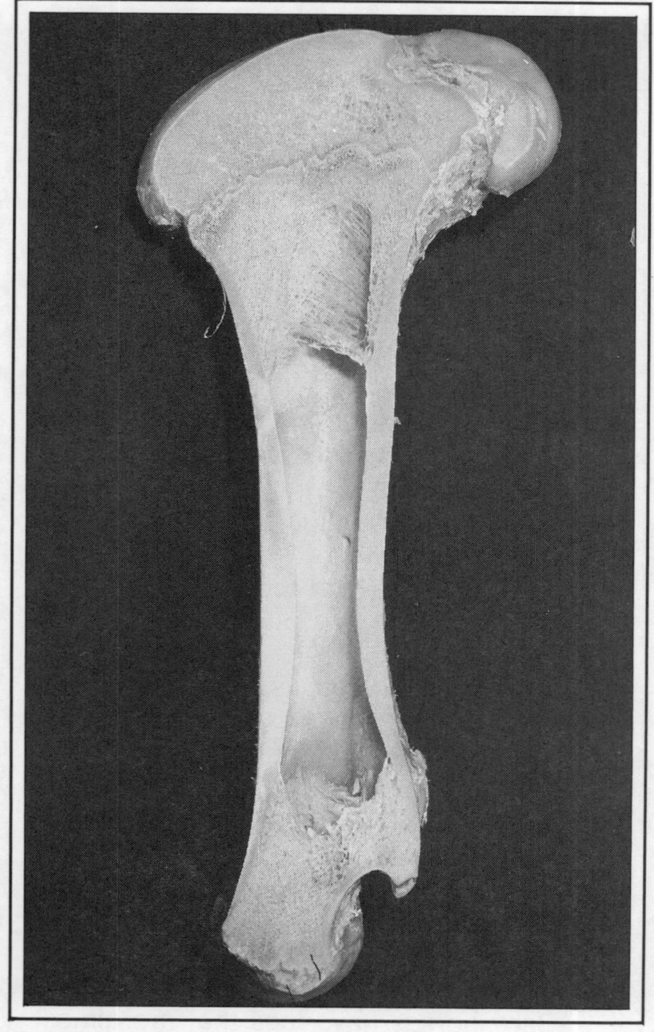

Fig. 2.30 A femur cut in half: this one is from a sheep

much the same way plants get some support as they grow. Think of the way a tree grows: as it grows taller, it also gets wider about the trunk. In the last section you learned how osmosis helps to support plants and how trees become woody to give them extra support. Their wide-at-the-bottom trunks are another part of the design to keep them upright. As you can see in Fig. 2.31, one of the biggest plants in the world, a redwood tree, has a thick, 'bottom-heavy' trunk.

In Unit 9 you will learn something more about stability, but if you stop to think about the way plant roots spread out underground in their search for water, you can see what a good 'anchor' they make. The larger a plant grows, the bigger its root system grows. That could explain why gardeners should not plant young trees too close to a house. Can you think why?

Fig. 2.31 A giant redwood tree makes people and even other trees look very small

WHAT! NO LEGS?

Some organisms do not have legs or roots to support their weight. They must have some other means of support. Very often such organisms use water to support them. Fig. 2.32 shows just what a difference water can make.

One of the animals to benefit from the support water can give is the whale. Whales are the largest animals on earth. Blue whales can be 30 m long and have a mass of 120 000 kg. However, the whale's skeleton cannot support the rest of its body out of water and a stranded whale will die. Its organs are crushed by the weight of its body.

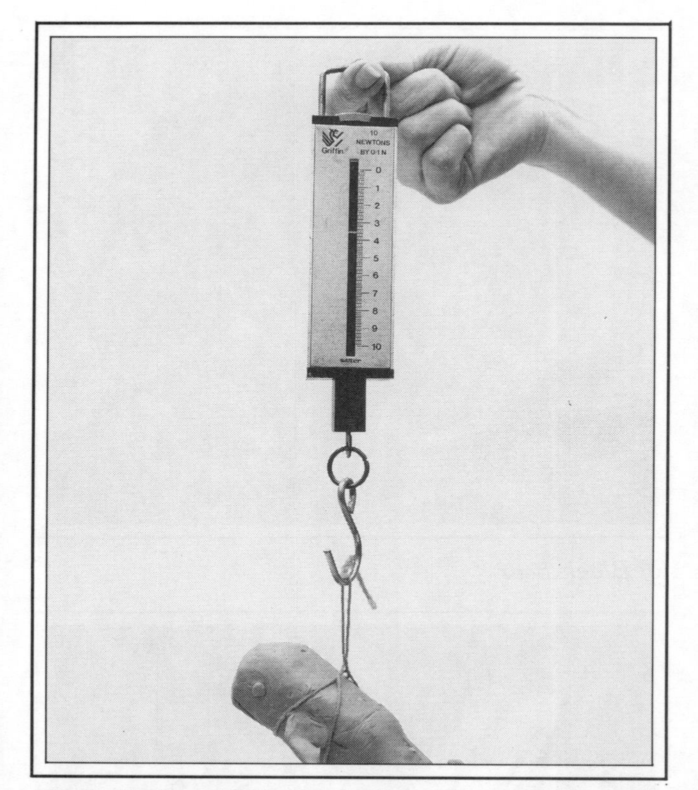

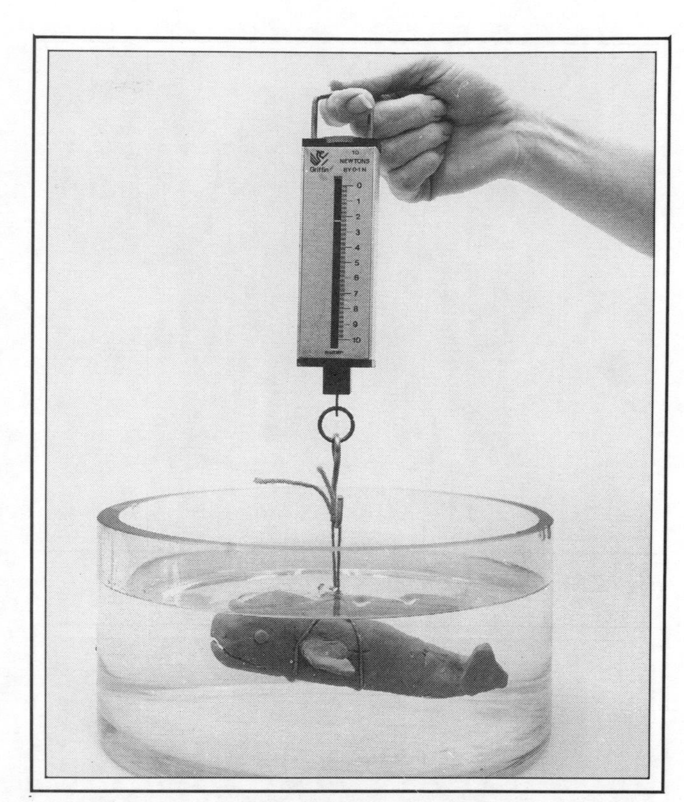

Fig. 2.32 An object weighs less in water than it does in air: the balance on the left reads 3.5 N; the one on the right reads 2.3 N

Fig. 2.33 Whales stranded on a beach in Orkney

SIZE, SHAPE AND CLIMATE

The size and shape of an organism affects the way it lives as well as supporting it. On page 86 you learned how surface area affects the way an organism loses water and heat. The bigger surface area becomes, the easier it is for an organism to lose energy as heat. All organisms have to control their temperature in some way, and Unit 9 will go into this in more detail. One of the design features of an organism that can help is its surface area.

Fig. 2.34 (a) Polar bear

(b) Black bear

(c) Snowshoe hare

(d) Jack rabbit

(e) Arctic fox

(f) European fox

Investigation 2.7
Built for the weather

Study the pictures of animals in Fig. 2.34.

1 For each pair of animals, make a list of the differences between them.
2 Which of the animals need to lose heat to keep their body temperature steady?
3 Which animals need to prevent heat loss?
4 (a) Describe the body shape of the animals that want to lose heat.
   (b) What other feature do they have to help them lose heat?

As you saw in the last investigation, organisms that need to lose heat have large surface areas. Those that do not want to lose heat have a shape that gives them a small surface area. If you look back to the photographs in Fig. 2.34 you can see that round, compact shapes give a smaller surface area. Animals from colder climates also have smaller ears and tails, to reduce the area exposed to the cold air. They often are covered in thick fur, too, which traps air and prevents heat escaping. Why is trapped air good at stopping heat escaping?

Not all organisms have the perfect design for the climate where they are found. Look at Fig. 2.35 which shows two animals that have about the same mass. The giraffe has the biggest surface area and is best designed to live in warm climates. The hippopotamus also lives in the same climate, but has a much smaller surface area. Its rounder shape means that it has less surface from which heat can escape. Hippos solve this problem by spending time in water, to keep themselves cooler.

Fig. 2.35 The hippopotamus (below) and the giraffe (above) cope with a hot climate in different ways

Investigation 2.8 Surface area and heat loss

In this experiment you will use flasks of hot water to represent organisms. If all the flasks are the same shape but different sizes, you can see how animals lose heat as they get bigger.

Collect

2 round flasks of different sizes (e.g. 50 cm^3 and 500 cm^3)
2 thermometers in bungs, one to fit each flask
hot water
graph paper

What to do

1 Set up the flasks as the diagram shows.

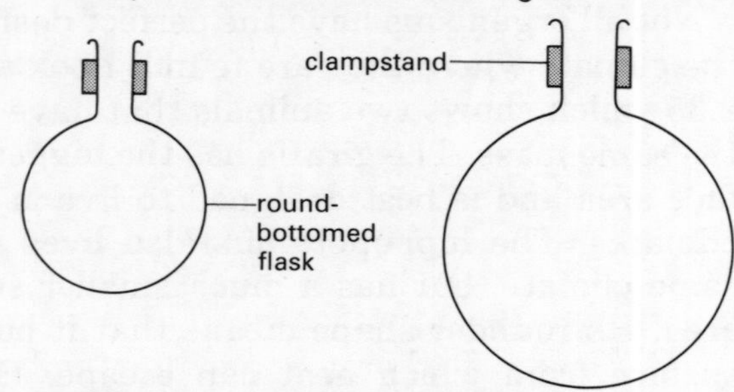

2 Collect some hot water and fill both flasks until water is half way up the necks.

3 Put the thermometers in and record the starting temperature.

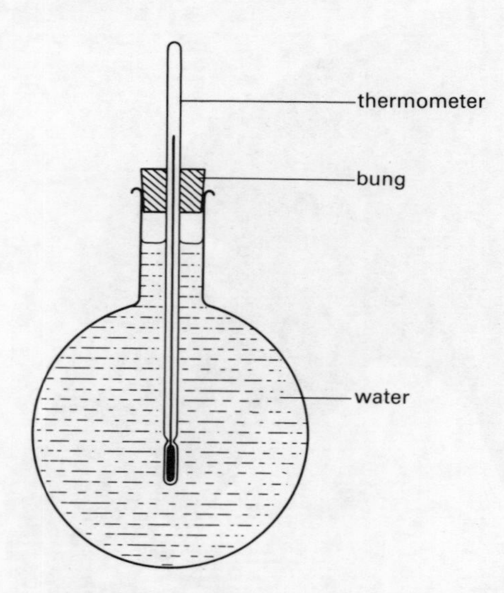

4 Take the temperature in both flasks every five minutes and record your results in a table like the one below.

5 After about half an hour, remove the thermometers, empty the flasks and use your results to draw two curves on graph paper. One graph should show how the water temperature in the small flask changed; the other should show how the water temperature in the bigger flask changed.

Questions

1 Which flask kept warm longest?

2 Which flask has the smallest surface area?

3 Young babies need to be wrapped warmly even when the weather is not very cold. Use the results of this experiment, and what you understand about surface area and heat loss, to explain why this is so.

4 A girl decided to go on a diet and lost weight easily. After a few months she could wear clothes several sizes smaller. She was pleased about that, but complained that she always seemed to feel cold. Can you explain the reason for this?

	Time/min.	0	5	10	15	20	25	30
Water tempera-ture /° C	Small flask							
	Large flask							

Complete organisms

WHAT YOU SHOULD KNOW

1 A living organism is a collection of organs.
2 Organisms vary in the way they are designed to survive in their surroundings.
3 All organisms need support. In humans and other vertebrate animals, this support comes from a skeleton made of bones.
4 The human skeleton also allows movement and protects internal organs.
5 Plants get some support from a combination of osmosis and transpiration. These keep stem cells firm, and the plant upright.
6 Plant that do not die at the end of the growing season develop strong, woody stems for extra support.
7 As animals become bigger they need thicker legs to support them. Leg bones are hollow because they give the same strength as solid bones for less mass.
8 As plants become bigger they grow thicker stems, which are wider at the bottom. They also get support from a larger root system below the ground.
9 Large water organisms get additional support from the water they live in. This means they can grow to a greater size than if they lived on land.
10 The design of an organism must also relate to the climate in which it lives.
11 As an organism grows, its surface area gets smaller in relation to its mass and volume. This means young animals have a comparatively large surface area.
12 A large surface area makes loss of heat energy easier.
13 Animals from cold climates tend to have smaller surface area than those from a warm climate.
14 Round, compact organisms tend to have a smaller surface area than any other design.
15 Other features of organisms in different climates include:
fur covering to reduce heat loss in cold conditions;
large ears to increase surface area for losing heat in hot conditions.

QUESTIONS

1 Pine trees have leaves that are shaped like needles.
(a) How will the surface area of a pine needle compare to that of an ordinary leaf of the same mass?
(b) What effect will this have on the amount of water that the pine tree will lose?
(c) Trees like the pine do not lose their leaves in winter. Suggest why they often have needle-shaped leaves.
(d) Suggest why other trees lose their leaves in winter.

2 As part of her homework a pupil wrote: 'Animals from cold climates have fur or a layer of fat under their skin. They are usually small and do not have large ears or other bits that stick out.'
(a) Explain why such animals have fur.
(b) Why may animals from cold areas have fatty layers under their skin?
(c) Explain why the pupil's teacher disagreed with her when she said that animals from cold climates are small.
(d) How would having 'bits that stick out' make the animals lose heat?

3. Energy 1

Fig. 3.1 Forces can cause changes in speed, direction, shape and size. What changes are being caused in these photographs?

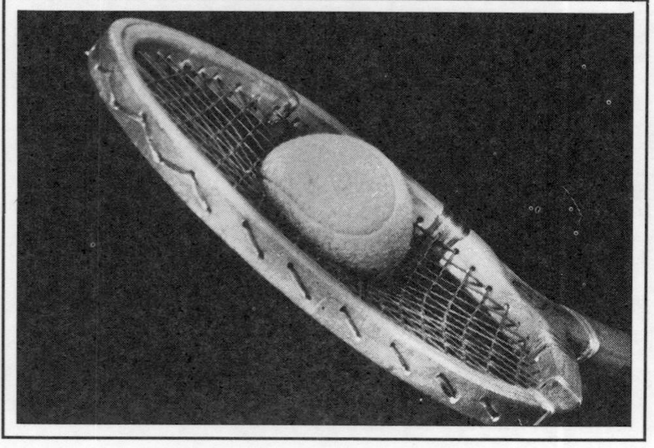

FORCE, WORK AND POWER

FORCE

You have already met forces in Unit 1. You will remember that forces produce changes. Figure 3.1 shows some of the changes that forces can produce. You will see that the only way that we know a force is acting is by the *change* it produces.

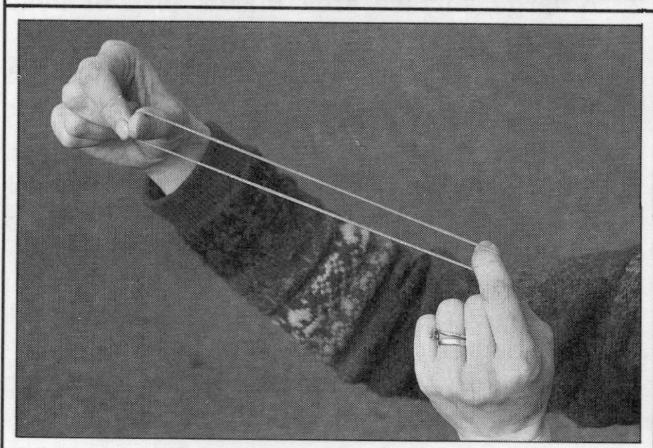

Fig. 3.2 Balanced forces

Balanced forces

Look at the photograph of the bridge in Fig. 3.2. Are there any forces acting? At first sight, the answer could be 'no' because there are no obvious changes. But the effects of all the different forces cancel out. The bridge remains stationary because all the forces on it are exactly balanced.

The parachutist in Fig. 3.2 is moving down at a constant speed. Moving at a constant speed is *not* a change (see Unit 1, p.27), so the forces acting must be balanced. The downward force (called the weight) of the parachutist and equipment is balanced by an upward force due to air resistance.

Directions of forces

With the parachutist, the directions in which the forces acted are shown by arrows. We cannot find out if the forces are balanced or not unless we know the directions in which the forces act. For any force, it is necessary to know both its size and the direction in which it acts.

In scientific measurements, there are two types of quantity: SCALAR and VECTOR. Figure 3.3 shows a few examples of scalar quantities.

These quantities have size (and units) but no direction. Can you think of other scalar quantities? But a vector quantity must have its direction stated as well as its size (and units). You

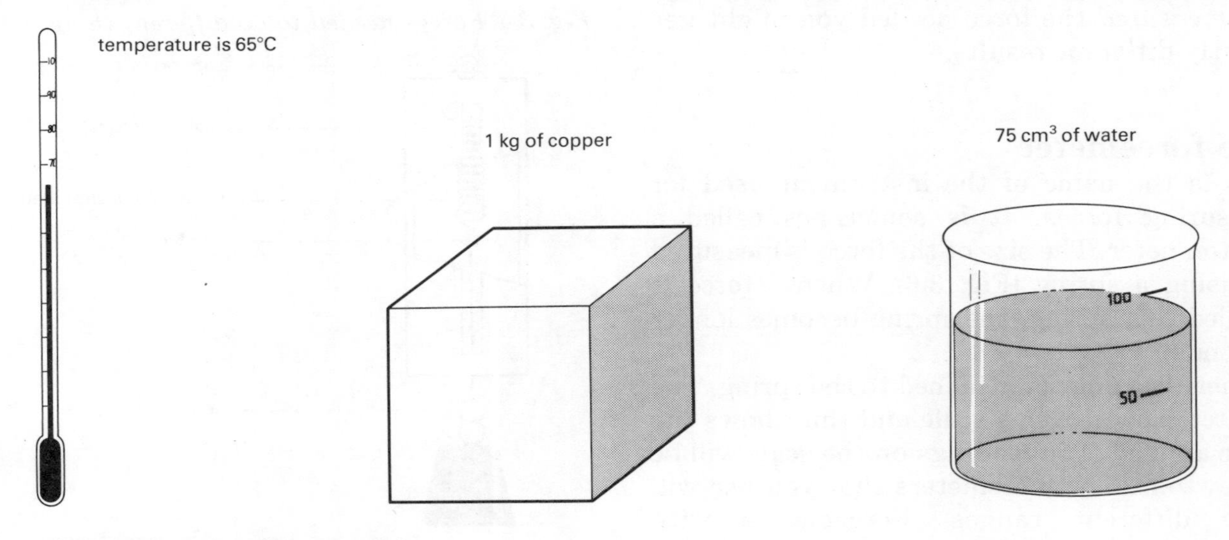

Fig. 3.3 Examples of scalar quantities

will meet a number of vector quantities while studying this course.

Here is an example of the importance of knowing directions. A person walks 30 m, stops and then walks 40 m. How far is that person from the starting position? There is no way of knowing *unless* you know the directions in which the person was moving. If you are told that the person walks 30 m due north, stops and then walks 40 m due east, could you solve the problem?

As all forces are vector quantities, it is essential to show the directions in which they act. In diagrams, the direction should always be shown by an arrow. The size (and units) of the force should be printed close to the arrow.

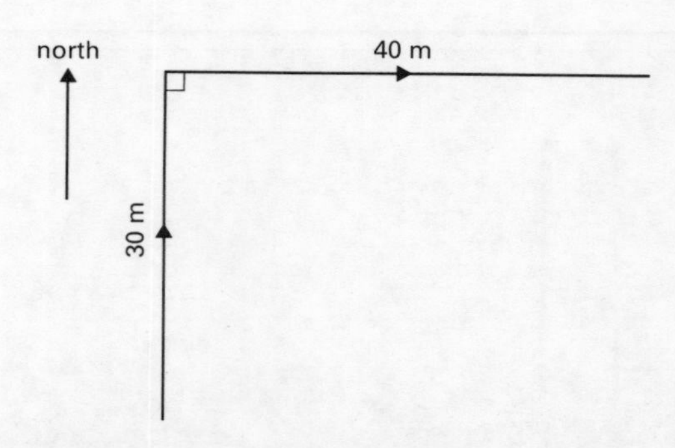

Fig. 3.4 How far away is the person?

Measuring forces

Apart from knowing the direction in which a force acts, it is necessary to know its size. All forces are measured in units called *newtons*. The symbol for the newton is N. The unit is named after Sir Isaac Newton (1642–1727). One of his discoveries was the law of gravitation. The force due to gravity pulls us (and everything else!) towards the surface of the earth. This force is usually called the weight of the object. Newton's law allows us to calculate the size of this force.

It is said that an apple falling from a tree started Newton thinking about gravity. The force needed to support an apple is about 1 N. In Fig. 3.5, you can see what force is needed to do certain things. These are typical values. If you *measured* the force needed you might get slightly different results.

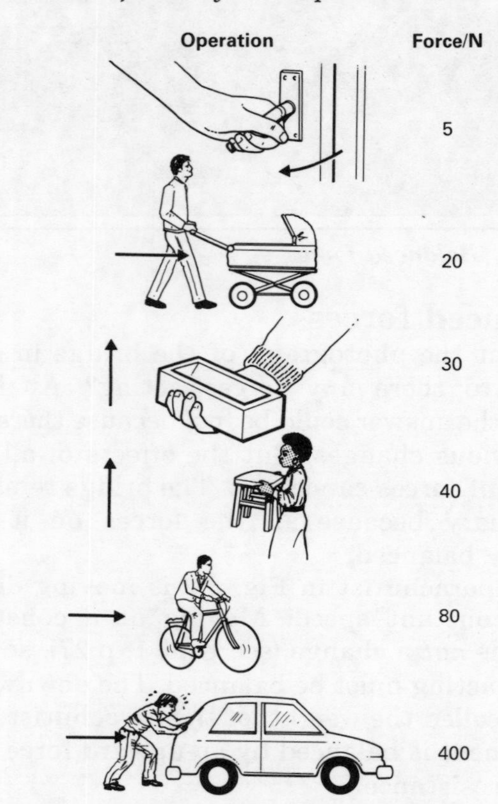

Fig. 3.5 Forces needed to do different things

The forcemeter

This is the name of the instrument used for measuring forces. It is sometimes called a newtonmeter. The size of the force is measured by using a spring (Fig. 3.6). When a force is applied to a spring, the spring becomes longer (extends).

There is a pointer attached to the spring. The pointer moves over a scale and this shows the force applied. The readings on the scale will be in newtons. The forcemeters that you use will have different ranges. Forcemeters with stronger springs will measure larger forces.

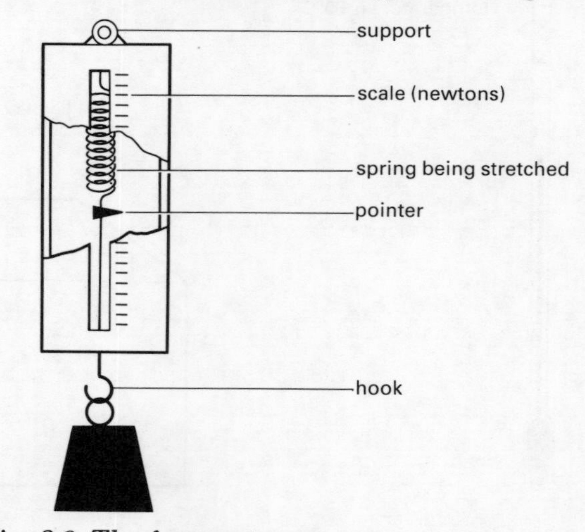

Fig. 3.6 The forcemeter

Assignment – Measuring forces

1 What are the readings recorded on the following forcemeters?

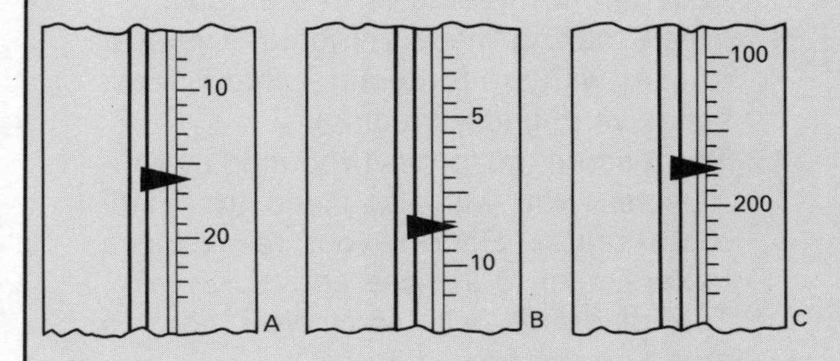

2 Assume that you had three forcemeters reading different ranges:
0–10 N; 0–100 N; 0–500 N.
Which forcemeter would you use to measure each of the forces mentioned in Fig. 3.5?

3 What do you think the reading will be on each forcemeter in the following examples?

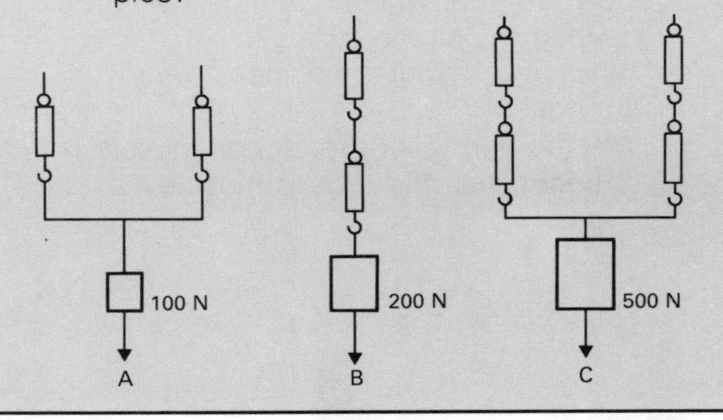

4 In an experiment to investigate the relationship between the extension of a spring and the force applied to it, the following results were obtained.

Load/N	0	1	2	3	4	5	6
Extension/cm	0	2.1	4.2	6.3	8.4	10.5	12.6

Plot these results on a graph. The horizontal axis should be the load and the vertical axis the extension.
(a) Is the graph a straight line or curved?
(b) If the load is 4.6 N, what is the extension?
(c) If the extension is 7.2 cm, what is the load?

5 You have been asked to design a cheap forcemeter that uses an elastic band instead of a spring. Describe how you would design the device. How would you make sure that it measured forces accurately?

6 Try to find out who discovered the connection between the load on a spring and the amount it stretches. When did he live? What *is* the connection he discovered?

Muscles and the body

All movements made by the body are produced by muscles. There are more than 600 muscles in your body. Muscles that enable us to produce forces are attached to bones by means of tendons. These muscles can only exert forces by contracting (becoming shorter). The muscle, as it contracts, becomes fatter and bulges. You can easily see this with your arm or leg muscles. Figure 3.7 shows the important muscle systems in the human leg. Fig. 2.13 on page 84 shows the muscles of the human arm.

To understand how these muscles work, look at the diagram of the thigh muscles. You can see the tendons that attach the muscles to the bones. To lift your lower leg, the front thigh

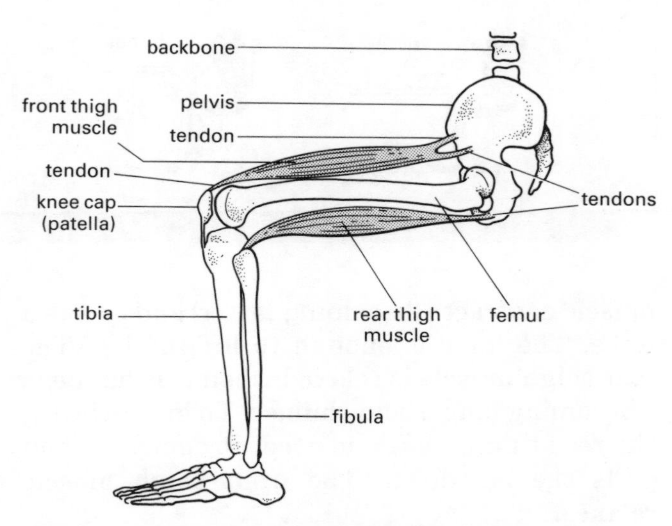

Fig. 3.7 Muscle system of the upper leg

Investigation 3.1 To measure the size of various forces

Collect

different range forcemeters
sandbag
trolley, with wheels
trolley, without wheels
brick
string
bathroom scales, reading in newtons

What to do

1 Draw a table in your book like the one shown below. Leave space for other measurements of force that you make.

Force to be measured	Size of force/N
Pulling a 10 N sandbag along bench	
Lifting a 10 N sandbag	
Pulling a trolley along bench (with wheels)	
Pulling a trolley along bench (without wheels)	
Lifting a brick	
Opening a door	

2 You may have to devise ways of connecting the forcemeter to the object.
3 When making measurements, it is best to start with the forcemeter that is least sensitive (highest reading).
4 To measure the forces produced by your muscles, you will also need the bathroom scales. Record your results in a table similar to the one above.
The diagrams will show you how to measure the forces.

Questions

1 When pulling the sandbag along the bench, how could you make the force smaller?
2 Why do you think that the force needed to pull the sandbag along the bench is smaller than the force needed to lift it?
3 How do the wheels affect the force needed to pull the trolley?
4 Which of your muscles was the strongest?
5 Why do you think the biceps muscle is stronger than the triceps muscle?

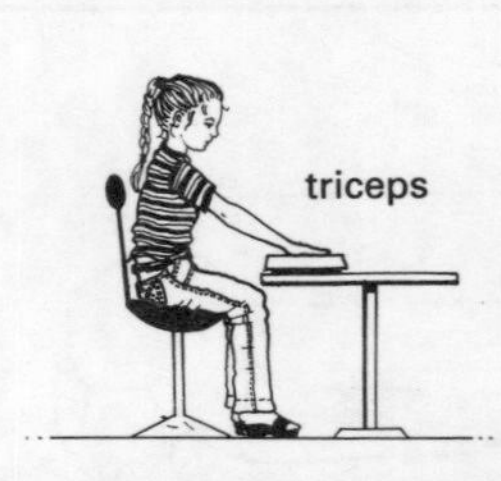

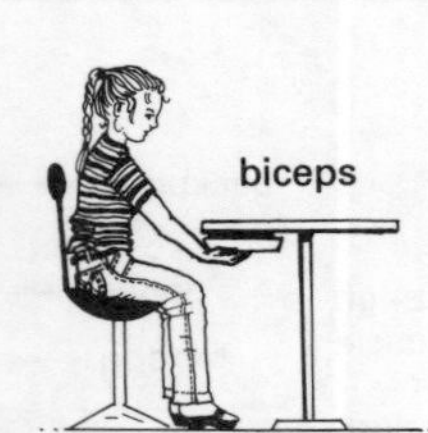

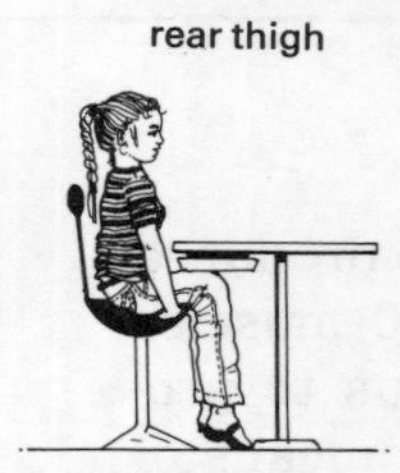

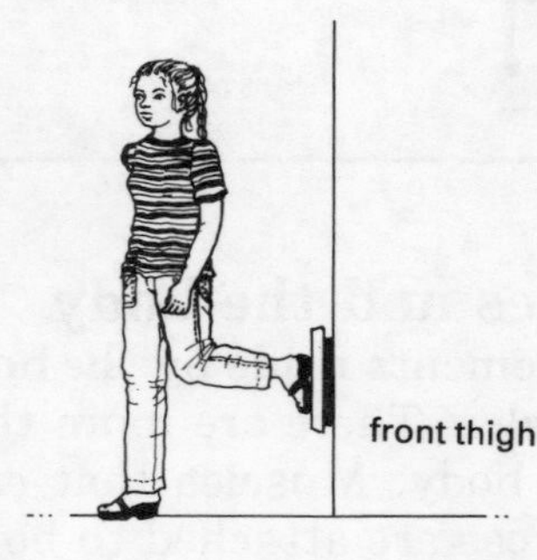

muscle contracts (becoming fatter) and exerts a force. The force is enough to lift the leg. The rear thigh muscle is relaxed when this happens – becoming longer and thinner. To lower the leg the rear thigh muscle is used. It contracts and pulls the leg down. The front thigh muscle relaxes.

For every movement of the body, there must be a pair of muscles. Why must they always be in pairs? You have seen that the forces exerted by different muscles are not the same. If the muscle has to exert a large force then that muscle is large. The thigh muscles have to help to move the whole body and so they are very large. The muscles that move the fingers do not have to be very strong. So they are much smaller.

Increasing the force available

From the last experiment, you will have seen that there is a limit to the forces that muscles can produce. If you have to exert a larger force than this, then you will need a device to multiply the size of your available force. For instance, breaking open a hazel nut using one's fingers would be impossible for most people. The nutcracker is a device that will increase the available force and break the shell of the nut.

The nutcracker is one example of a whole set of devices called LEVERS. Some of these are shown in Fig. 3.8. In each case, a force (called the EFFORT) has to be applied to produce a change in the LOAD. Sometimes the forces are in pairs (as with the nutcracker). The lever has to turn about a fixed point. This is called the FULCRUM (or pivot). You will notice that the effort, load and fulcrum are not always in the same order. How many different ways can you find of arranging the effort, load and fulcrum?

In most of the examples above, the lever increases the size of the effort applied. Such levers are called *'force multipliers'*. But look at the last example. If the effort (produced by the muscle) moves through a small distance, then the load moves through a much larger distance. This is called a *'distance multiplier'*. Why do you think a 'distance multiplier' lever is useful for the movement of limbs in the human body?

A lever can act as a 'force multiplier' or a 'distance multiplier'. It cannot be both at the same time. With a 'force multiplier' lever, the effort moves through a larger distance than the load. The opposite is true for the 'distance multiplier': the load moves through a larger distance than the effort but the size of the effort is greater than that of the load. Look at Fig. 3.8 and decide which of the devices are 'force multipliers' and which are 'distance multipliers'.

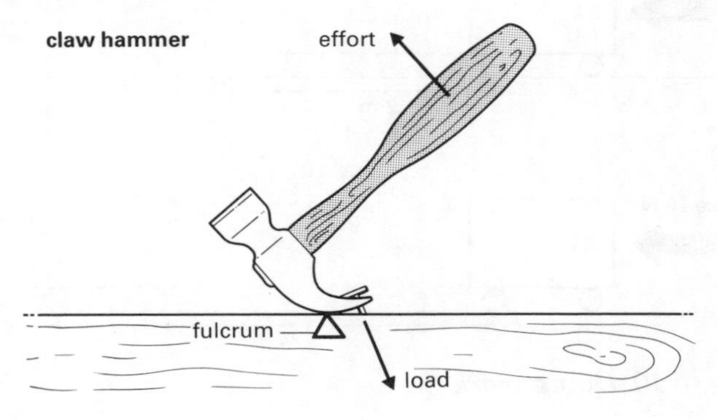

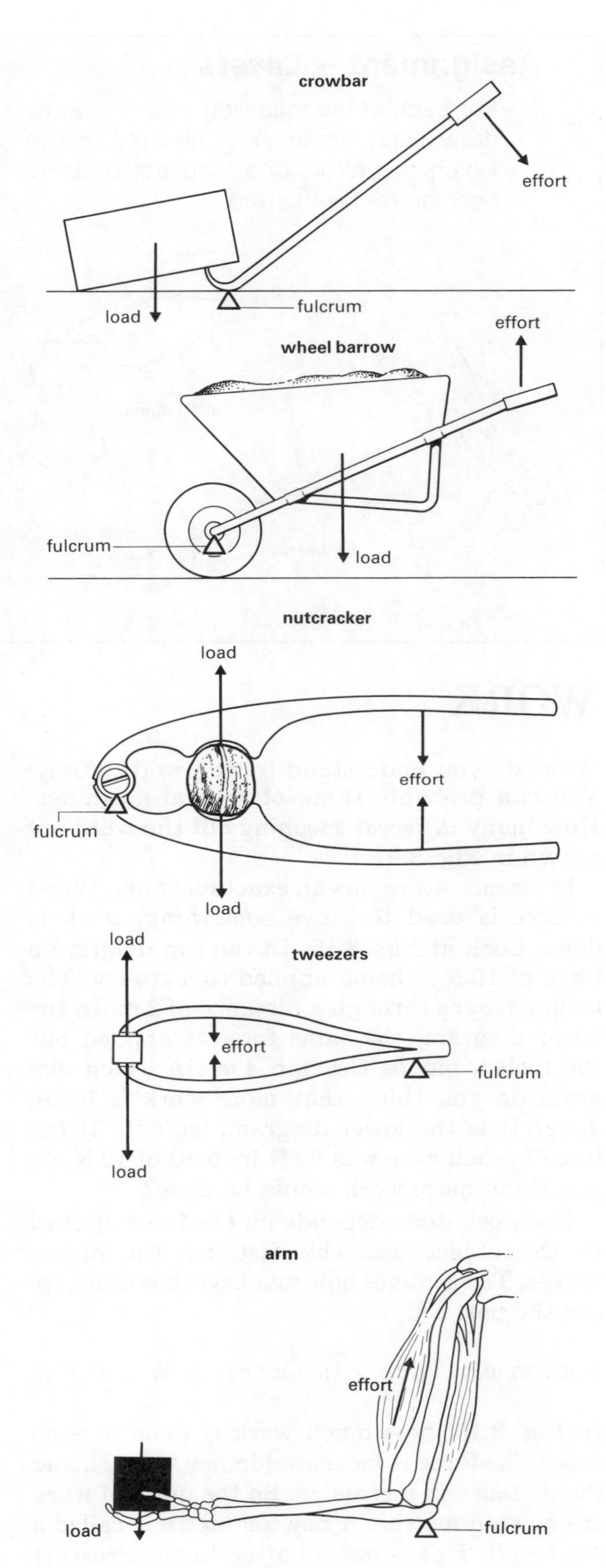

Fig. 3.8 Some examples of levers

Assignment – Levers

1 For each of the following lever systems, draw a diagram to show the directions in which the effort and load act and the position of the fulcrum.

2 In each case, say if it is a 'force multiplier' or a 'distance multiplier'.

3 Try and think of other levers that you know. Draw diagrams to show how they work. Say if they are 'force multipliers' or 'distance multipliers'.

4 The human body has many lever systems involving the muscles. We have already looked at the thigh muscle example. Sketch and describe two other lever systems.

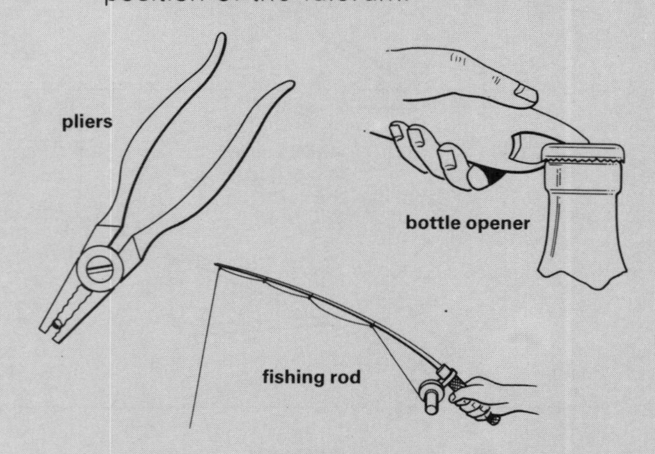

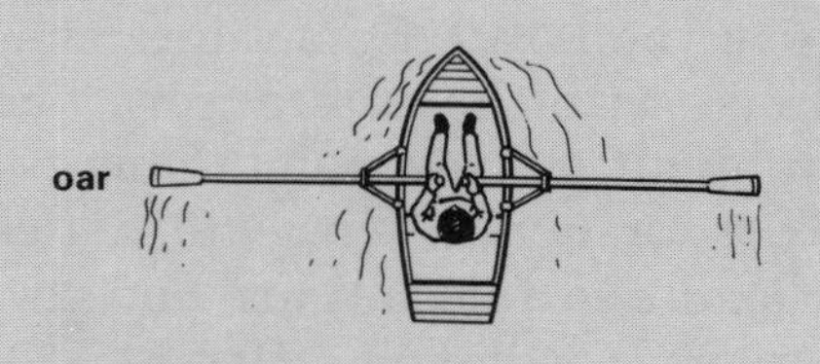

WORK

What do you understand by the word WORK? You can probably think of several meanings. How many *different* meanings of the word are shown in Fig. 3.9?

In science, *work* has an exact meaning. When a force is used to move something, work is done. Look at Fig. 3.10. In the top diagram a force of 10 N is being applied to a trolley. The trolley moves through a distance of 2 m. In the lower diagram, the same force is applied but the trolley moves through 4 m. In which diagram do you think that more work is being done? It is the lower diagram, isn't it? If the force in each case was 20 N instead of 10 N, do you think more work would be done?

The work done depends on the force applied to the object and the distance the object moves. To calculate how much work is done, we use the pattern:

work done = force × distance or $W = F \times d$.

In Fig. 3.10, how much work is done in each case? The force is measured in newtons, N, and the distance in metres, m. So the units of work are newton metres. A newton metre is called a JOULE (J). This is named after James Prescott Joule (1818–1889).

Fig. 3.9 Everyday meanings of the word 'work'

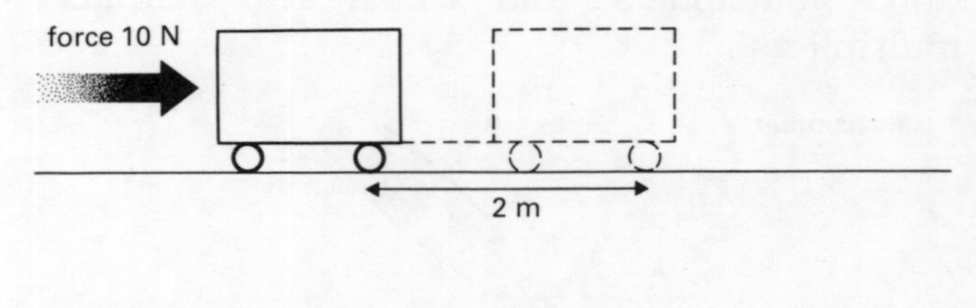

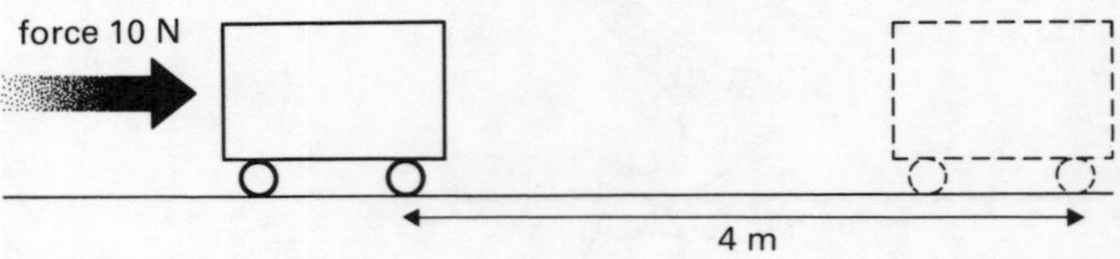

Fig. 3.10 Doing work

Using the work formula

Making calculations about work can be done using the triangle method. If you know two of the quantities in the formula W = F × d, you can always calculate the third one (Fig. 3.11).

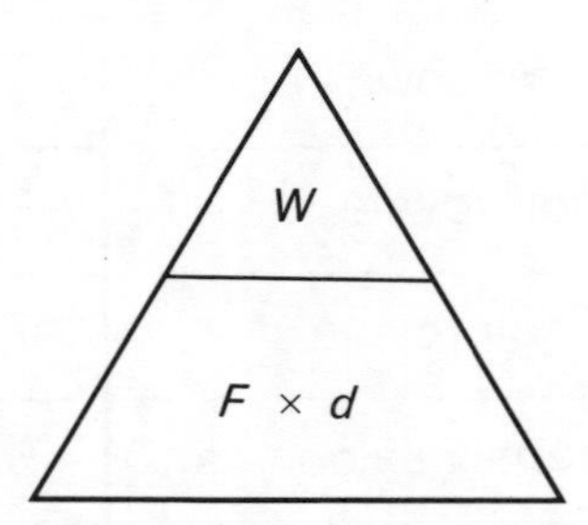

Fig. 3.11 Calculating the work done

For example: If 500 J of work is done moving a body through a distance of 10 m, find the force acting. (Cover *F* with your finger, then the triangle tells you that $F = \frac{W}{d}$.)

$$F = \frac{W}{d} = \frac{500}{10} = 50 \text{ N}$$

If a 1000 N force moves a body and 5000 J of work is done, what distance does the body move? (Cover *d* with your finger, then the triangle tells you that $d = \frac{W}{F}$.)

$$d = \frac{W}{F} = \frac{5000}{1000} = 5 \text{ m}$$

Assignment – Working out work

What quantities should be in the blank spaces below? What are the units in each answer?

Work	Units	Force	Units	Distance	Units
		80	N	5	m
		200	N	0.2	m
500	J			5	m
1 000	J	200	N		
10 000	J			0.8	m
7 500	J	250	N		

POWER

The word 'power', like the word 'work', has a number of different meanings in everyday life. Can you write down some of its different meanings? In science, the word 'power' has a very exact meaning.

Fig. 3.12 How fast is the work done?

In Fig. 3.12, the man is lifting boxes, each of weight 100 N, onto the back of the lorry 1 m high. When he starts work, he is able to load 20 of the boxes in 200 seconds. Later on he takes 500 seconds to lift 20 of the boxes onto the lorry. He has done the same amount of work in each case (100 × 20 × 1 = 2 000 J) but the time taken is *different*. His rate of doing work has decreased as he gets tired. The rate of doing work is called POWER.

Using the power formula

To calculate the power, you need to know both the work done and the time taken to do it.

$$\text{power} = \frac{\text{work done}}{\text{time taken}}$$

$$\text{or} \quad P = \frac{W}{t}$$

In the above example, the man does 2 000 J of work in 200 s:

$$\text{his power is } \frac{2000}{200} = 10 \text{ J s}^{-1}.$$

Later on, he does 2000 J of work in 500 s:

$$\text{his power is } \frac{2000}{500} = 4 \text{ J s}^{-1}.$$

Because we have divided joules by seconds, we measure power in joules per second (J s^{-1}). A joule per second is usually called a WATT (W).

This is named after James Watt (1736–1819) who developed one of the first steam engines.

It is possible to use the triangle method again. But to do this we must write the power formula in a different way:

$$P = \frac{W}{t}$$

becomes $$W = P \times t.$$

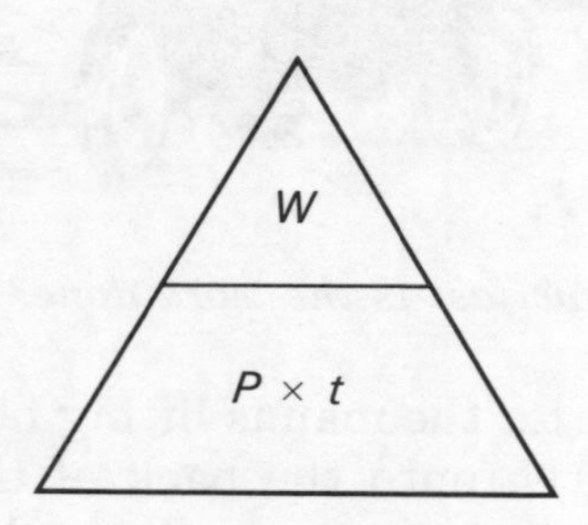

Fig. 3.13 Calculating the power

Putting this into the triangle we get Fig. 3.13. *For example*: An electric motor has a power of 500 W. How much work can it do in 20 seconds?

(Cover W with your finger, then the triangle tells you $W = P \times t$).

$$W = P \times t = 500 \times 20 = 10\,000\text{ J}$$

Assignment – Working out power

What quantities should be in the blank spaces below?

What are the units in each answer?

Power	Units	Work done	Units	Time taken	Units
		5 000	J	100	s
		2 500	J	80	s
100	W			10	s
500	W	10 000	J		
1 500	W	9 000	J		
40 000	W			20	s

Assignment – Power output

Here are some figures for power output:

Human body	100 W
Horse ('horse-power')	750 W
250 cm³ motorcycle	25 000 W
1000 cm³ motorcycle	75 000 W
1600 cm³ family saloon car	67 000 W

Here are some more figures comparing in various ways the performances of cars and motorcycles:

	125 cm³ motor-cycle	1000 cm³ motor-cycle	1600 cm³ family car
Top speed/ $km\ h^{-1}$	140	210	180
Annual insurance cost/£	£150	£350	£210
Carrying capacity/ persons	2	2	5
Fuel con-sumption/ $km\ l^{-1}$	17.5	14.0	14.0
Average tyre life/ km	14 000	9 500	40 000

Using the above figures and any other information that you know or can find out, write an article for a newspaper which tells its readers the various advantages and disadvantages of owning each of these three vehicles. End your article with a conclusion which says which one of them *you* would like to own and give reasons for your choice.

Investigation 3.2 The power of the human body

In this investigation you will try to find the power output of your own body.

Collect

bathroom scales
sandbags or bricks
stopwatch
metre rule
ball of string
convenient low bench

What to do

1 Work with a partner
2 Copy this table into your notebook

Activity	Force /N	Distance /m	Work done/J	Time taken/s	Power /W
1 Lifting sandbags					
2 Running upstairs					
3 Carrying bricks upstairs					
4 Step-ups					
5 Press-ups					

3 Follow the instructions given for the different activities.
4 After you have finished an activity, fill in your results in the table.
5 Calculate the work done and the power.

1 *Lifting Sandbags* Lift the sandbags, one at a time, from the floor to the bench top. Get your partner to time how long it takes you to lift them all. The force is equal to the total weight of all the sandbags. What is the distance moved?

2 *Running upstairs* Use the bathroom scales to find your weight. Run upstairs at a pace you can manage easily and ask your partner to time you. The distance moved will be equal to the vertical height of the stairs. Use the string to help you measure this.

3 *Carrying bricks* You can choose whether to carry the bricks one at a time or more than one at a time. (Is there any difference?) The total force here is the total weight of all the bricks.

4 *Step-ups* Use a low gym bench. Step on to the bench (with both feet!) and see how many you can do in, say, 60 s. Measure the height of the bench. What force is being exerted here?

5 *Press-ups* This is more difficult. Do a few trial press-ups with your hands on the scales. The weight reading will change but make an estimate of the average value. The distance moved is equal to the vertical height moved by your shoulders. Your partner will be busy measuring this and also timing you to see how many press-ups you can do in, say, 20 s.

Questions

1 From the results of the investigation, in which activity were you producing (a) most power; (b) least power? Suggest reasons for your answers.
2 Compare the figure for your power output when climbing stairs with the figure for your power output when carrying bricks upstairs. Give reasons for the differences you find.
3 The average power output of the human body is about 100 W. Suggest reasons why this figure is smaller than your figure from the above investigation. Do you think that you have a power output when you are asleep? Give reasons for your answer.

MACHINES

Here are photographs of some of the many machines used in everyday life. You should be able to see in each case where the load and the effort are. The force required for the effort can come from a person's muscles or from an engine or motor.

Machines are designed to make work 'easier'. This means that they can do one of two things. They can increase the size of the effort so as to move a larger load. Then the machine is a 'force multiplier'. Or the machine can increase the distance moved by the effort so as to move the load through a larger distance. Then the machine is a 'distance multiplier'. Why can a machine not be both of these at the same time? Practically every machine is a 'force multiplier', so we will consider only this type. Levers are the simplest type of machine. They need a fulcrum to operate. All machines need something similar.

The wheel

Throughout history, humans have tried to increase the force available from their muscles. The wheel was probably the earliest machine in common use. It probably started as a number of tree trunks that were placed under heavy loads (Fig. 3.15). As the load was hauled forward, the tree trunks were removed from behind the load and placed in front. What type of tree trunks would you have selected to make the effort needed as small as possible? For example, each trunk should be completely straight. Can you think of any other good properties?

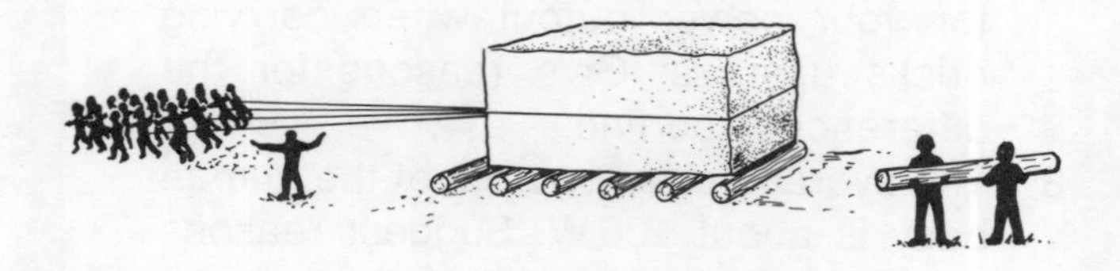

Fig. 3.15 The earliest machine?

Later on, it was found that two thin sections of wood joined by an axle would be more satisfactory. What advantages would this arrangement have?

Fig. 3.14 Examples of machines

Assignment – Building Stonehenge

Stonehenge stands on Salisbury Plain. It took over 1000 years to construct and is believed to have been started in about 2150 BC (over 4000 years ago). Some of the stones used were brought from the Preseli Mountains in Wales, more than 160 km (100 miles) away. These stones weigh about 40 000 N each. This is roughly the weight of two Rolls-Royces. Using a contour map of England and Wales, work out a route that could have been used to transport these stones. Remember that you want the route to be as flat as possible. Can you guess how long it took to transport each stone?

Fig. 3.16 Pulley systems: (a) (top) single pulley for lifting water from a well in India, (b) (centre) double pulley system for raising a medieval portcullis, (c) (bottom) a complex pulley system on a crane

The pulley

The pulley is a wheel with a groove in it. Pulleys have been in use for over 3000 years. Originally they were made of wood. Nowadays most pulleys are made of steel. Ropes or chains (for very heavy loads) are used to connect the pulleys. The single pulley (Fig. 3.16(a)) allows the user to apply a force more easily. Would you call the single pulley a machine? Is it a 'force' or 'distance' multiplier?

Most practical pulley systems have a number of pulleys arranged in two blocks. The more pulleys there are, the more the effort is multiplied. This allows us to lift heavier loads. But the effort must be exerted through a greater distance.

Investigation 3.3 The pulley system

In this investigation, we will try to prove that the pulley is a 'force multiplier'. Also, the amount of work got out of the machine will be compared with the amount of work put in.

Collect

pulleys and string
retort stand or suitable support
1 N and 0.1 N weights
2 weight holders
ruler

What to do

1 Set up the pulley system. The diagram shows a four-pulley system. The pulleys are drawn in this way so that you can see how to connect them. Your pulley system might have a different number of pulleys.
2 Check that it works, with a suitable load on the lower pulley.
3 Draw a table like the one below.
4 With the smallest load, find the smallest effort needed to lift it. Write down this value.
5 For a certain distance moved by the load, measure the distance moved by the effort *at the same time*. Record these values in *metres*.
6 Repeat instructions 4 and 5 for each value of the load.
7 Calculate the work output (load × distance moved by load) and the work input (effort × distance moved by effort).

Questions

1 Is the machine a 'force multiplier'?
2 The work output is never greater than the work input. Is this true for all your results?
3 Why do you think more work has to be put into the machine than can be got out of the machine?
4 Is there any pattern between the distance moved by the load and the distance moved by the effort in the same time?

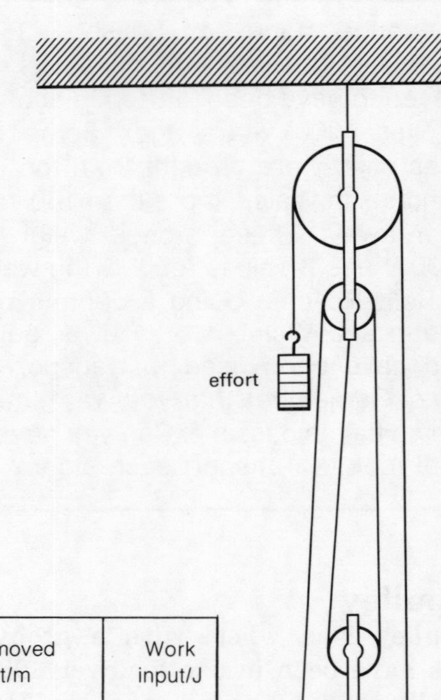

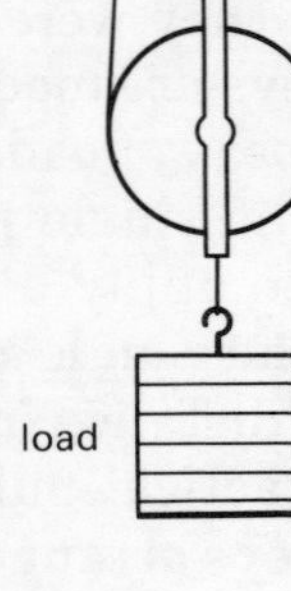

Load/N	Distance moved by load/m	Work output/J	Effort/N	Distance moved by effort/m	Work input/J
1					
2					
3					
4					
5					
etc					

Fig. 3.17 Examples of inclined planes

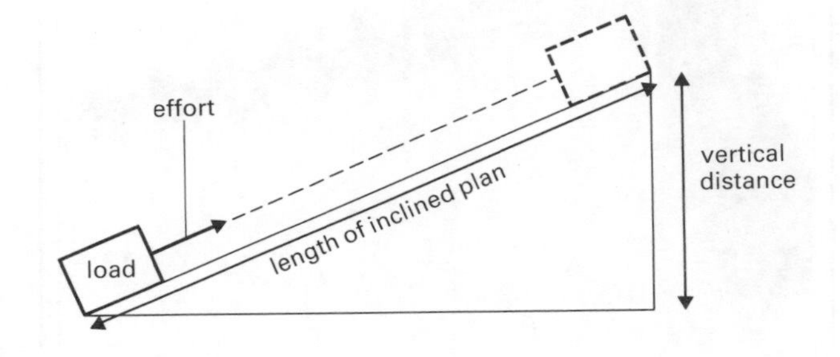

Fig. 3.18 An inclined plane

The inclined plane

Figure 3.17 shows examples of another type of machine called the inclined plane. This type of machine has been in use for thousands of years. The Egyptians used inclined planes for building the Pyramids. Thousands of slaves were used to pull the heavy stones (on rollers) up the inclined planes to get the stones into their correct positions.

Inclined planes are normally used to lift heavy loads vertically upwards (i.e. against the force of gravity). Look at Fig. 3.18. When the load has been moved to the upper end of the inclined plane, it has been moved the vertical distance shown. But the effort applied has to move the load along the length of the inclined plane. This means that:

work output = load × vertical distance lifted
work input = effort × length of inclined plane

Do you think the output will be greater, the same or smaller than the input for a practical inclined plane?

Assignment – The inclined plane

In an experiment on an inclined plane in the laboratory, the following results were obtained. The inclined plane remained the same length but the vertical distance moved by the load was changed.

Load/N	10	20	30	40
Vertical distance/m	0.2	0.4	0.6	0.8
Work output/J				
Effort/N	5	6.5	15	24.5
Length of inclined plane/m	2.0	2.0	2.0	2.0
Work input/J				

Copy the table and fill in the spaces for the work output and the work input.

1 Is the machine a 'force multiplier'?
2 How does the work output compare with the work input?

The screw thread

There are other machines that use the inclined plane but look completely different. A screw is an inclined plane that is wrapped round itself. You can see this with a triangular piece of paper and a pencil. Wrapping the paper round the pencil will produce something looking like a screw thread (Fig. 3.19).

Figure 3.20 shows some machines that use the screw thread form of the inclined plane. Both these machines multiply the effort by a large amount. It would be impossible for most people to lift (and hold!) the side of a car so that the wheel could be changed without a car jack.

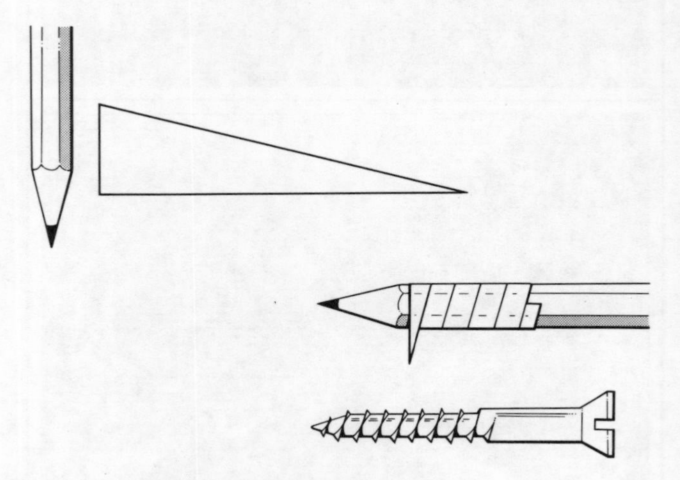

Fig. 3.19 The screw thread

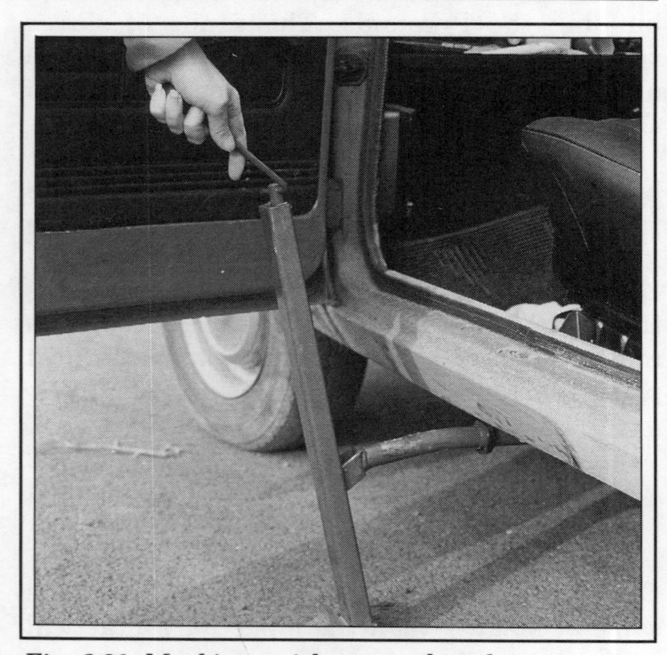

Fig. 3.20 Machines with screw threads

THE EFFICIENCY OF MACHINES

Efficiency is another word that has many meanings in everyday life. In science, the meaning is very precise. Look back at the results that were obtained in the pulley investigation (p.116) or the inclined plane assignment (p.118). The results show that the work output is always less than the work input. Some of the work appears to have 'disappeared'. With some machines, moving parts have to be lifted as well as the load. So part of the effort applied is used for this. For example, in a pulley system the lower block has always to be lifted with the load. But, with every machine, work also has to be done against friction. This is discussed in the next section.

The efficiency is defined using the following formula:

$$\text{efficiency} = \frac{\text{work output}}{\text{work input}} \times 100\%.$$

The work output and work input must be for the *same* period of time. Multiplying by 100 means that the efficiency is written down as a percentage.

For example: What is the efficiency of a machine if a work input of 4000 J produces a work output of 3200 J?

$$\begin{aligned}\text{efficiency} &= \frac{\text{work output}}{\text{work input}} \times 100\% \\ &= \frac{3200}{4000} \times 100\% \\ &= \frac{4}{5} \times 100\% \\ &= 80\%\end{aligned}$$

In all practical machines, the efficiency is less than 100%. The work output is less than the work input, as we have seen. In the past, a number of people have tried to produce a machine that has an efficiency of more than 100%. This means that the work output is greater than the work input. Such devices are called PERPETUAL MOTION MACHINES. Nobody has yet managed to make such a device. Do you think a perpetual motion machine is possible?

Since the work input and output for the same time must be considered, they can be replaced by the work done per second, i.e. the power. The formula now becomes:

$$\text{efficiency} = \frac{\text{power output}}{\text{power input}} \times 100\%.$$

Friction

In all practical machines, we have said that work done has to be done against a force called FRICTION. A frictional force comes into effect when one solid surface moves over another. Friction occurs because no surface is completely smooth. However, the roughness can only be observed by viewing the surface under a very powerful microscope (Fig. 3.21).

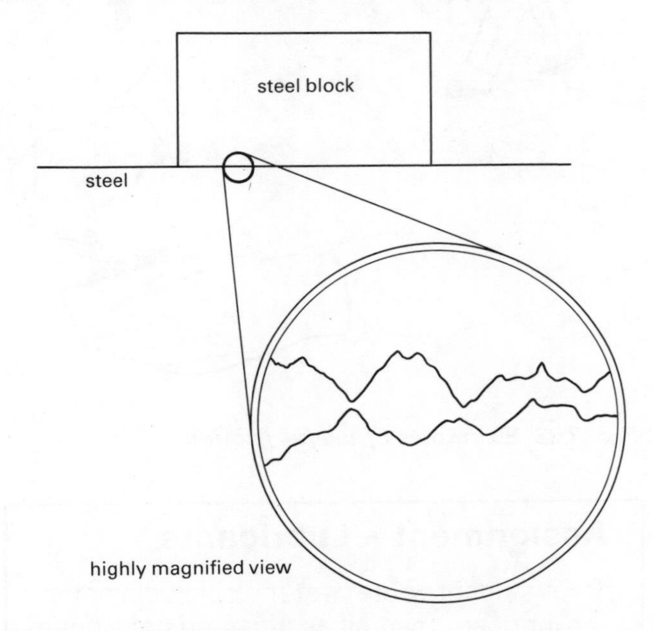

Fig. 3.21 How friction occurs

When one surface moves over another, the frictional forces cause wear. To reduce friction and wear of the surfaces, materials called lubricants have been produced. Most of these are made from oil. In a car engine the piston inside a cylinder might be moving up and down 100 times a second. Without a lubricant the engine would become so hot that the piston would weld itself to the cylinder. Try rubbing your hands together to feel the heat produced due to friction.

From what has been said, it appears that friction is a nuisance. But there are occasions when friction can be very useful. Figure 3.22 shows some of these. Can you think of any others?

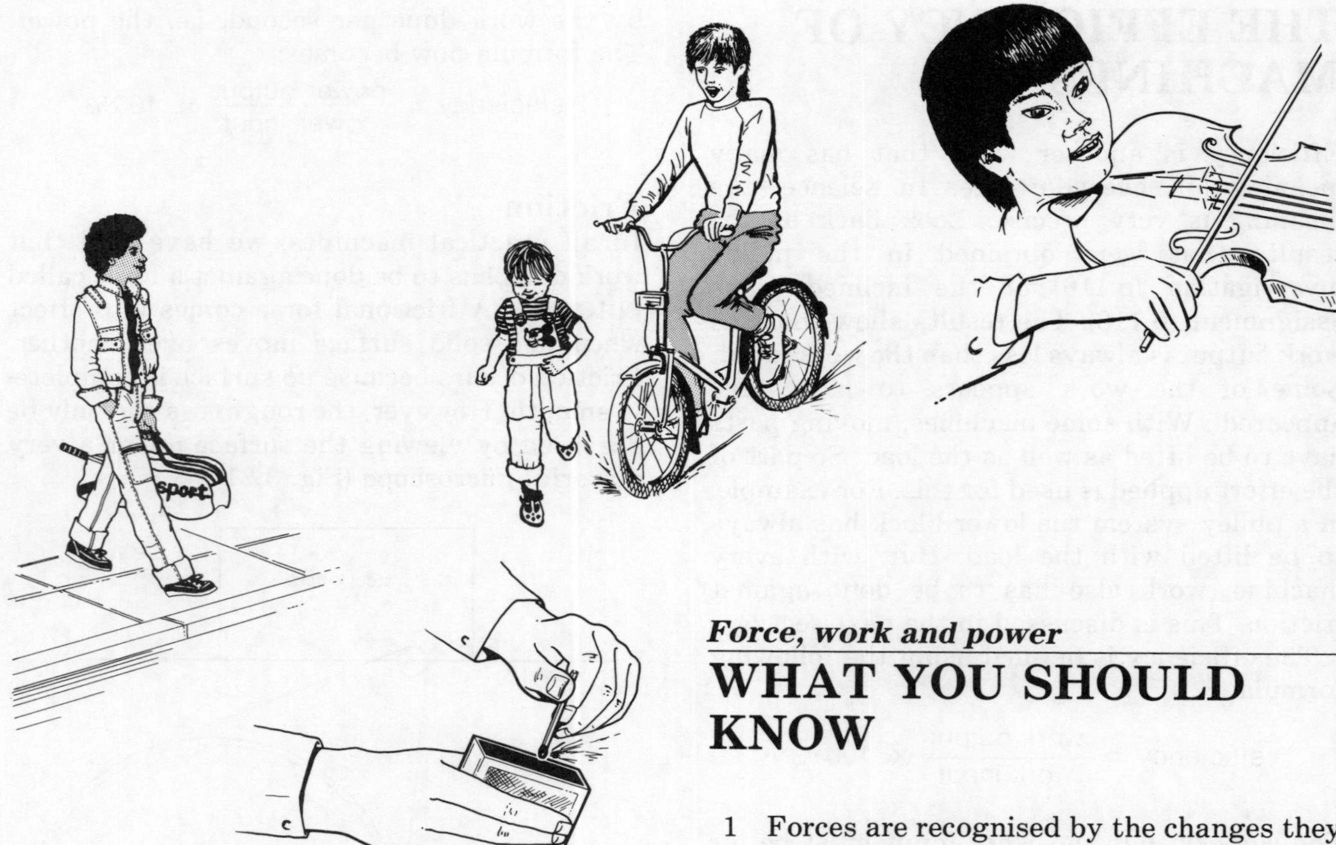

Fig. 3.22 Examples of useful friction

Assignment - Lubricants

1 A student wrote 'water would be a better lubricant than oil as it can be got from a tap'.

However, water is unsuitable as a lubricant. What properties does water have that normally makes it useless as a lubricant?

2 Friction also occurs when solid objects move through liquids or gases. Describe how the effects of friction in these cases can be made as small as possible.

3 A certain type of motor car is said to have a 'drag coefficient of 0.28'. Try to find out what this means and why it is important.

4 The first hovercraft was launched in 1959. This uses air as a lubricant to minimise friction. Try to find out how this machine works.

Force, work and power

WHAT YOU SHOULD KNOW

1. Forces are recognised by the changes they produce.
2. If all the forces on a body are balanced, no change is observed.
3. Forces are vector quantities - they have size *and* direction.
4. Forces are measured in newtons, N.
5. Muscles produce the forces in the body.
6. The available force can be increased by using a lever.
7. Levers can be classed as 'force multipliers' or 'distance multipliers'.
8. Work done = force × distance, and is measured in joules, J.
9. Power $= \dfrac{\text{work done}}{\text{time taken}}$ and is measured in watts, W.
10. Machines make work 'easier', usually by acting as 'force multipliers'.
11. For a machine,
$$\text{efficiency} = \frac{\text{work output}}{\text{work input}} \times 100\ \%.$$
12. This can be expressed as
$$\text{efficiency} = \frac{\text{power output}}{\text{power input}} \times 100\ \%.$$
13. Friction is a force that appears when one surface moves over another.

QUESTIONS

1 The table shows the results obtained when a wire was stretched by different forces.

Load/N	0	20	40	60	80	100
Extension/cm	0	1.9	3.8	5.7	7.9	10.5

(a) Plot a graph of load (horizontal axis) against extension.
(b) Up to what load does the wire appear to obey Hooke's Law?
(c) What load would produce an extension of 3.5 cm?
(d) What extension would be produced by a load of 50 N?

2 A simple lever starts in a horizontal position and moves to the position shown in the diagram. Calculate:
(a) the work input;
(b) the work output;
(c) the efficiency.

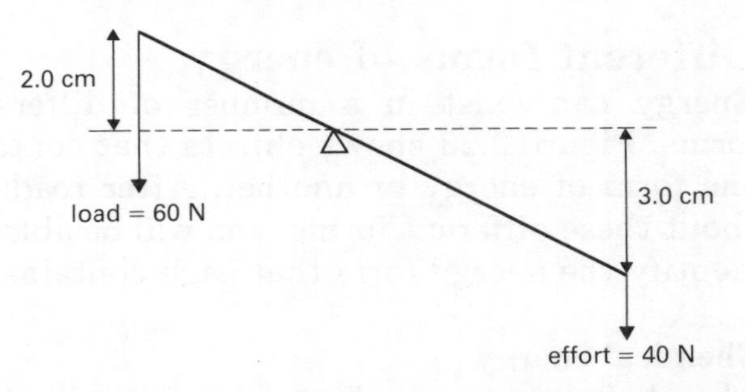

3 (a) A force of 30 N is needed to push a trolley along a horizontal surface. If the trolley is pushed through a distance of 10 m in 8 s, calculate (i) the work done and (ii) the power.
(b) An electric motor is used to raise bricks on a building site. If the motor can raise 300 bricks, each of weight 20 N, through a vertical distance of 5 m in 30 s, calculate (i) the work done by the motor and (ii) its power output. If the efficiency of the motor is 40%, what must be its electrical power input?

4 A person of weight 600 N runs up a flight of 20 steps, each 0.2 m in height, in 6 s.
(a) Calculate the work done running up the stairs.
(b) Calculate the power developed during this run.
(c) If the person then runs up the stairs carrying a 150 N load at the same power rate, how long would it take?

5 A student was investigating how the angle of an inclined plane affected its efficiency. The following results were obtained.

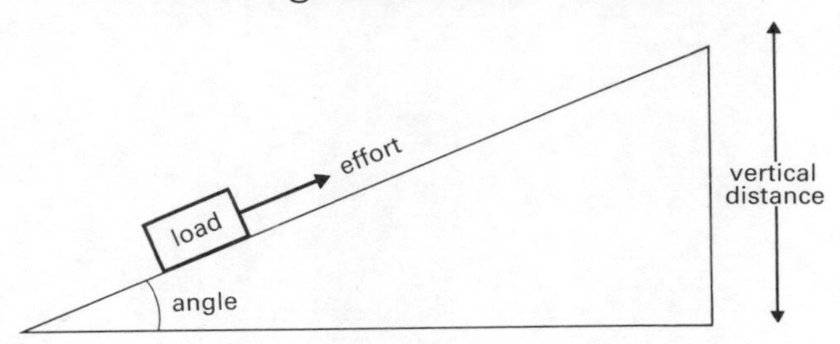

Angle/ degrees	Load/ N	Vertical distance/m	Effort/ N	Length of inclined plane/m
10	10	0.17	4.5	1.0
15	10	0.26	5.6	1.0
20	10	0.34	6.6	1.0
25	10	0.42	7.4	1.0
30	10	0.50	8.2	1.0

The conclusion drawn from the results was that 'as the inclined plane becomes steeper, the effort increases and so the efficiency must decrease'. Point out the error in this conclusion and substitute a more correct conclusion.

Energy 1

ENERGY

FORMS OF ENERGY

In the last section we looked at the work done on machines. In many cases, the work was done using the muscles of the body. What is it in the body that allows work to be done? We know that the more active a person is, the more food that person requires. If your supply of food is stopped, firstly you would feel hungry. But over a period of time you would become weaker and weaker. We say that food provides ENERGY.

If you have got energy then you can do work. This does not apply just to human beings. Every object that does work has to have a supply of energy. The motor car will not keep running unless it is kept supplied with petrol. Petrol provides energy that the internal combustion engine can use to do work. The next section will look at the different forms that energy can take.

Fig. 3.23 All these objects contain energy

Different forms of energy

Energy can exist in a number of different forms. Figure 3.23 shows objects that contain one form of energy or another. After reading about these different forms, you will be able to identify the energy form that each contains.

Chemical energy

All substances are made up of small particles. By changing the way in which the particles are arranged, energy can be released. Food is one example of a substance that contains chemical energy. Other examples are: petrol, explosives, wood, batteries. You would not know that such substances contained chemical energy just by looking at them. How could you show that wood contained chemical energy?

Heat energy

In Unit 5, you will find out that molecules are continually moving. And the higher the temperature, the faster they move. This form of energy is called heat energy. So, 1 kg of water at 100° C would have more heat energy than 1 kg of water at 10° C. Measurement of heat energy will be explained on page 136.

Mechanical energy
There are two different types of mechanical energy. One type is called KINETIC ENERGY. This is the energy of a moving object, e.g. a falling book or a car travelling at speed. The second type is called POTENTIAL ENERGY. The word 'potential' means 'stored'. So potential energy is stored mechanical energy. The potential energy of an object can be due to its position (e.g. the water behind a dam) or to its shape (e.g. a wound-up clockspring).

Electrical energy
This is the energy that moving electrical charges have. It will be studied in more detail in Unit 4. This is probably the most useful form of energy today. Make a list of all the devices in your home that use electrical energy and you will begin to see how much your life revolves around electrical energy.

Wave energy
Waves will be studied in more detail later in this unit. One of their important properties is that they carry energy from one place to another. Perhaps the commonest wave is the water wave. How would you show that water waves carry energy? Two other important types of wave are sound waves and light waves. Light is, in fact, one example of a whole family of similar waves called electromagnetic waves.

Nuclear energy
This is the energy available from the nucleus of certain types of atom. How this energy is obtained is described in Unit 5 (page 268).

Assignment – Types of energy

1. Look at Fig. 3.23 on page 122. Write down the energy form that you think each object shown has.
2. For each energy form mentioned in the text, give an example of an object that has this type of energy. Try *not* to use an example already mentioned.
3. Sometimes energy forms are classified as 'stored energy' or 'moving energy' Put the energy forms mentioned in the text under these two headings.

How do we measure energy?

The food that we eat contains chemical energy that allows us to do work. Assume that you lift a 100 N weight through a vertical distance of 2 m. The work done is $100 \times 2 = 200$ J. We say that the work done is equal to the energy changed. In this case, it means that 200 J of chemical energy has been changed.

Into what has the chemical energy been changed? Lifting the weight gives it potential energy. So, 200 J of chemical energy has been changed into 200 J of potential energy. We will find that, whenever work is done, one form of energy is changed into another. This may be represented by the following formula:

$$\text{energy changed} = \text{work done.}$$

For example: A person weighing 800 N runs up a flight of stairs of vertical height 5 m. How much chemical energy has been changed into potential energy?

$$\begin{aligned}\text{energy change} &= \text{work done}\\ &= 800 \times 5\\ &= 4000\ \text{J.}\end{aligned}$$

ENERGY CHANGES

We have said that when one form of energy is changed into another then work is done. If the work done can be measured, then we know how much energy has been changed. In many cases, however, the work done cannot be measured exactly. But we can usually recognise the initial energy form and the final energy form. Usually, when we change energy from one form to another, it is to produce a form of energy that is more useful.

In some respects, energy is rather like money. Money consists of metal coins or paper notes. These items are no use as they are. But they can be exchanged for other items that are of much greater use (e.g. food, bus tickets, a radio, a holiday, etc.). Energy is similar to this. Its original form is usually of little use to us. It has to be changed into another form of energy before it becomes useful.

Examples of energy changes

The Bunsen burner uses natural gas (Fig. 3.24(a)). Natural gas contains chemical energy. By burning the gas in air, heat energy is produced. (A small amount of light energy is produced as well, but this will be ignored here.) So the energy change is from chemical energy to heat energy. Figure 3.24(b) shows a solar cell. This device uses light (wave energy) from the sun and turns it into electrical energy. The electrical energy is shown by the movement of the needle of the ammeter when the solar cell is illuminated. The energy change here is from wave energy to electrical energy.

Assignment – Energy changes

In the diagrams below, there are a number of energy changes shown. In each case you have to name the energy form at the start and at the end. Some of these energy changes may be set up for you to see. The first one has been done for you. Electrical energy has been supplied to the kettle. The electrical energy has been changed into heat energy. This heat energy boils the water.

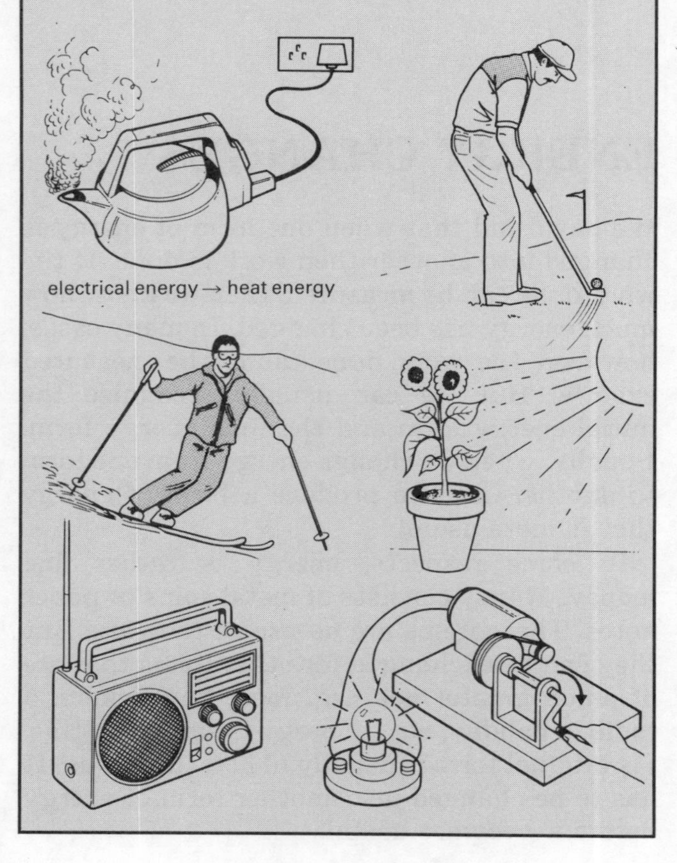

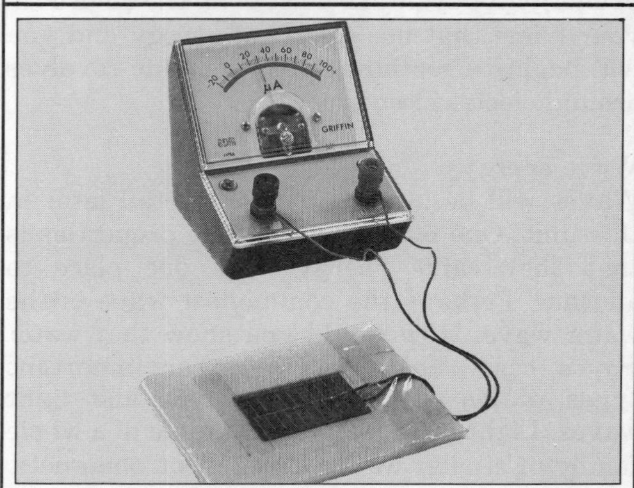

Fig. 3.24 Two examples of energy changes:
(a) chemical energy → heat energy
(b) wave energy → electrical energy

Energy chains

The examples of energy changes in the assignment show some of the ways in which energy can be changed. But some of the changes are not as simple as you might think. Only very rarely is there a single energy change. Let us look at the last example. The kinetic energy to turn the generator comes from the chemical energy in the body. This is changed to electrical energy. An electrical current flows through the bulb filament, producing heat energy. The filament gets very hot and produces the energy form that is needed - light energy. At each energy conversion, some of the original energy is changed to unwanted heat energy. How could this energy be detected? The changes are shown on the next page.

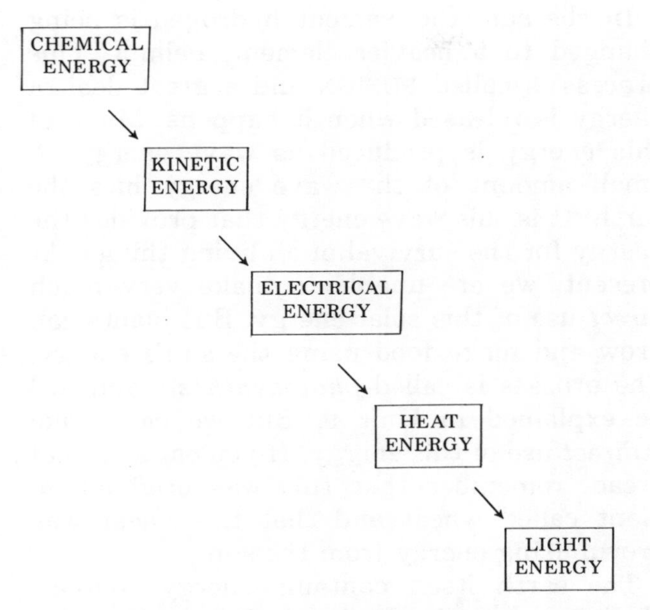

A series of energy conversions is called an ENERGY CHAIN.

Law of Conservation of Energy

In order to understand the efficiency of energy changes, we must first understand the rule called the LAW OF CONSERVATION OF ENERGY which says that

new energy cannot be created

existing energy cannot be destroyed

energy *can* be changed from one form to another

For example, what are the energy changes that take place when the switch in Fig. 3.25 is pressed? They can be shown as an energy chain:

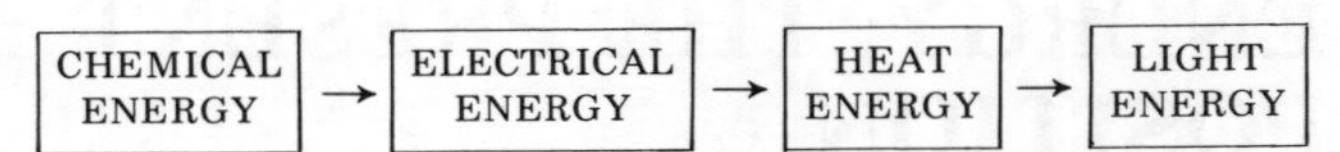

The bulb is designed to produce *light* energy. However, at every step in the chain some energy is lost to the surroundings so that less than 5% of the original energy becomes light energy. The total amount has remained the same (i.e. it is conserved). We find that whenever energy is changed from one form to another, some of the energy is turned into a form that we do not want. Usually, the energy form that is of little use is heat energy. Can you think of any other examples where the heat energy produced is wasted?

Assignment – The pendulum

Set up a pendulum as shown below. Then try to answer the questions.

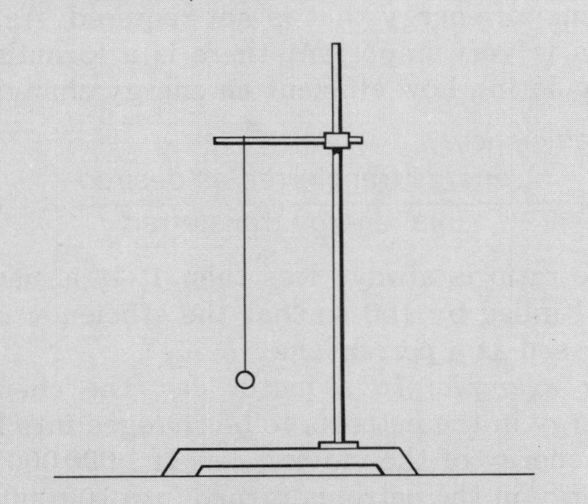

1 The time for one complete swing is called the *period*. How is the period affected by
   (a) the length of the pendulum,
   (b) the size of the swing,
   (c) the mass of the pendulum bob?
   You will have to make measurements, if you do not know the answers.
2 What are the energy changes that happen when the pendulum swings?
3 What happens to the period of the pendulum as the size of the swings gets smaller?
4 Eventually, the pendulum will come to rest. Explain, in terms of energy changes, why this happens.

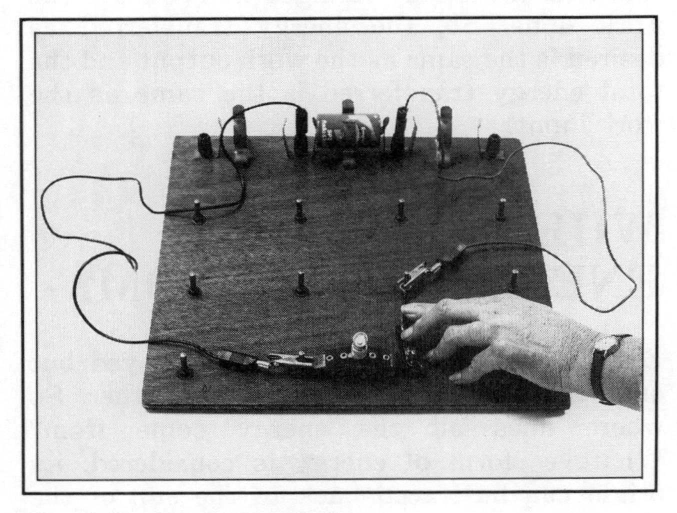

Fig. 3.25 An energy chain

The efficiency of energy changes

Whenever an energy change occurs, the total quantity of energy remains constant. But the change always produces another form (or forms) of energy that is not required. As this idea is very important there is a formula for calculating how efficient an energy change is.

$$\text{efficiency} = \frac{\text{energy transferred as desired}}{\text{total energy transferred}} \times 100\%$$

The ratio is always less than 1. It is usually multiplied by 100 so that the efficiency is expressed as a percentage.

For example: In a motor car, the chemical energy in the petrol is to be changed into kinetic energy of the moving car. If 1 000 000 J of energy in the petrol is turned into 100 000 J of kinetic energy, what is the efficiency?

$$\text{efficiency} = \frac{\text{energy transferred as desired}}{\text{total energy transferred}} \times 100$$

$$= \frac{100\,000}{1\,000\,000} \times 100$$

$$= \frac{1}{10} \times 100$$

$$= 10\%.$$

Where does the rest of the energy go? If you are not sure, think why a car has to have a cooling system.

The formula for efficiency looks very similar to the formula that was used to calculate the efficiency of a machine (on p.119). This is not just a coincidence. You will remember that we said that the energy changed was equal to the work done. So, the energy transferred as desired is the same as the work output and the total energy transferred is the same as the work input.

WHERE DOES OUR ENERGY COME FROM?

Energy cannot be created nor destroyed but only changed from one form to another. So where does all the energy come from? Whatever form of energy is considered, its origin can be traced back to the sun or the earth.

In the sun, the element hydrogen is being changed to a heavier element, helium. This process is called FUSION and a great deal of energy is released when it happens. Much of this energy is produced as wave energy. A small amount of the wave energy hits the earth. It is this wave energy that provides the energy for the survival of all living things. At present, we are unable to make very much *direct* use of this solar energy. But plants can grow and make food using the sun's energy. The process is called *photosynthesis* and will be explained in Unit 9. But we can make *indirect* use of this energy. If you eat a slice of bread, remember that this was originally a plant called wheat and that the wheat was grown using energy from the sun.

The earth itself contains energy sources. Nuclear energy is obtainable from radioactive elements such as uranium. Our most important energy sources at the present time are the fossil fuels - oil, coal and natural gas.

Assignment – The black cloud

In a science fiction story by Fred Hoyle called *The Black Cloud*, most of the radiation from the sun reaching the earth was absorbed by a black cloud from outer space. If such a thing really happened, describe how life on earth would change. Reading the book might give you some more ideas.

ENERGY: THE PRESENT POSITION

Figure 3.26 shows the percentages of different energy sources and the way they are used in the UK.

The largest percentage is that supplied by oil; the next largest percentage is that of coal. Natural gas has become of increasing importance due to the amounts found under the North Sea. Oil, coal and natural gas are called 'high grade' energy sources because a lot of energy can be extracted from them quickly. Remember that the original energy form is always

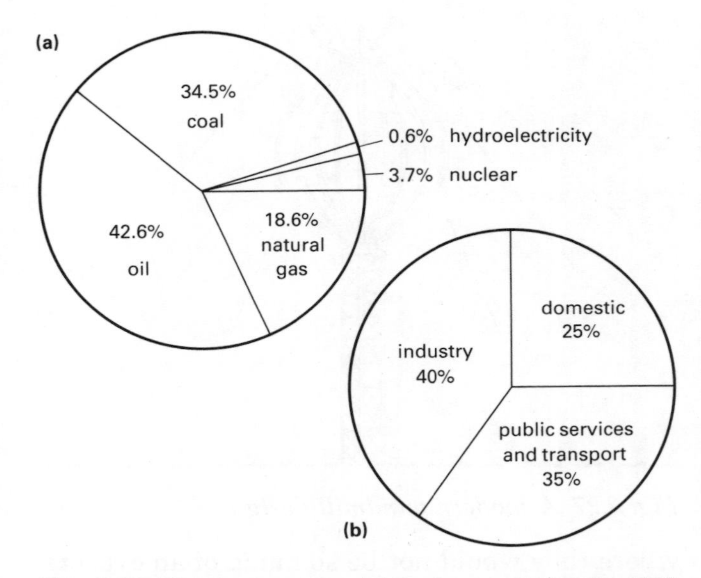

Fig. 3.26 Energy sources (a) and uses (b) in the UK

being changed into a more useful form.

But oil, coal and natural gas are three of the energy sources that are classed as 'non-renewable'. This means that after the present reserves are used up, there will be no more available. By the year 2050, most of the present sources of oil and natural gas will have run 'dry'. However, coal reserves could last another 300 years. More reserves may be found but it is becoming increasingly important to find 'energy sources' to replace the present ones.

There are two other energy sources used in the UK at present: nuclear and hydroelectric. The use of nuclear energy is very controversial for reasons you will find out about in Unit 5. The 'fuel' for reactors is uranium. This is in short supply and will run out in the foreseeable future. There is another type of nuclear reactor called a breeder reactor. This produces more nuclear fuel than is present at the start. This seems a good idea but there are many problems associated with its development. Hydroelectric energy provides about 0.6% of our energy needs. Although this energy form produces no pollution, the number of suitable sites is very limited. All nuclear and hydroelectric energy is converted into electrical energy.

Assignment – Patterns in energy use

The graphs show the changing patterns in the use of energy sources over a period of 30 years. Use the graphs to answer the following questions.

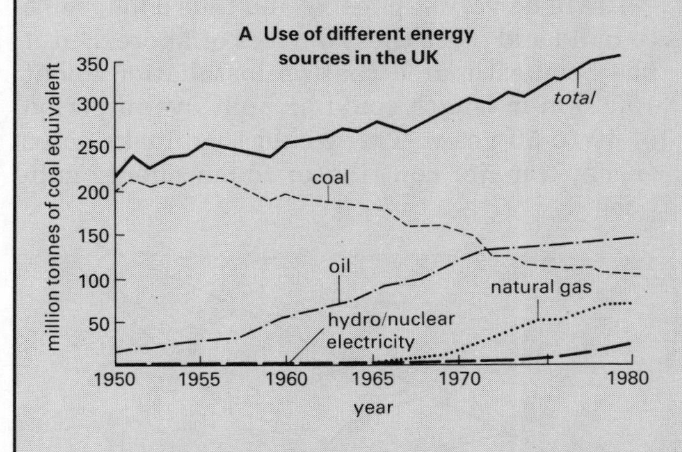

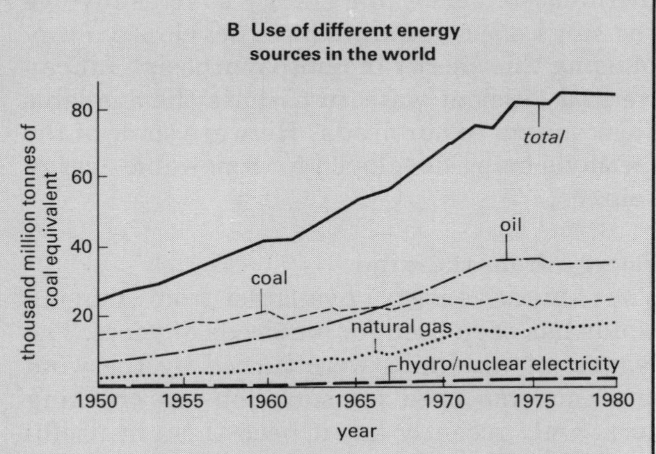

1. What was the main energy source in the UK in (a) 1950 (b) 1980?
2. (a) Why do you think coal has declined as an energy source?
   (b) Why do you think oil has increased as an energy source?
3. Do you think that you could predict how much energy the UK will use in the year 2000, using graph A? Explain your answer.
4. (a) Look at graph B. List as many differences as you can between graphs A and B. Note that the vertical scale on graph B is 1000 times larger than that on A.
   (b) Suggest reasons for each of the differences you can find.

Assignment – Careful with oil!

At the beginning of this section, on page 127, we talked about the need to use oil, gas and coal as carefully as possible. Find out about other uses of these fuels which make it even more important for us to be careful.

FUTURE ENERGY SOURCES

The non-renewable energy forms, e.g. fossil fuels, must be used as carefully as possible. Reserves of coal will last much longer than those of oil. Coal can be turned into a form of oil, though this has only been done on a large scale in South Africa. But the fact is that, sometime in the near future, these energy sources will be exhausted.

We will then have to rely on renewable energy sources. The largest energy source available to us is the sun. Most of the proposed methods for renewable energy sources involve the sun's energy. Plants have developed a way of using this energy in photosynthesis. But can we find efficient ways to harness the available solar energy to our needs? Here are some of the methods being developed for renewable energy sources.

Energy from the wind

The kinetic energy available from moving winds has been used for hundreds of years. The 'vanes' of windmills were turned by the wind and this was used for such jobs as crushing corn. Only recently has it been thought useful to produce electrical energy by using the vanes of a windmill to drive a generator. The main aim of the design of a windmill is to extract as much kinetic energy from the air as possible. Some of these designs do not look much like traditional windmills (Fig. 3.27).

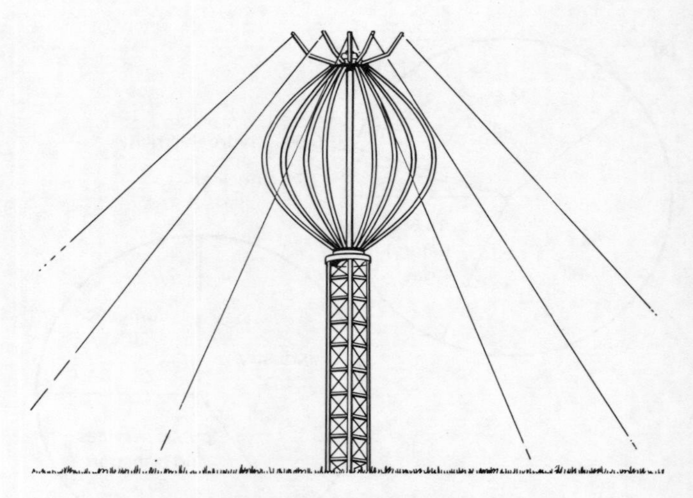

Fig. 3.27 A modern windmill design

The problems with windmills are twofold. One is that the wind does not blow all the time and the second is that a lot of windmills are needed to match the output of a power station. There have been suggestions for building large numbers of these at sea close to the shore where they would not be so much of an eyesore and the strength of the prevailing winds would be more constant.

Energy from the waves

There is a vast amount of kinetic energy in water waves. A number of devices have been developed to extract this energy. One important type is called Salter's duck (Fig. 3.28).

As the waves pass the 'duck' it rotates backwards and forwards. This rocking motion can be used to drive a generator to produce electrical energy. A number of the ducks would be connected together and their electrical output sent by cable to the shore.

It will be very expensive and take a long time to build and place these devices offshore. But it has been estimated that an installation about 1000 km in length could be built over a period of 40 to 50 years. This would then make wave energy a major contributor to our energy supplies.

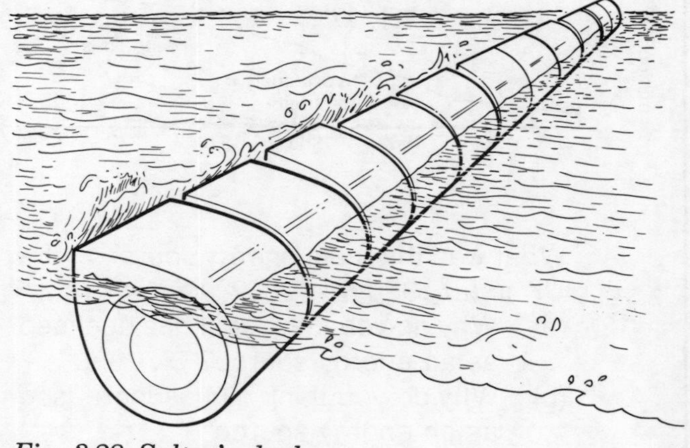

Fig. 3.28 Salter's ducks

Solar energy

It is possible to use the sun's energy directly. There are two ways of doing this. One is by using solar panels (Fig. 3.29). These absorb the radiation and use it to heat water that circulates through the panels. The panels must be set up to catch as much of the available radiation as possible. The main problem is that when maximum energy is needed (in the winter), the sun's radiation is at its weakest.

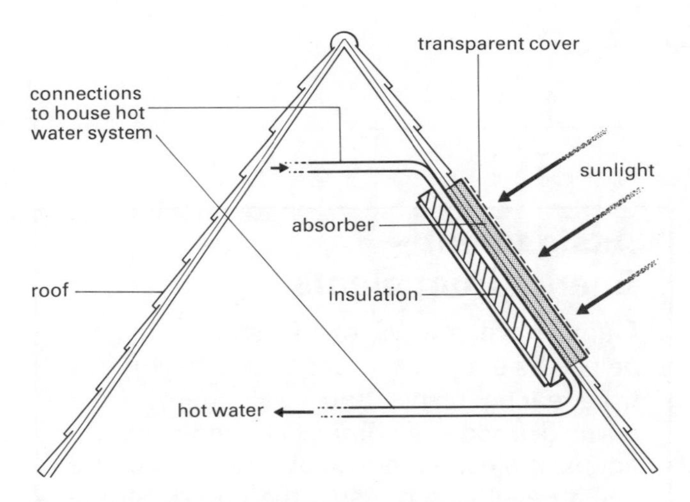

Fig. 3.29 Solar panel on the roof of a house

Although solar panels are expensive at the moment, the increasing price of fossil fuels could make solar panels more economic.

The other main method is using solar cells. These devices convert the sun's radiation directly into electrical energy. However, at present they are so expensive that they are only used in extreme cases when no other method is possible. For example, all the satellites launched now have large numbers of solar cells to produce their electrical energy. It is possible that in the future we may be able to use satellites to 'collect' solar energy and beam it to earth.

Hydroelectric energy

This method of using a renewable energy form is widely used. Water is evaporated by the sun, forms clouds and falls on high ground. It then has potential energy. If this water is collected in a reservoir, the potential energy is changed into kinetic energy as it falls from the reservoir to lower ground. The kinetic energy is changed to electrical energy by a generator. The common name for the energy produced by this process is hydroelectricity. Although non-polluting, hydroelectric schemes can flood a lot

Fig. 3.30 An array of solar cells at Marchwood Power station, Southampton

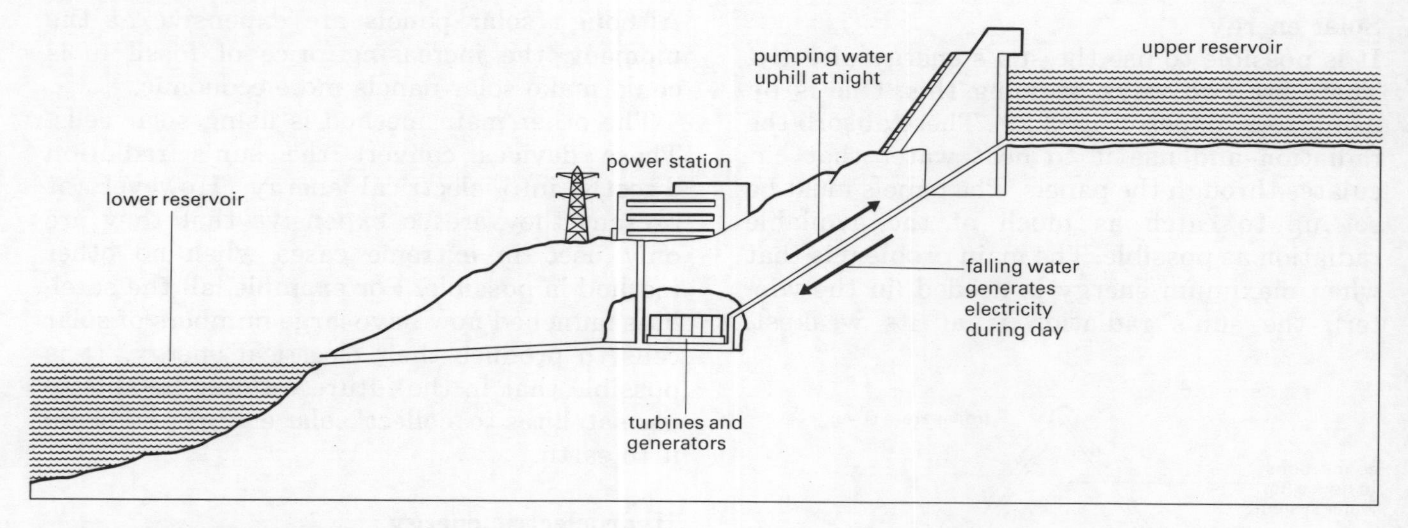

Fig. 3.31 The pumped storage system

of useful land (as the reservoir fills up) and disturb the balance of nature in the surrounding areas.

Because of the variation during a 24-hour cycle of the demand for electrical energy, pumped storage systems are being built (Fig. 3.31). During the day, electrical energy is obtained from the potential energy stored in the water above the generating station. At night, the demands for electrical energy are less but power stations are impossible to turn 'on' or 'off' as needed. So these schemes use the electrical energy produced by power stations at night to pump water 'uphill'. This energy can be recovered during the day. There are only two of these pumped storage systems in operation in Great Britain, one in Wales and one in Scotland.

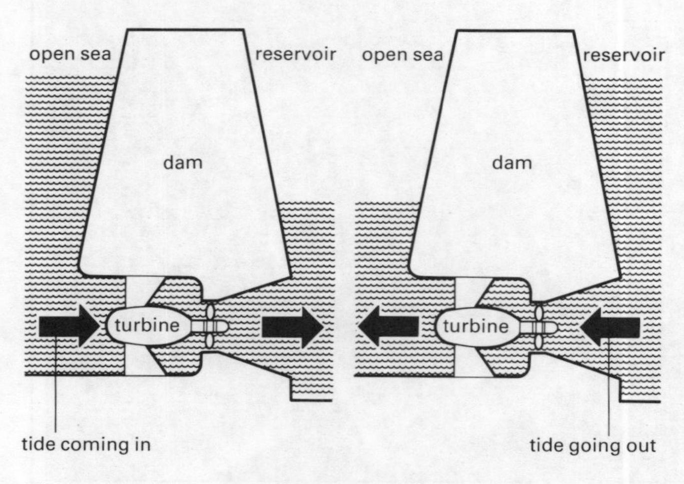

Fig. 3.32 Using tidal energy

Assignment – Energy from plants

Plants have always been used by human beings as energy sources. Chemical energy from eating plants and heat energy from burning wood are perhaps the most important examples. But nowadays many ways are being developed of using the energy stored in plants and other organic materials. Find out about some of the ways in which this is being done.

Energy from the tides

Tides occur because of the gravitational force between the earth and the moon, and, to a smaller extent, between the earth and the sun. The changing positions of these bodies produce high and low tides (see page 6). In many places the difference in water level between high and low tide can be considerable, perhaps more than 10 metres.

Figure 3.32 shows a scheme for using tides to generate electricity. The tidal flow is allowed to fill a reservoir, driving a turbine connected to a generator. As the tidal level outside the reservoir falls, the water can be allowed to drive the turbine in the opposite direction.

The cost of building such a scheme is very high. To date, there is only one large-scale system in operation. This is at La Rance in France and has been producing electrical energy in this way since 1968 (Fig. 3.33).

Fig. 3.33 The barrage across the estuary at La Rance

In the United Kingdom, there is a plan to use the Severn Estuary for such a tidal energy store. Look at the map on page 133 and see what might be the problems with this suggestion. Are there any possible benefits apart from harnessing the energy?

Nuclear fusion

The main difficulty with present *renewable* energy sources is that they will never produce more than a small fraction of our energy needs. For example, the amount of electrical energy we use doubles every ten years! If nuclear fusion can be made to work then it would solve all our energy requirements at a stroke. The main way the the sun produces energy is to turn hydrogen atoms into helium atoms. This can take place only at very high temperatures but the reaction releases a large quantity of energy. The aim is to reproduce this solar reaction on earth. At Culham, in Oxfordshire, the European nations have joined together to try and do this.

Even if all the problems can be solved, it will be some years before sufficient energy for our needs can be produced. But our supply of hydrogen (from water) is almost unlimited and the system produces very little in the form of radioactive wastes that have to be disposed of (unlike the conventional nuclear power station).

Assignment – Everyday energy

Newspapers and other periodicals regularly have articles on energy, such as those here. Collect any that you see. You will be surprised at how often the topic of energy is mentioned. You will also become aware of why it is of such great importance.

WHAT YOU SHOULD KNOW

1 Energy is necessary for doing work.
2 Energy exists in a number of different forms. The forms of energy are: chemical, heat, mechanical (kinetic and potential), electrical, wave and nuclear.
3 To do work, one form of energy has to be changed into another.
4 Energy change = work done; energy is measured in joules.
5 A series of energy conversions is called an energy chain.
6 The law of conservation of energy states that energy cannot be created or destroyed but only changed from one form to another.
7 The efficiency of an energy change =

$$\frac{\text{energy transferred as desired}}{\text{total energy transferred}} \times 100\%.$$

8 All our energy sources can be traced back to the earth or the sun.
9 Most of our present energy sources are non-renewable and will become exhausted in the foreseeable future.
10 Future energy sources will have to be based on renewable forms of energy.

QUESTIONS

1 In each of the following situations, describe the energy changes that take place.
(a) A ball thrown upwards.
(b) A torch that is switched on.
(c) A meteor entering the atmosphere around the earth.
(d) A gas fire that is alight.
(e) A car hitting a crash barrier on a motorway.
(f) A spring-wound alarm clock.

2 Read the following article and then answer the questions. You may need access to reference books to help you answer all the questions.

Developing countries outside OPEC: energy balances

Oil currently supplies over one-half, and coal one-quarter of all commercial energy in developing countries outside OPEC. Not included in this balance are traditional household fuels of firewood, dung and crop residues whose contribution is difficult to quantify, but which may be at least as great as that of coal to the total energy balance.

The energy balance in the year 2000 is based on a projection in which the rate of economic growth varies but is lower than in the 1970s. Higher, steadier economic growth would not greatly alter the fuel balance, although oil might retain a slightly larger share.

These projections do not accord a great role to new and renewable sources of energy. Although there may be some interesting possibilities for using such things as solar power or biomass conversion to meet some rural energy needs, the capital investment required to make a substantial contribution to energy supply is very large and the power obtained often too unreliable for industrial or even many household needs. A better option may be hydro-electric power, of which there is still a lot to be tapped in developing countries. Much simpler technology, like enclosed, wood-burning stoves, could multiply many times the useful energy obtained from traditional fuels.

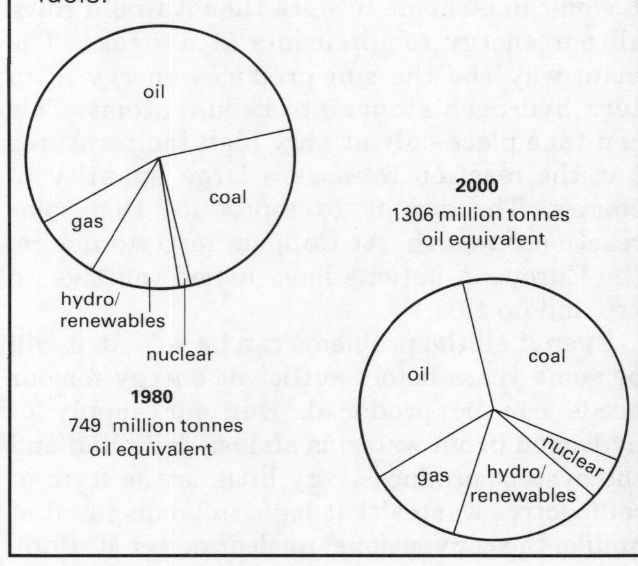

(a) What is meant by the phrase 'developing countries'? Give some examples.
(b) Why do you think the contribution of traditional household fuels is difficult to quantify?
(c) Why do you think it is not possible to make accurate forecasts about economic growth?
(d) Why do you think that 'new and renewable sources of energy' will not make much of a contribution to the energy balance by the year 2000?
(e) What is meant by 'biomass conversion'?
(f) Why do wood-burning stoves increase the useful energy obtained from traditional fuels?
(g) Which fuels (i) increase (ii) decrease in their percentage contribution to the energy balance over the 20 year period?

3 Read the article on the Severn Barrage. Assume that you are a resident of Weston-Super-Mare and that it has just been announced that the building of it will commence within one year. Write an article for the local newspaper either *defending* or *attacking* the proposed building of this scheme.

4 The graph represents the past and predicted future production of the fossil fuels oil and coals.

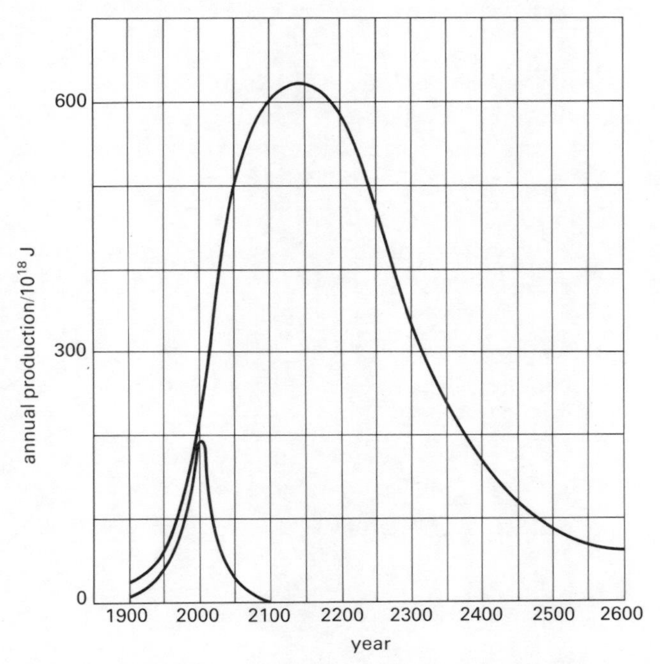

(a) In which year is oil production expected to be a maximum?
(b) Which fossil fuel has the greater reserves? Explain your answers.
(c) What assumptions do you think have been made in predicting the future production of oil and coal?

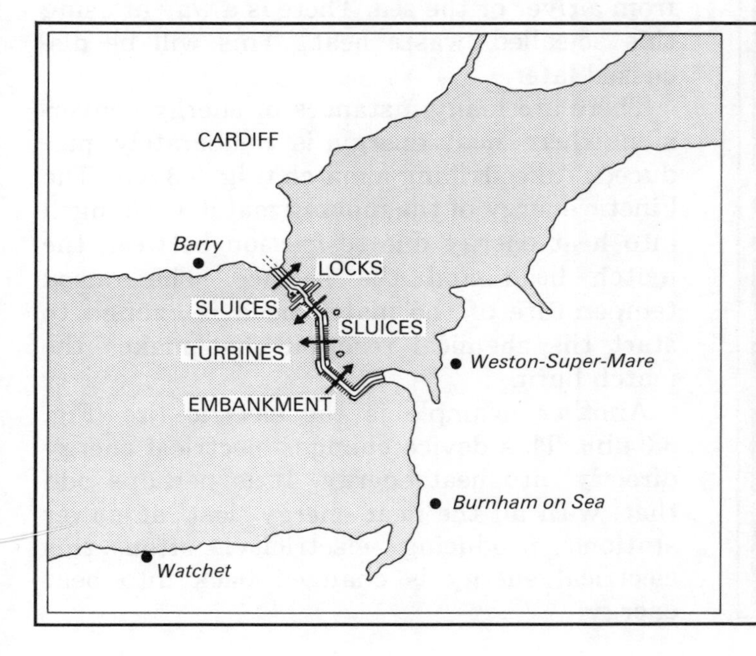

The Severn Barrage

This is the scheme recommended to harness the tidal power of the Severn Estuary. It envisages an inner barrage, costing about £5600 million, which would use the ebb tide to generate power. The system would allow a rising tide to flow through sluices and turbines, which idle in reverse. Both passageways are then closed at high tide and are kept closed until the tide has ebbed sufficiently for the difference in water level between the barrage and the sea to drive the turbines and their generators. The water is then allowed to flow through the turbines until the difference in level is too low to turn them efficiently.

HEAT ENERGY

OBTAINING HEAT ENERGY

We have seen that heat energy is often produced when energy changes occur. For example, to stop a car its brakes are applied. The kinetic energy of the car is changed into heat energy (in the brakes themselves). Various ways of storing the kinetic energy of a car have been suggested, such as spinning flywheels. As the car comes to rest its kinetic energy would be given to the flywheel and could then be used to move the car again.

In many instances the heat energy produced is a nuisance. Again, the motor car provides a useful example. Changing the chemical energy of the petrol into kinetic energy of the vehicle produces a great deal of heat energy that has to be removed. An elaborate cooling system is provided to carry away this energy. Most of this (a small amount is used to heat the inside of the car) is lost into the atmosphere.

On a larger scale, the power stations that generate electricity using fossil fuels produce a vast amount of heat energy as a by-product. Normally this is released into the atmosphere via the cooling towers or used to heat water from a river or the sea. There is a way of using this so-called 'waste heat'. This will be discussed later.

There are many instances of energy conversion where heat energy is deliberately produced. Take striking a match (Fig. 3.34(a)). The kinetic energy of the moving match is changed into heat energy due to friction between the match head and the surface. The raised temperature of the match head is enough to start the chemical reaction that makes the match burn.

Another example is the electric fire (Fig. 3.34(b)). This device changes electrical energy directly into heat energy. It is perhaps odd that, with all the heat energy 'lost' at power stations producing electricity, often this electrical energy is changed back into heat energy.

Fig. 3.34 Producing heat energy

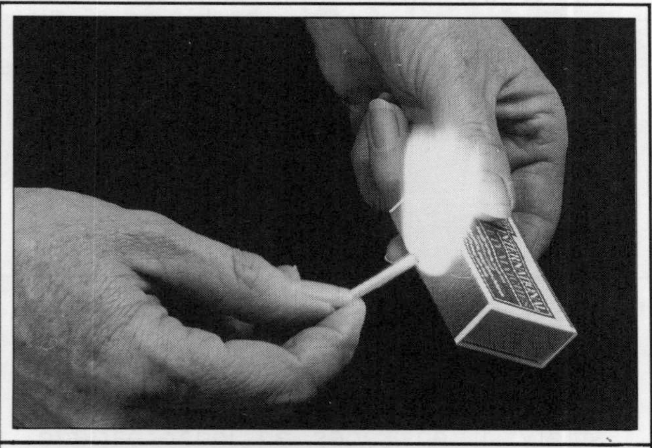

Importance of heat energy

The most plentiful source of heat energy that we have is the sun. Without this heat energy being continually delivered to the earth, there would be no life on earth.

Heat energy is very important to humans, as we are warm-blooded animals. Our bodies have to remain at a reasonably steady temperature (about 37 °C). If the temperature changes from this value, the body feels uncomfortable. The main source of heat energy is food. But a low external temperature can easily make us feel uncomfortable. Many buildings have some form of central heating. This is usually adjusted to keep the body temperature at a reasonable level if the temperature outside is low.

What is heat energy?

Up until the 19th century, little was known about the nature of heat. It was thought to be a mysterious substance called 'caloric'. A hotter body contained more 'caloric' than a colder one. But then a series of experiments showed that heat was produced by energy changes.

But the nature of heat energy is rather more difficult to understand. You will learn in Unit 5 about a theory which says that all particles are in a state of continual motion (the 'kinetic theory'). For example, in a solid the particles are fixed in position (this is why a solid keeps its shape). But they are vibrating backwards and forwards about a fixed point. The heat energy is, in fact, the total kinetic energy of all the individual particles in the solid. If the substance is made hotter then the speed of these vibrations increases.

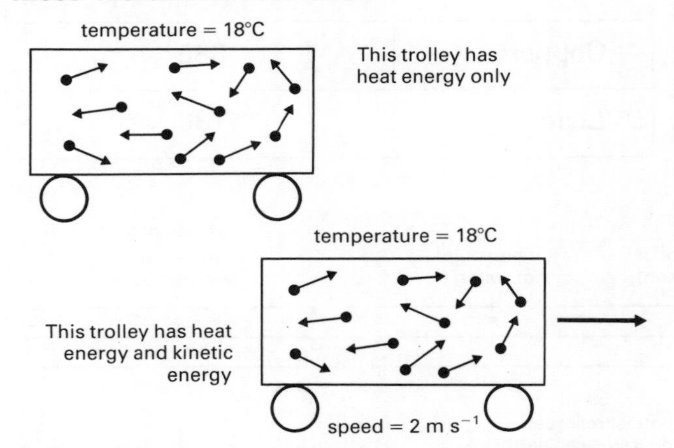

Fig. 3.35 Difference between heat and kinetic energy

You may think that kinetic energy and heat energy are the same. They are, in one sense, because the energy is due to motion. But with heat energy the motion of the individual particles is entirely random. There is no particular direction to their movement. With kinetic energy, all the particles are moving in the same direction. Any object that has kinetic energy will have some heat energy. But the amount of heat energy will be the same whether it is moving or still (Fig. 3.35).

MEASURING HEAT ENERGY

The difference between heat and temperature

These terms have precise meanings in science. (In everyday life, there is little difference in their meanings.) When we say heat we mean 'heat energy'. That is, the total kinetic energy of all the particles in the body. It is possible to measure changes in this (see page 136). The word 'temperature' means how hot (or cold) a body is.

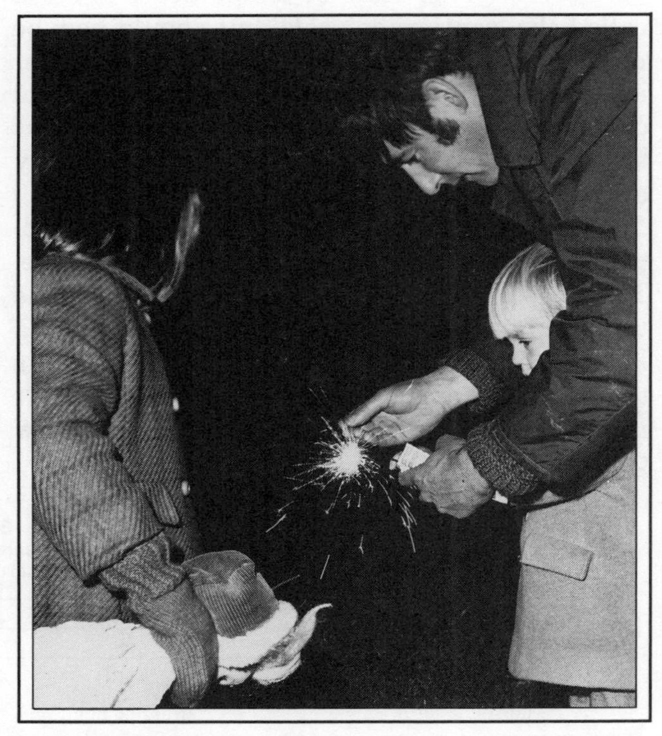

Fig. 3.36 The temperature of each spark may be 800 °C

You have probably lit 'sparklers' on Bonfire Night. When the 'sparkler' is lit, large numbers of sparks are produced. If they fall on your hand, you do not even feel them. Yet the temperature of each spark (actually it is a tiny piece of white-hot metal) may be as high as 800 °C (the temperature of boiling water is 100 °C). Why do you not feel the sparks? The answer is that the actual heat energy of each spark is extremely tiny. As the spark cools down, it transfers this tiny amount of heat energy to the skin. But the temperature rise it causes is so small that the skin cannot detect it.

If 1 g water (at 100 °C) fell on your left hand and 1 kg of water (at 100 °C) fell on your right hand, which would be worse?

Measuring temperature

When measuring lengths, a scale is needed. With the SI system of units, this unit is the metre. Likewise with temperature, there must be a scale. But the way in which the scale is devised is rather different. Certain temperatures are given fixed values. For example, the temperature at which ice changes to water is given the value '0' and the temperature at which water changes to steam is given the value '100'. This scale of temperature is often called the 'centigrade scale' but should now be more correctly called the 'Celsius scale' after its inventor. The scale is divided into 100 equal divisions. There are other temperatures that are given fixed values but you need not worry about these at this stage.

Figure 3.37 shows that the range of temperatures that is possible is extremely large. There is, in fact, no upper limit to the temperature that can be reached. But there is a lower limit. It is called ABSOLUTE ZERO. At this temperature, molecules would stop moving. It is not possible to cool anything down below this value.

Measuring heat energy

We cannot actually measure the total amount of heat energy that a body contains. But what we can do is measure how much energy is supplied to (or removed from) a body. This is what will happen in Investigation 3.4, which you should do now.

Specific heat capacity

The amount of heat energy needed to raise the temperature of 1 kg of the substance by 1 °C is given a special name. It is called the SPECIFIC HEAT CAPACITY of the substance. Table 3.1 gives the values of specific heat capacity for the liquids you may have used in Investigation 3.4. Why do you think your answers did not completely agree with the values quoted here? The table also gives values for some solids.

Table 3.1 Specific heat capacities

Substance	Specific heat capacity/ $J\,kg^{-1}\,°C^{-1}$
Water	4200
Brine	3000
Methanol	2300
Paraffin	2200
Aluminium	900
Iron	480
Copper	385
Lead	130

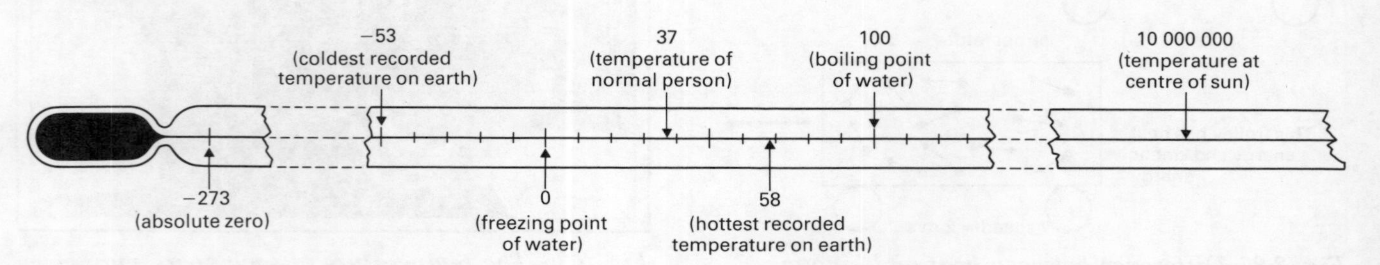

Fig. 3.37 Temperatures (°C)

Investigation 3.4 To compare the heat energy supplied to different liquids

You are going to see the effect of heating equal masses of *different* liquids with the same amount of heat energy. The heat energy is provided by an electrical heater.

Collect

heater and power supply
thermometer (0–100°C)
stopclock
boiling tube
beakers each containing 25 g of the liquids used (e.g. water, oil, paraffin, brine, methanol, glycerol etc.)

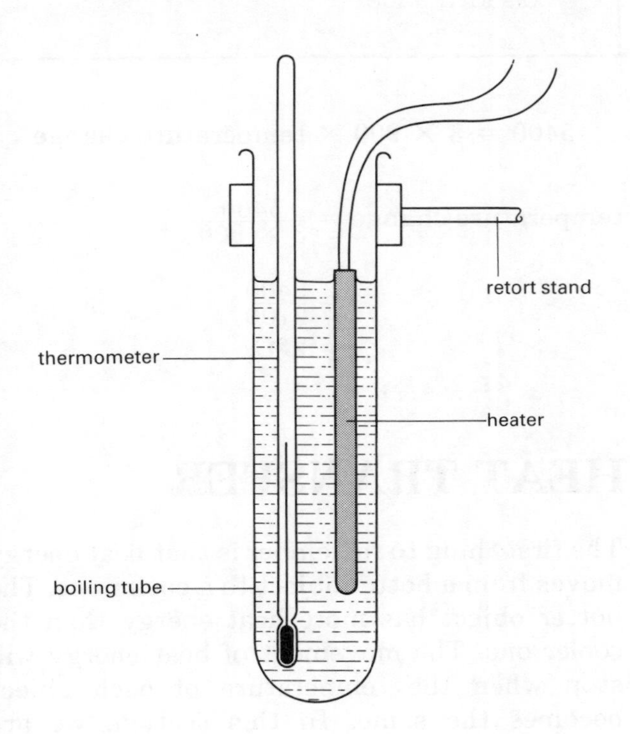

What to do

1 Connect up the heat to the power supply. Do not switch it on.
2 Set up the apparatus as in the diagram.
3 You will be told the voltage setting and the heater power.
4 Check that the stopclock works and then, for each liquid, heat the liquid for, say, 200 s. Start the stopclock at the same moment that you switch the heater on.
5 When you switch off the heater, remove it from the liquid at *once*. Stir the liquid before taking the temperature.

Copy and complete the table below.

heat energy = heater power × time (in seconds)

Questions

1 For the energy supplied, are all the results the same?
2 Which liquid needed most energy?
3 Which liquid needed least energy?
4 For each of the liquids, calculate how much energy is needed to raise the temperature of 1 kg by 1°C.
This is done as follows:
The mass must be in kg, i.e. 0.025 kg.

$$\text{Energy needed} = \frac{\text{energy supplied}}{\text{mass} \times \text{temp. rise}}$$

Liquid	Initial temperature	Final temperature	Time taken	Energy supplied

You will notice that the specific heat capacities of the solids are much less than those of the liquids. You may be given a chance to measure the specific heat capacity of a metal. But, again, your results will probably not be in agreement with those quoted here. The units for specific heat capacity are $J\,kg^{-1}\,°C^{-1}$. This is pronounced as joules per kilogram per degree Celsius.

Calculations on specific heat capacity
The specific heat capacity tells us how much energy is needed to raise the temperature of 1 kg by 1 °C. In practice, it is very unlikely that the mass would be 1 kg or the temperature rise 1 °C. We have talked about temperature rise but it should be remembered that, if the temperature is raised, heat energy is taken in. But if the temperature falls, heat energy is given out. For example, 1 kg of aluminium would need 900 J of energy to raise its temperature from 27° C to 28° C. If the temperature of the aluminium falls from 28° C to 27° C, it will give out exactly 900 J of energy. So we need a general formula for use with heat energy changes and it should apply to bodies receiving heat energy and giving out heat energy. Here is the formula:

$$\text{heat energy} = \text{mass} \times \text{specific heat capacity} \times \text{temperature change}$$

Examples:

1 How much heat energy must be supplied to raise the temperature of 2 kg of water by 20 °C?

The specific heat capacity of water is $4200\,J\,kg^{-1}\,°C^{-1}$ (from Table 3.1).

$$\text{heat energy} = \text{mass} \times \text{specific heat capacity} \times \text{temperature change}$$
$$= 2 \times 4200 \times 20$$
$$= 168\,000\text{ J.}$$

2 If 5400 J of energy is supplied to 3 kg of aluminium, by how much will the temperature rise?

The specific heat capacity of aluminium is $900\,J\,kg^{-1}\,°C^{-1}$.

$$5400 = 3 \times 900 \times \text{temperature change}$$

$$\text{temperature change} = \frac{5400}{3 \times 900}$$
$$= \frac{5400}{2700}$$
$$= 2°\text{C.}$$

> **Assignment – Heat energy in practice**
>
> 1 How much heat energy is needed to raise the temperature of
> (a) 10 kg of lead by 5 °C?
> (b) 2.5 kg of brine by 10 °C?
>
> 2 If 5600 J of energy raises the temperature of 2 kg of copper by 8 °C, what value does this give for the specific heat capacity of copper?
>
> 3 Water has a much higher specific heat capacity than rock. Try to find out how this fact affects the climate of an island such as Britain.
>
> 4 Find out how night storage heaters work. What are their advantages and disadvantages?

HEAT TRANSFER

The first thing to remember is that heat energy moves from a hotter object to a cooler one. The hotter object has more heat energy than the cooler one. The movement of heat energy will stop when the temperature of each object becomes the same. In this section, we are interested in how this heat energy moves from one place to another. The three methods by which heat moves are called conduction, convection and radiation.

Conduction
Look at Fig. 3.38. One end of the metal rod is being heated by the Bunsen burner. After a period of time, the wax melts and the drawing pin falls. Heat energy has travelled along the rod. This method of heat transfer (or movement) is called CONDUCTION. The only group of

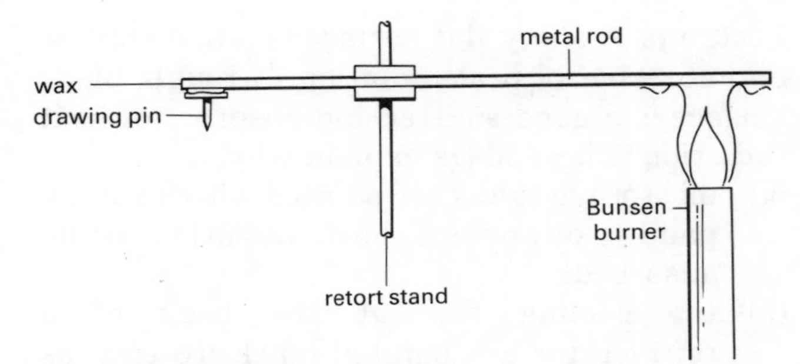

Fig. 3.38 Conduction of heat

substances that can be called *good* conductors of heat are metals. All other substances are classed as INSULATORS (bad conductors of heat).

> ## Assignment – Good conductors
> Not all metals conduct heat at the same rate. Some conduct better than others. Silver is, in fact, the best conductor. Using Fig. 3.38 as a start, describe how you could set up an experiment to find out which is the best conductor among the following metals: iron, copper and aluminium.

In general a good conductor is used when we want heat energy to pass through it and a bad conductor when we do not. For example, a saucepan is normally made of aluminium (good conductor) so that heat energy from the gas or electric ring gets through to heat the contents. It has a plastic covered handle (bad conductor) so that it can be moved safely. The handle does not heat up even when the contents of the saucepan are boiling.

Convection

Look at Fig. 3.39. In both diagrams, CONVECTION is occurring. The left hand one shows convection in a liquid. As the water is heated, it starts to rise (shown by the coloured stream). It rises because it expands and becomes less dense than the water around it. Cooler water moves in to take its place and this, in turn, is heated. Figure 3.39(b) shows convection in a gas (air). The heated air rises (together with the smoke) and draws in cooler air that is in turn heated. With convection, heat energy is carried by the moving medium. Convection can occur only in liquids and gases. Why do you think this is?

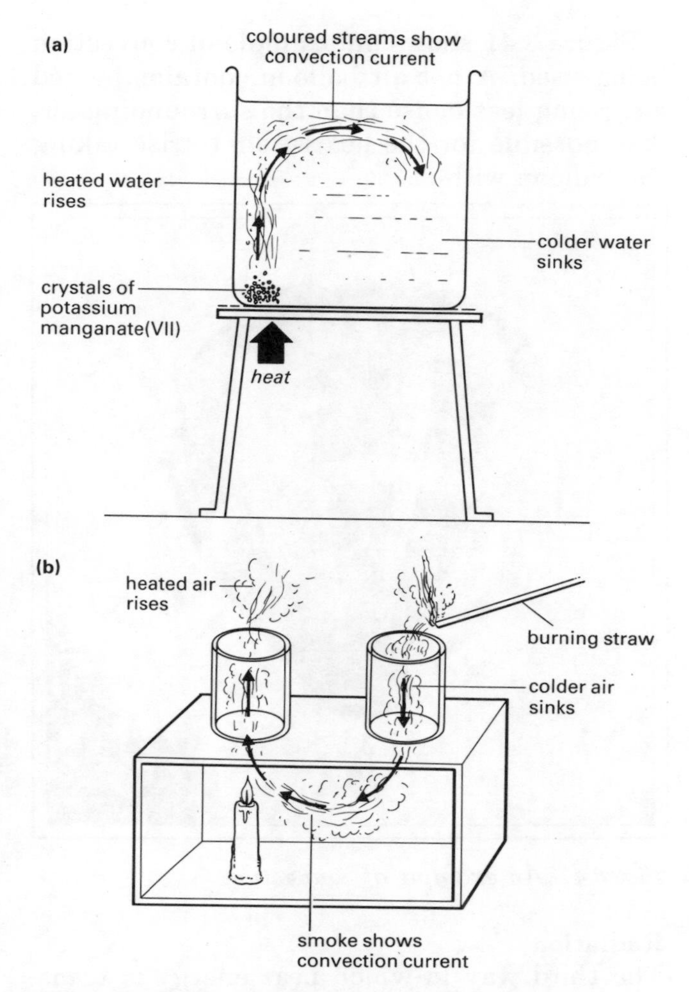

Fig. 3.39 Convection in a liquid and a gas

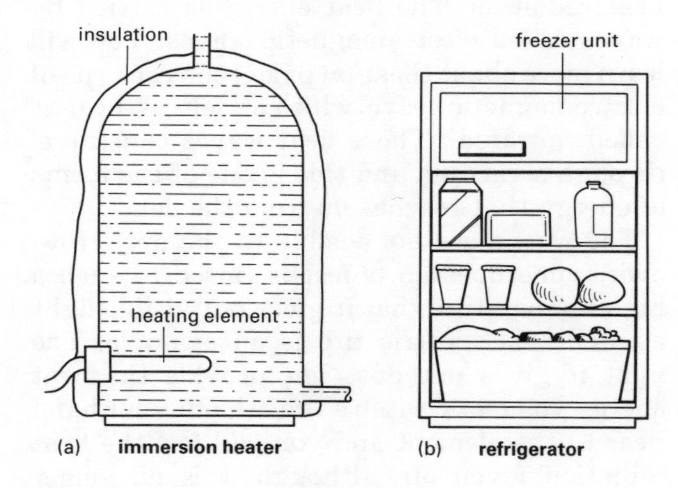

Fig. 3.40 Convection in the home

In a domestic hot water tank, the heating element is near the bottom (Fig. 3.40(a)). Why do you think this is? Why do you think the freezer unit in a refrigerator (Fig. 3.40(b)) is at the top?

Figure 3.41 shows an example of convection being used. A hot air balloon contains heated air. Being less dense than the surrounding air, it is possible for the heated air to rise taking the balloon with it.

Fig. 3.41 An example of convection

Radiation
The third way in which heat energy is transferred is called RADIATION or, more properly, heat radiation. The heat energy is carried by waves called electromagnetic waves. You will learn more about these on page 156. The type of electromagnetic wave which is felt as heat is called 'infrared'. These heat waves can travel through a vacuum and this is one of the forms of energy that reaches us from the sun.

The waves are not visible to the eye. However, sometimes an object is raised to such a high temperature that it gives out visible light as well. The electric fire is an example. The light it gives out does *not* provide the heat energy you feel. Switch a fire off; put your hand near to the element and you will feel the heat radiation given off, although it is no longer giving off light.

If an object is hotter than its surroundings, then it will give out heat radiation. In doing so it will lose heat energy and cool down. If the object is cooler than the surroundings, then the opposite happens. The type of surface will affect how quickly the object cools down or heats up. A shiny flat surface is a bad emitter and absorber of heat radiation. A rough, black surface is a good emitter and absorber of heat radiation. These ideas explain why

(a) oil storage tanks are painted with a silvery paint (to prevent heat radiation being absorbed);
(b) the cooling fins at the back of a refrigerator are painted black (to emit as much heat radiation as possible);
(c) white clothes are more comfortable to wear in hot climates (they do not absorb as much heat radiation as dark colours).

You can probably think of other examples.

Assignment – Vacuum flasks

1 Try to find out how a vacuum (or Thermos) flask works.
2 A student wrote: 'The Thermos flask is a fantastic invention. It can keep hot things hot and cold things cold. But how does it know which is which?'
Can you help the student?

'Saving' heat energy

In a house, it has to be remembered that most of the energy used has to be paid for. Much of the energy expenditure goes towards heating the house. It is estimated that about one third of the heat energy produced is 'lost' directly to the atmosphere outside. Figure 3.42 shows the main sources of heat loss from a house.

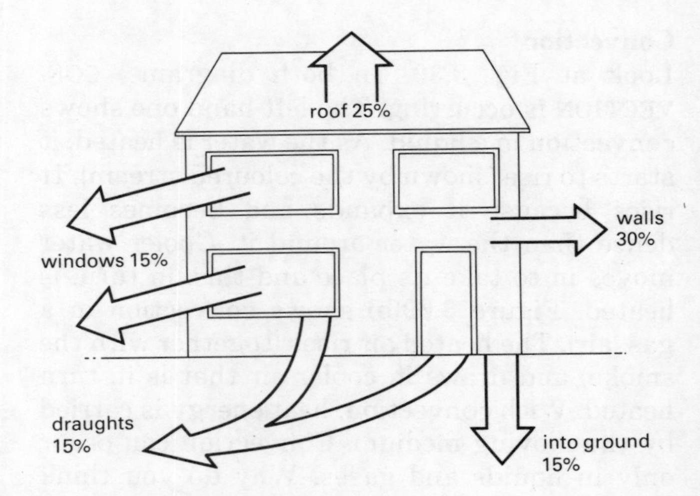

Fig. 3.42 Heat losses from a house

Assignment – Save it!

1 Heat energy can be lost from a house through the roof, the windows, the walls and the ground. For each of these, find out ways in which the heat losses can be reduced.
2 You have been given two plastic materials and have been told that they would be suitable for lagging a hot water tank to prevent heat losses. What tests could you do to find out which one is better?

Developments in the 'saving' of heat energy
Apart from providing insulation around a house, what other methods are available for conserving heat energy? One of the more important methods at the moment is the attempt to use the heat energy wasted from conventional power stations.

In a power station there is a great deal of heat energy that is simply lost to the surroundings. This heat energy either goes straight into the atmosphere or is used to heat adjacent water in a river or the sea. A few schemes have been developed by which the 'surplus' energy is used to heat houses and other buildings close to the power station. It is called 'district heating'. However, as the cost of land in urban areas is so great, most power stations are now being built in rural areas and nuclear power stations are always built well away from urban areas. So this method will not be very attractive.

It has also been suggested that large industrial consumers could have their own small power station. Besides the electrical energy produced, the surplus heat energy could heat the surrounding buildings and perhaps the neighbourhood houses.

Assignment – Heat pump

A 'heat pump' works rather like a refrigerator. Find out how this device works and why it would be useful in 'saving' heat energy.

WHAT YOU SHOULD KNOW

1 When one form of energy is changed into another, heat energy is usually produced.
2 Heat energy is the total kinetic energy of the particles of a substance.
3 If the substance is made hotter, then the total kinetic energy of the particles increases.
4 Temperatures are measured using the Celsius scale.
5 The specific heat capacity of a substance is the amount of heat energy needed to raise the temperature of 1 kg of it by 1 °C. The units of specific heat capacity are $\mathrm{J\,kg^{-1}\,{}^\circ C^{-1}}$.
6 With heat energy changes:
heat energy

$$= \text{mass} \times \text{specific heat capacity} \times \text{temperature change}$$

7 There are three methods of heat transfer: conduction, convection and radiation.

QUESTIONS

1 (a) If 2 kg of water (specific heat capacity $= 4200\,\mathrm{J\,kg^{-1}\,{}^\circ C^{-1}}$) is heated through 40 °C, how much heat energy must be supplied?
(b) How much heat energy is given out when 2 kg of iron (specific heat capacity $= 480\,\mathrm{J\,kg^{-1}\,{}^\circ C^{-1}}$) cools from 500 °C to 200 °C?
(c) An electrical heater is rated at 200 W. If it supplies heat energy to 0.5 kg of water for 84 s, calculate
(i) the electrical energy supplied,
(ii) the temperature rise, of the water, assuming no heat energy loss.

Time/min	0	1	2	3	4	5
Temperature/°C	19.8	21.4	23.0	24.6	25.9	26.9

2 An electrical heater rated at 50 W is used to heat 0.5 kg of a liquid in a container. The table shows the variation in temperature with time.
   (a) Explain the shape of the graph.
   (b) Describe how the graph would differ if
      (i) 1.0 kg of the liquid had been used instead,
      (ii) the container had been lagged,
      (iii) the container had been painted black.
   (c) Calculate the heat energy supplied in the first three minutes.
   (d) Use this result to calculate a value for the specific heat capacity of the liquid. Assume there is no heat energy loss.

3 The three ways in which heat energy can be transferred are by conduction, convection and radiation. Use these ideas to help you explain the following statements.
   (a) Many teapots and kettles have shiny metal surfaces.
   (b) In winter, birds fluff up their feathers when stationary.
   (c) A woollen sweater is usually warmer than one made from synthetic materials.
   (d) Aluminium foil is often used to cover meat when it is roasted in an oven.
   (e) Double glazing is more effective for heat insulation in a house than single glazing.
   (f) Water moves through a car radiator from top to bottom.
   (g) The reflector behind the element of an electric fire should be kept as clean as possible.
   (h) A motor cycle engine has fins on the outside and is often painted black.
   (i) The heating element of a kettle is always placed at the bottom.
   (j) When the weather is cold, frost is more likely to form on clear nights than on cloudy nights.

Energy 1

WAVES

VISIBLE WAVES

The commonest visible waves are water waves. They could be the waves seen on the seashore. Or they could be the waves you make when you sit down in the bath. Both are called water waves. They travel along the surface of the water. These waves will be studied in detail as they can be seen easily and do not travel very fast.

Making waves

If a wave travels along a water surface, then something has to start the wave moving. At sea it is the wind that produces the original waves. In the bath it is your body that starts the wave. To make things a little simpler, use a piece of wood and a small tank or beaker containing water covered with oil (this slows the wave down) (Fig. 3.43). Dip the wood into the water. What happens to the water surface? Now remove the wood. What happens now? This is called a PULSE.

A wave is a series of continuous pulses. What happens when the pulse hits the side? How could you set up a continuous train of pulses? This is how a wave is produced. One thing moves up and down (the wood). This sets something else moving (the water surface). The thing moving up and down is called an OSCILLATOR. An oscillator repeats the same motion time and time again. All waves are produced by something oscillating.

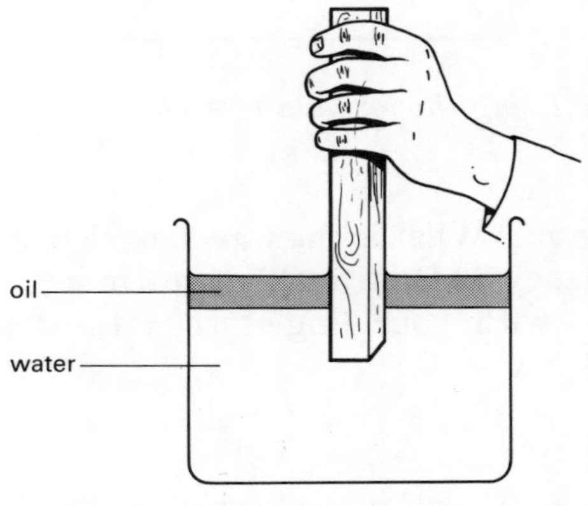

Fig. 3.43 Making water waves

Why study waves?

At first sight, waves do not appear to be very interesting (or useful) to study. However, they turn out to be very important for the following reasons.

(a) Waves carry energy. You have only to look at the effects of a rough sea on the coast to see that energy is carried by waves. Other types of wave can be shown to carry energy. All the energy that reaches us from the sun is carried by waves. They are known as electromagnetic waves.

(b) Waves can carry information. The eye receives light waves from our surroundings. These gives the brain a great deal of information on which to act. Sound waves are a most important way for individuals close to each other to communicate. But it is also possible to produce other waves to carry information. Radio waves are used to transmit speech and music over large distances. Most methods of communication use waves of one type or another.

(c) Waves can produce a great deal of information about our surroundings. All that is known about the sun, the stars and the different galaxies in the universe has been learned by studying waves. Much of what is known about the structure of matter has been learned through the behaviour of waves. We have also learned a lot about the structure of the earth by studying the waves produced by earthquakes or explosions.

Parts of the wave

Before looking at the properties of waves, there are a few names and labels that must be learned. Figure 3.44 shows a 'frozen' picture of a water wave. (The broken line represents the undisturbed surface.) The largest distance from the undisturbed surface to the wave is called the AMPLITUDE. This is labelled a in the diagram.

The distance from one crest to the next (or one trough to the next) is called the WAVELENGTH. The symbol for the wavelength is λ (the Greek letter 'l', called lamda).

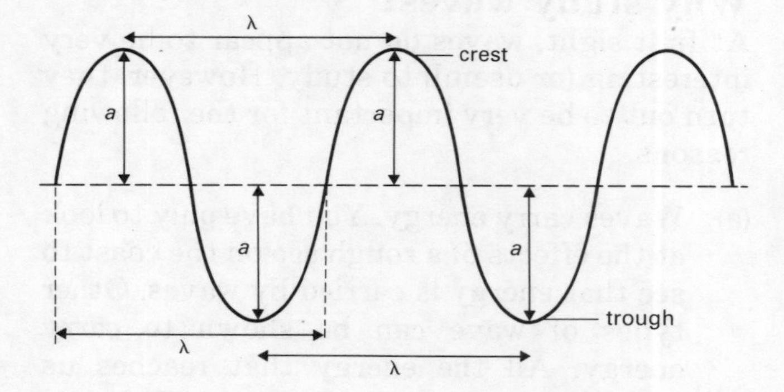

Fig. 3.44 Parts of a wave

The number of waves produced per second is called the FREQUENCY. The symbol is f. The unit for frequency is the hertz (abbreviation: Hz). For example, if you were using your finger to generate water waves and your finger moved up and down 3 times each second, then the wave frequency would be 3 Hz.

The *speed* (symbol: v) is the last quantity. It is defined as the distance travelled by the wave in 1 second. The speed, as will be seen, is related to the frequency and the wavelength. The units of speed are either metres per second ($m\,s^{-1}$) or centimetres per second ($cm\,s^{-1}$).

The wave formula

Assume that an oscillator produces 10 waves each second (the frequency, f, is then 10 Hz) and the wavelength, λ, is 5 cm. In one second, the first wave will have travelled a distance of $10 \times 5 = 50$ cm. This is the distance travelled in unit time, i.e. the speed. So the relationship between speed, frequency and wavelength can be written as

$$\text{speed} = \text{frequency} \times \text{wavelength}$$

or

$$v = f\lambda.$$

Again a triangle method can be used for solving problems using this formula (Fig. 3.45).

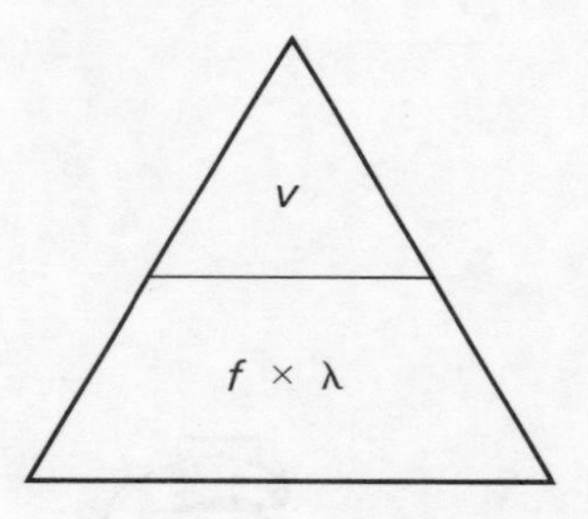

Fig. 3.45 Using the formula $v = f\lambda$

For example: What is the wavelength of a wave of frequency 20 Hz travelling at 3 $m\,s^{-1}$? (Cover λ with your finger, then the triangle tells you: $\lambda = \frac{v}{f}$)

$$\lambda = \frac{v}{f} = \frac{3}{20} = 0.15 \text{ m or } 15 \text{ cm.}$$

Investigation 3.5 Studying the properties of waves

In this investigation, you will look at waves travelling along a spring and on a water surface. By sending pulses it is possible to learn a lot about how waves travel. With both systems, the pulses travel slowly and can be directly observed.

Collect

stopwatch
long spring
tank and plunger
string

A. Spring waves

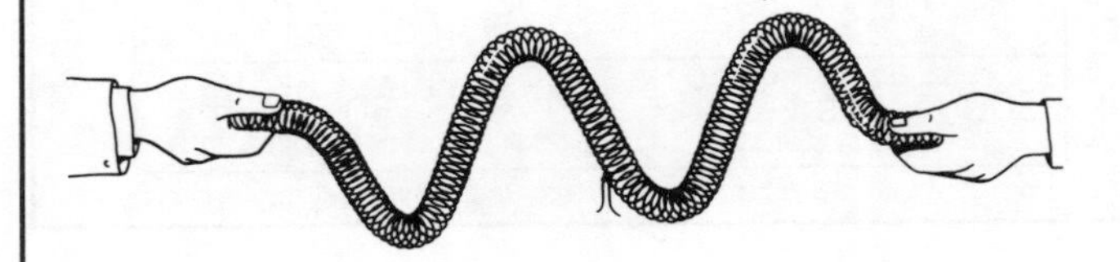

What to do

1 Lay the long spring on the floor or benchtop. Tie a piece of string to one of the coils.
2 One end of the spring must be securely fixed: your partner could hold one end so that it does not move.
3 Stretch the spring slightly and then send a pulse along the spring by moving your hand quickly one way and then the other.
4 Try to answer the following questions.

Questions

1 What happens to the size of the pulse as it travels backwards and forwards?
2 What happens to the shape of the pulse as it travels backwards and forwards?
3 Eventually the pulse disappears. Why do you think this happens?
4 What happens to the pulse when it is reflected from the fixed end?
5 Use the stopwatch and try to calculate a value for the speed of the pulse (speed $= \frac{\text{distance}}{\text{time}}$).
If the pulse moves very quickly, do not stretch the spring so much.
6 If you stretch the spring more, what happens to the speed? Does it get greater or smaller?
7 Does the spring move *along* with the pulse? Think of a way to use the string to check this.
8 Try producing continuous waves. Would a continuous wave make answering questions 1 to 7 easier or harder?

B. Water waves

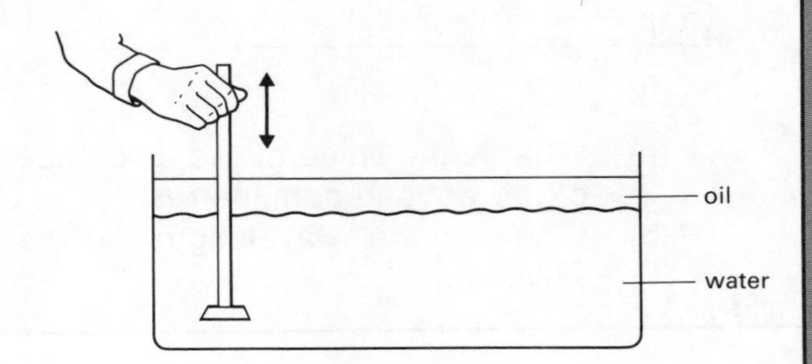

What to do

1 Move plunger vertically up and down to produce the water waves. (The layer of oil on top of the water is to make the speed of the water waves smaller.)
2 Using pulses produced by the plunger, try to answer the following questions.

Questions

1 What happens to the size of the pulse as it travels backwards and forwards?
2 What happens to the shape of the pulse as it travels backwards and forwards?
3 Eventually the pulse disappears. Why do you think this happens?
4 What happens to the pulse when it is reflected from the fixed end?
5 Use the stopwatch and try to calculate a value for the speed of the pulse (speed $= \frac{\text{distance}}{\text{time}}$).
6 Does the water move *along* with the pulse? Try to devise a way of checking your answer to this question.
7 Try producing continuous waves. Would a continuous wave make answering questions 1 to 6 easier or harder?

Assignment - Water waves

1 The diagram below shows the appearance of waves on a water surface at a given instant of time. The scales shown in centimetres are to enable the wavelength and amplitude of the wave to be measured.

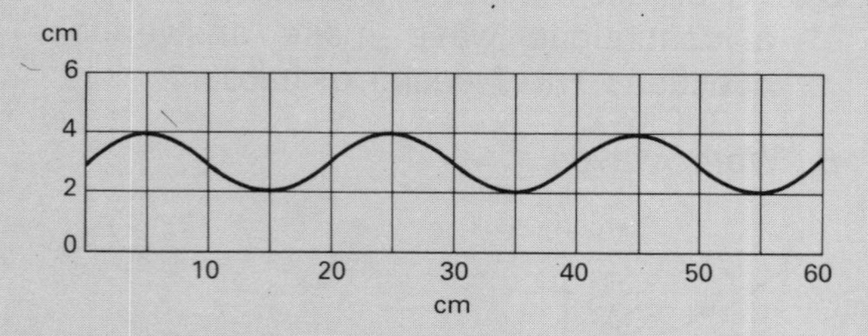

(a) What is the value of the amplitude of the wave in centimetres?

(b) What is the wavelength of the wave?

(c) If the frequency at which these waves are produced is 5 Hz, what is their speed?

2 What quantities should be in the blank spaces below? State the units of each answer.

Speed	Units	Frequency	Units	Wavelength	Units
		50	Hz	2	cm
50	$cm\,s^{-1}$			5	cm
2.5	$m\,s^{-1}$	20	Hz		
500	$m\,s^{-1}$			50	m

Other properties of waves

In your investigation, the waves travelled in only one direction. To investigate the other properties of waves, we need to look at waves that travel in more than one direction. The apparatus used is called a ripple tank (Fig. 3.46).

The wooden bar vibrates up and down continuously, producing water waves. The waves travel at a speed that is difficult to observe directly. It is possible to 'freeze' the waves so that they appear to stand still and can be photographed. Figure 3.47 shows some of the wave patterns that are produced. The diagrams show four important properties of waves: *reflection, refraction, diffraction* and *interference*. The lines show the positions of the crests (i.e. the tops) of the waves. How would you measures the wavelength of these waves?

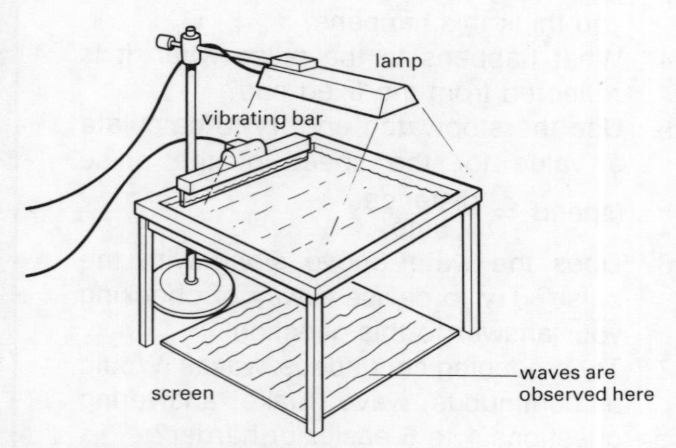

Fig. 3.46 The ripple tank

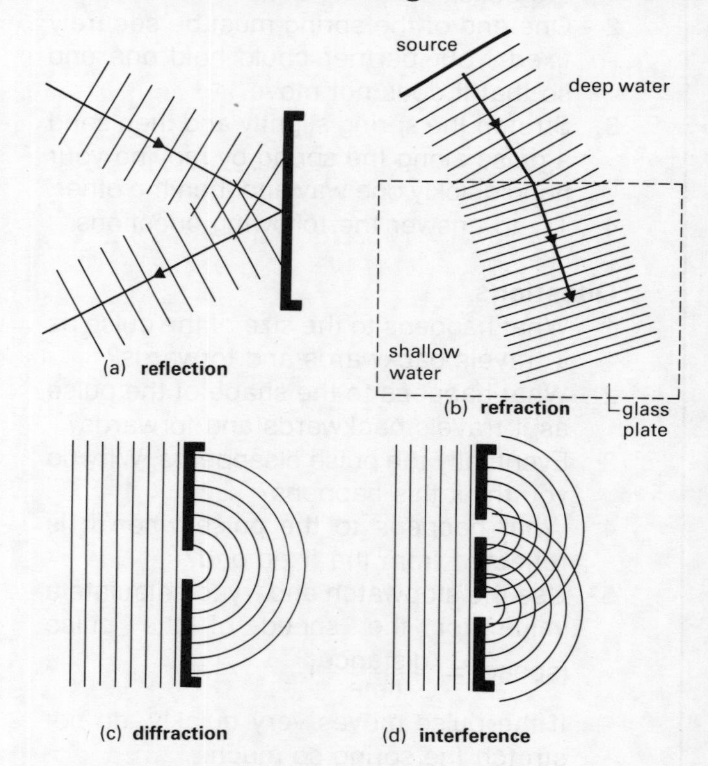

Fig. 3.47 The important properties of waves

Reflection

When waves hit a solid object, they may be reflected. The solid object in the ripple tank (Fig. 3.47 (a)) is a metal barrier. The direction in which the waves travel before and after REFLECTION is shown by an arrow. Can you see any connection between the angle at which the waves hit the metal barrier and the angle at which they are reflected?

Refraction

Instead of being reflected, waves can sometimes pass from one material into another material. When this happens, the speed changes and also the direction in which the waves are travelling. This effect is called REFRACTION. The speed of the water waves in the ripple tank depends on the depth of the water. By making the water shallower, the speed of the water waves is decreased. As the waves travel into the shallower water (shown by the dotted outline) the speed decreases and the waves change direction (Fig. 3.47(b)). The arrows show the change in direction.

Diffraction

The waves can only pass through the gap between the metal barriers in Fig. 3.47 (c). But they spread out after they have passed through the gap. This effect is called DIFFRACTION. You can see from the diagram that the waves have become circular but the wavelength remains the same.

Interference

Two gaps in the metal barriers are needed to show this effect (Fig. 3.47 (d)). As the waves pass through the gaps they are diffracted. But there are now two sets of waves. You can see that the waves from one gap can pass right through the waves from the other gap. What happens when one wave meets another is very important. The effect is called INTERFERENCE.

When waves interfere, you can see regions where the water appears to be flat. In these positions, the crest of one wave is in the same position as the trough of another wave. As you might think, they cancel each other out. So the water appears flat. These are called positions of *destructive* interference (Fig. 3.48).

The diagram also shows what happens when the crests of two waves or the troughs of two waves meet. These are called positions of *constructive* interference. When this happens, the effects add. If two crests meet at the same place, the result is a crest twice as high. A similar thing happens when two troughs meet at the same place.

To find out if something is a wave or not, the last two properties of waves are the best to use. If it shows diffraction and interference, then it must be a wave. Since both sound and light show these effects they must both be waves.

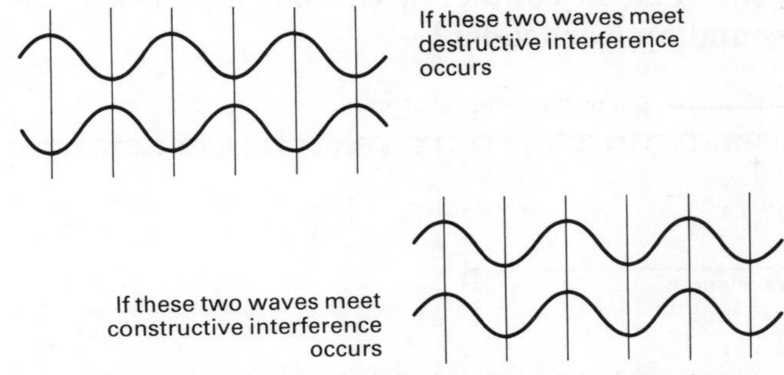

Fig. 3.48 Destructive and constructive interference

Assignment – Wave properties

1 Copy these diagrams into your book. Draw the appearance of the waves after they have been reflected (diagram (a)) and refracted (diagram (b)).

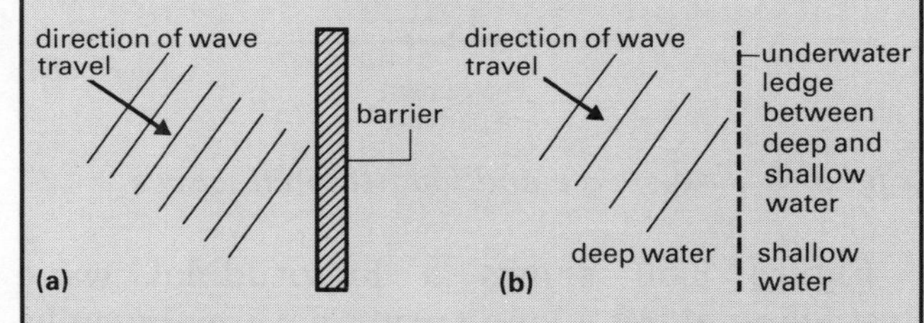

2 The diagrams below show an aerial view of water waves moving towards a harbour wall. In diagram (a), there is one entrance to the harbour and, in diagram (b), there are two entrances.

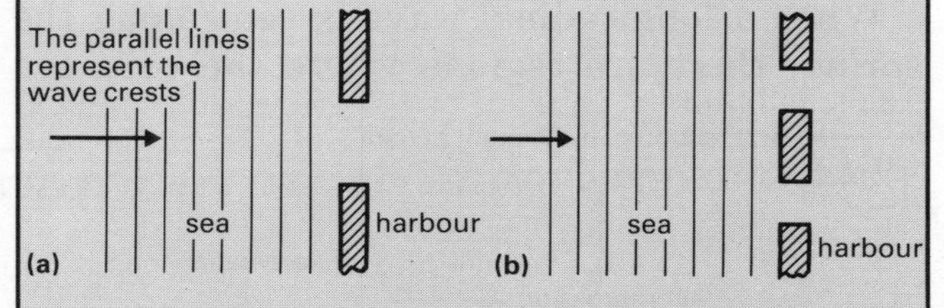

Copy these diagrams into your book. Draw the appearance of the waves after they have passed into the harbours.

SOUND: A DIFFERENT TYPE OF WAVE

The waves we have been studying are called TRANSVERSE WAVES. With water waves, the waves travel across the surface. But the water itself moves vertically up and down. There is another type of wave. This is called a LONGITUDINAL WAVE (Fig. 3.49).

With longitudinal waves, the movement of the particles in the medium is parallel to the direction in which the wave travels. Look at the lower diagram in Fig. 3.49. The wave travels from left to right. The particles move backwards and forwards. Compare this with the transverse wave where the movement of the particles is at right angles to the movement of the wave.

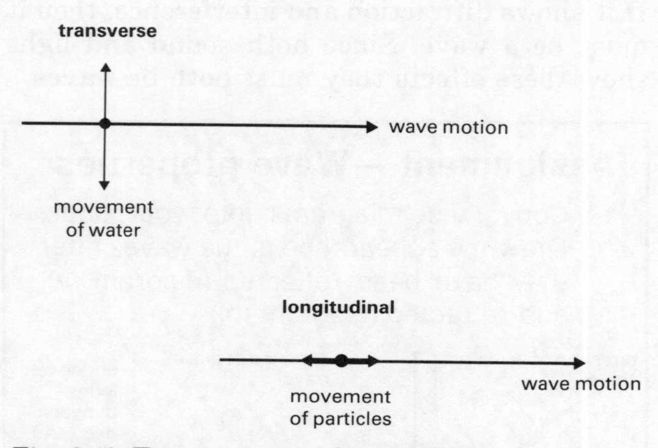

Fig. 3.49 Transverse and longitudinal waves

Figure 3.50 shows a longitudinal wave travelling along a long spring with many coils. The wave was produced by moving the left hand end of the spring backwards and forwards. You can see that the movement of the coils is in the same direction as that in which the wave travels.

When a longitudinal wave is sent along the spring, there are regions where the coils are closer together. These are called *compressions*. Midway between them are regions where the coils are more spread out. These are called *rarefactions*. The parts of the wave are similar to those explained for the transverse wave (page 144). But the meaning of the wavelength is slightly different. It is the distance between successive compressions (or successive rarefactions). The formula $v = f\lambda$ can be used in exactly the same way as with transverse waves.

Producing sounds

A sound is something that can be heard by the ear. All sounds are produced by vibrating objects. These vibrations are usually very small in amplitude and very rapid.

How do sounds reach the ear? This can be illustrated using a tuning fork as the sound source. If the tuning fork is struck against a hard object, the prongs vibrate (or oscillate) backwards and forwards. The frequency of these vibrations is usually stamped on the tuning fork. Different tuning forks vibrate at different frequencies.

The vibrations of the tuning fork set the surrounding air into vibration. These vibrations travel away from the tuning fork as longitudinal waves. If the ear is in the path of the waves produced, then a sound will be heard. Like all waves, sound waves lose energy as they travel through a medium. So the further the waves travel, the more energy they lose and the quieter the sound becomes.

Our ears are designed to pick up sounds transmitted through the air. It is also possible to detect sounds underwater, e.g. in a swimming pool. Sound waves will travel through solid materials as well. Try listening to the ticking of a watch with a wooden ruler. Place one end of the ruler on the watch and put the other end in contact with your ear. Does the sound appear louder?

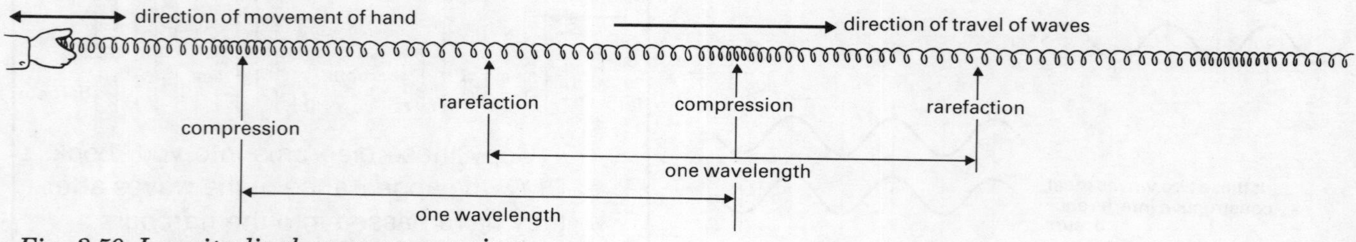

Fig. 3.50 Longitudinal wave on a spring

Fig. 3.51 Making sounds

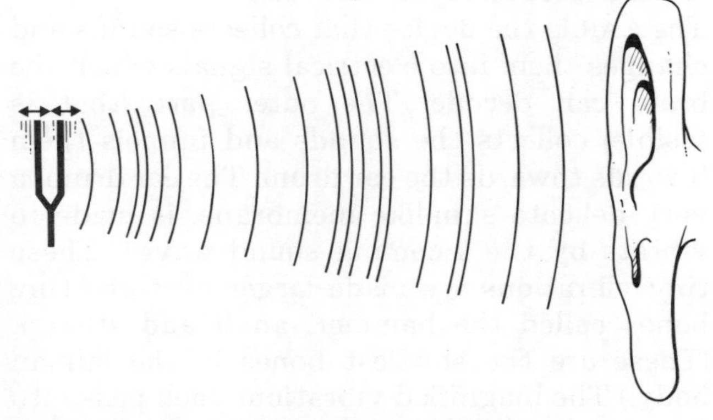

Fig. 3.52 How sound waves reach the ear

The speed of sound waves is lowest in gases, higher in liquids and highest of all in solids. For example, the speed of sound in air is $340\,\mathrm{m\,s^{-1}}$, in water $1500\,\mathrm{m\,s^{-1}}$ and in steel $5000\,\mathrm{m\,s^{-1}}$.

Assignment - The speed of sound waves

1 One of the first experiments to measure the speed of sound used a cannon. It was known that light travelled extremely fast. By seeing the flash produced as the cannon was fired, the observer could find out how long the sound of the shot took to reach him. If the distance was 1500 m and the sound took 5 s to reach the observer, what value does this give for the speed of sound?

2 How would you try to improve the accuracy of this result? Give a reason for each suggestion you make.

3 A violin sounds a note of 512 Hz (middle C); what will be the wavelength of the sound emitted? (Remember the formula $V = f\lambda$.)

4 A note of wavelength 0.5 m is sounded in air; what will be its frequency?

The structure of the ear

The EAR is the device that collects sounds and changes them into electrical signals which the brain can decode. The outer part (that is visible) collects the sounds and funnels them inwards towards the ear drum. The ear drum, a very delicate skin-like membrane, is made to vibrate by the incoming sound waves. These tiny vibrations are made larger by three tiny bones called the hammer, anvil and stirrup. (These are the smallest bones in the human body.) The magnified vibrations then pass into the COCHLEA. This is the shape of a snail's shell and is filled with liquid. Here the vibrations are changed into electrical signals that travel along the auditory nerve to the brain.

You will see in Fig. 3.53 two other parts of the ear: the SEMICIRCULAR CANALS that help us keep our balance and the EUSTACHIAN TUBE. The Eustachian tube connects the middle ear to the back of the nose. This enables the pressure on both sides of the ear drum to be kept equal. You may have experienced the change in pressure on your ear drum when flying in an aeroplane. As the plane increases its height above the ground, the air pressure in the cabin gets less. This feels uncomfortable. A similar thing can happen when going through a deep tunnel in a train or car. The solution is to swallow a few times. This makes the pressure inside the Eustachian tube equal to the pressure outside the ear drum.

What frequencies can the ear detect?

The ear is found to have frequency limits. Below a certain frequency, a sound will not be heard as continuous. It will sound like a series of 'blips'. Above a certain frequency, no sound will be detected by the ear. This range of frequencies is sometimes quoted as '20–20 000 Hz'. But the range of frequencies varies from individual to individual.

The upper frequency limit of the human ear can be investigated with a frequency generator and loud-speaker. The frequency generator produces electrical signals of varying frequencies. These are changed into sound waves with the *same* frequency by the loudspeaker. The person being tested faces away from the frequency generator (so as not

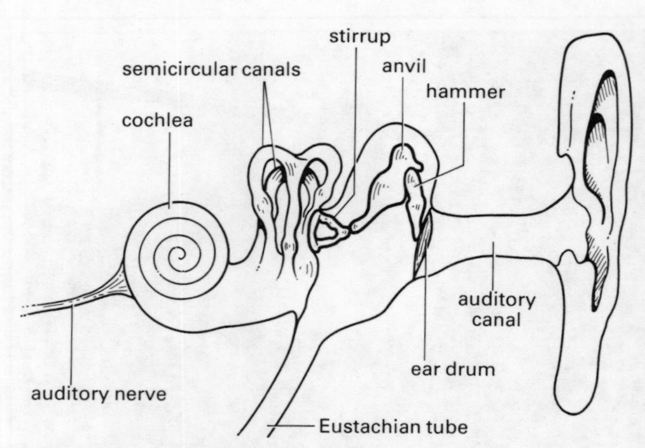

Fig. 3.53 Structure of the ear

to see the frequency being changed).

The frequency is increased until the person indicates that sound can no longer be heard. The frequency at this point is noted. The frequency is increased further and then decreased. When the person indicates that sound can be heard, the frequency is again noted. The average (or mean) of these two frequencies is taken as the upper frequency limit.

As you get older, the upper frequency limit decreases. This does not really affect your hearing, as most sounds in everyday life have frequencies lower than 10 000 Hz.

Looking at sounds

It is not possible to see sound waves directly. But the sound waves can be seen if they are changed into electrical currents. The apparatus shown in Fig. 3.54 does just this. A tuning fork is struck and placed close to the microphone. The microphone produces electrical currents that flow backwards and forwards in the same way as the sound waves. The oscilloscope displays these changing electrical currents as transverse waves on the television-like screen. We can then study sound waves. You will learn more about how the oscilloscope works in Unit 4.

Tuning forks were used to produce the traces shown in Fig. 3.54(b). What is the difference between a loud sound and a soft sound? What is the difference between a low frequency sound and a high frequency sound?

The tuning fork was used because it produces a pure sound with only a single frequency. The frequency of a particular fork is usually stamped on the fork. In practice, sounds usually contain a mixture of frequencies. Figure 3.55 shows some of the complex waveforms that sounds can have. They contain more than one frequency.

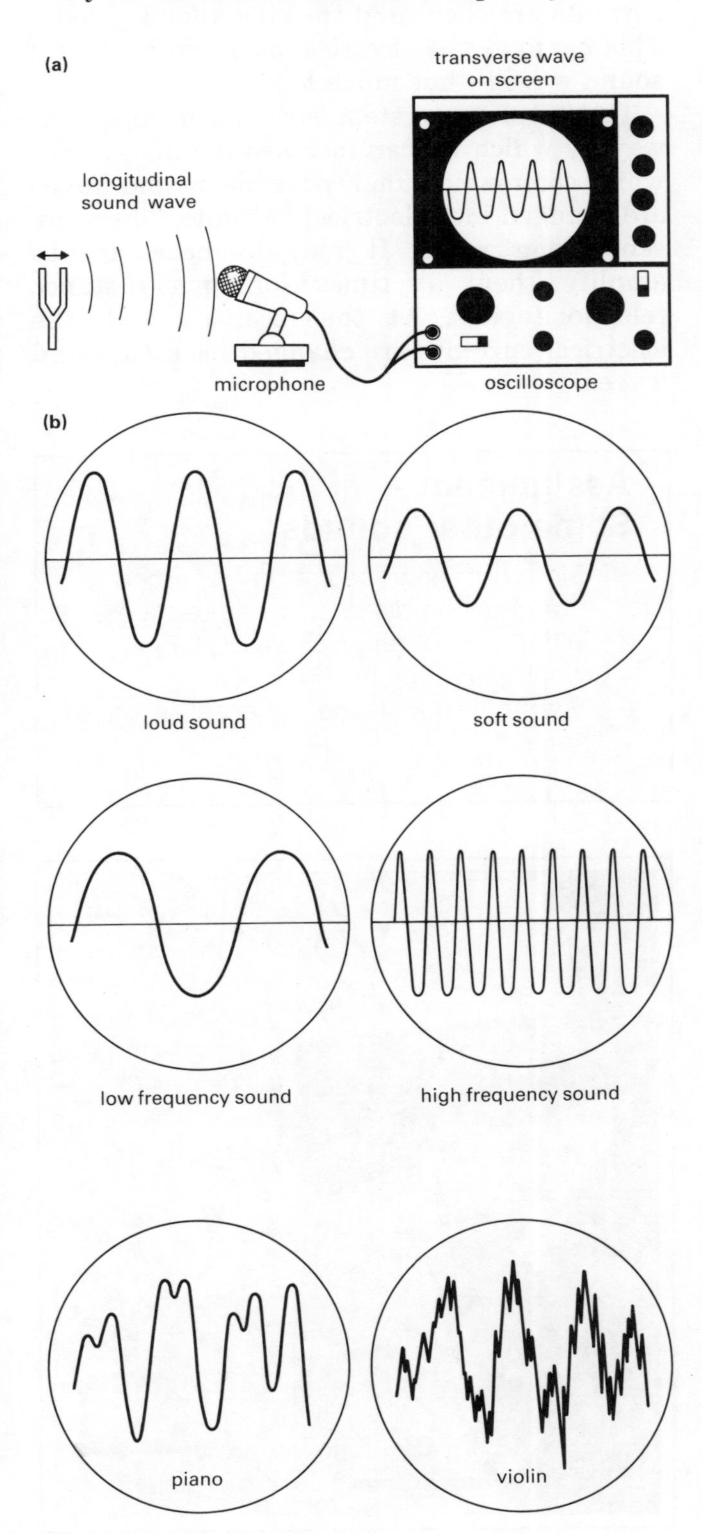

Fig. 3.55 The waveforms of some sounds

Sounds and noise

In the last section, we said that most sounds contain more than one frequency. It is the mixture of the frequencies present in the sound that will determine whether it is pleasant or not. There are certain sounds that everybody would agree are unpleasant. Such unpleasant sounds we call noise. But there are sounds which can be both pleasant and upleasant. It depends on the person hearing them. For example, what one person calls 'music' might be called 'noise' by someone else.

The loudness of noise is also a problem. Our society is becoming noisier and noisier. A special name has been produced for the effect this causes - noise pollution. Noise has a number of effects on our health. Besides being unpleasant, noise can cause stress and, perhaps most important, can directly affect our hearing. This can be temporary but constant exposure to sounds above a certain loudness level can produce permanent loss of hearing. It can sometimes lead to complete deafness. The first sign of hearing damage is the inability of the ear to hear high-frequency sounds. This can be measured by a device called an *audiometer*. People working in very noisy conditions are regularly tested to see if they are losing their ability to hear high-frequency sounds. Protective devices must be worn by people working where the sound levels are extremely high, e.g. near moving jet aeroplanes or using pneumatic drills.

Measurement of sound intensity

The loudness of a sound that we hear is determined by the energy that the sound wave carries. It is usually called the SOUND INTENSITY. The higher the sound intensity, the louder the sound will appear. Sound intensity is measured in units called decibels, dB. Some values of sound intensities, in decibels, are given in Table 3.2. The quietest sound that a person with normal hearing can detect is given the value 0 dB. The maximum safe sound intensity is usually taken as 85 dB. Sound intensities greater than this can produce a loss in hearing. Anybody exposed to such sound intensities for any length of time should wear protective devices over their ears.

Table 3.2 **Sound intensities**

Sound	Sound intensity/dB
Threshold of hearing	0
Whispering	20
Normal talking	60
Heavy traffic	80
Pneumatic drill	95
Rock concert	110
Jet plane taking off	120

Extending our ability to hear

Usually we want to make quiet sounds (small amplitude vibrations) louder (larger amplitude vibrations). This could be because the sound source is a long distance away or the person receiving the sound is suffering from deafness.

There are a number of simple devices that have been produced to amplify quiet sounds (Fig. 3.56). The stethoscope amplifies the very quiet sound produced by the pumping of the heart. An ear trumpet will help a partially deaf person. It collects the sound over a wide area (much larger than the ear) and passes this into the ear of the deaf person. The megaphone in Fig. 3.56(b) amplifies the boy's voice so that the rowers can hear him cheering them on from the bank.

Today, electronic methods are usually employed to make sounds louder (Fig. 3.57). The incoming sound is changed into an electrical current by a microphone. It is amplified by an electronic circuit. The much larger electrical currents are then used to drive a loudspeaker. This changes the electrical currents back into sound waves - but much louder.

The telephone system is the most important way by which we can increase the distance at which communication is possible. Sound waves are changed into electrical currents. These are sent along wires. It may be necessary to amplify them at times for long distance telephone calls. At the receiving end, the electrical currents are changed back to sound waves.

Assignment – Some other sounds

1. Find out how sound is recorded on gramophone discs or cassette tapes.
2. What are ultrasonic waves? Find out what uses they have.
3. Find out about the different uses of sonar.

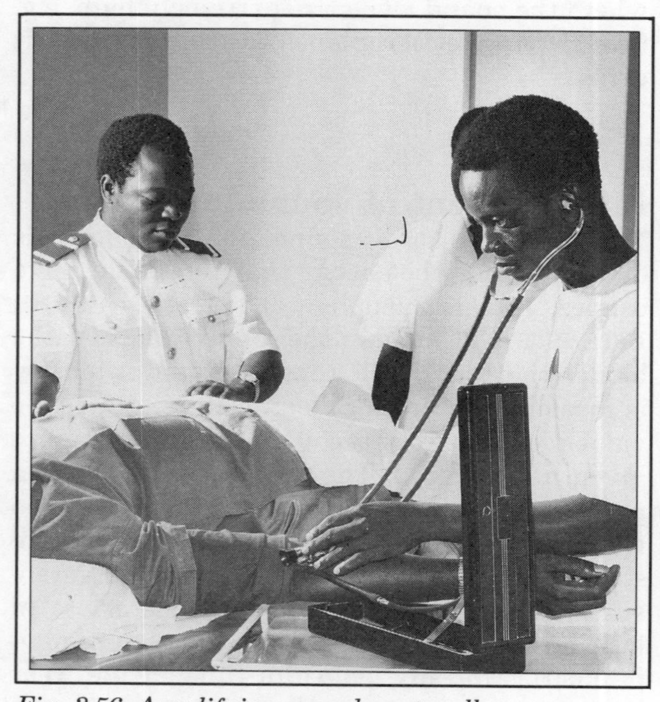

Fig. 3.56 Amplifying sounds naturally

Fig. 3.57 Amplifying sounds electronically

LIGHT

Figure 3.58 shows a dramatic use of a device called a laser. The light given out by this device is being used to cut through steel. The laser has become very important since its invention in 1960. However, its use is not limited to cutting through steel. Lasers with much smaller power outputs can be used in many different ways.

Fig. 3.58 Laser drilling holes in turbine blades

Assignment – Using lasers

Find out about how the laser has become so useful a device.

Not very many objects emit light. How many can you think of? In fact, practically everything that we see is seen by means of reflected light. Artificial illumination is extremely important to us. What changes would there be if we had no light sources to use when it became dark?

What is light?

Light travels as a transverse wave. It does not appear to have much resemblance to the water waves travelling along the surface of the ripple tank. One of the differences between them is that the wavelength of light is very much smaller than the wavelength of the water waves.

It is easy to show that light can be reflected and refracted (Fig. 3.59). But diffraction and interference are more difficult to observe because the wavelength of light is so small.

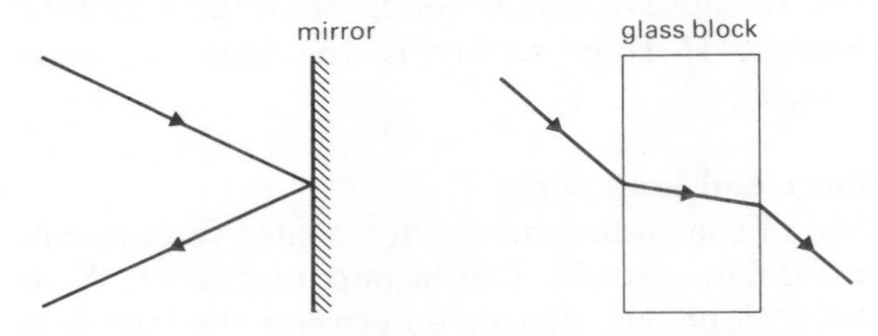

Fig. 3.59 Reflection and refraction of light

The eye

Information about our surroundings cames from the light emitted or reflected from all the objects around us. Light enters the eye through the transparent CORNEA (Fig. 3.60). The light is focused by the LENS onto the back of the eye. This is called the RETINA and is a light-sensitive surface. The retina changes the light received into electrical pulses that are transmitted via the OPTIC NERVE to the brain. The brain interprets these electrical signals as visual images.

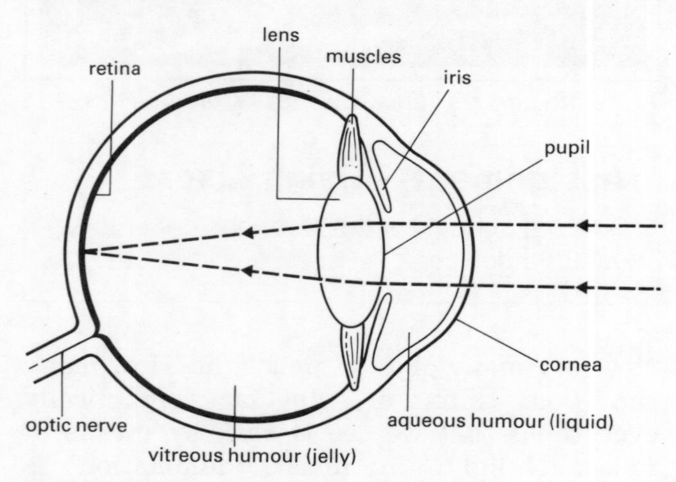

Fig. 3.60 Structure of the eye

The IRIS, in front of the eye, controls the amount of light that enters the lens. On a bright day, the iris will expand (i.e. the pupil will contract) and less light reaches the lens. The opposite will happen when it is dim or dark.

For an object to be seen clearly, the rays of light must focus on the retina. The lens of the eye has the ability to change its shape. If the object being viewed is a long distance away, the lens is thin (less 'powerful'). But, if the object being viewed is close to the eye then the lens becomes thicker (more 'powerful'). Changing the thickness of the lens is done automatically by the muscles surrounding the eye. If these muscles relax, the lens becomes thinner. If they contract, the lens becomes thicker.

Short and long sight

Two of the commonest defects that the eye can suffer from are short sight and long sight. With short sight, the distance between the lens and the retina is too great for the light to be properly focused. Light from a distant object is focused in front of the retina, producing a blurred image. It can be corrected by using a diverging (concave) lens in front of the eye. (Fig. 3.61).

Long sight occurs when the distance between the lens and the retina is too small for the light to be properly focused. This can be corrected by using a converging (convex) lens in front of the eye.

Instead of the spectacles that many people have to wear, it is possible to have contact lenses. These act in the same way as the lenses in spectacles but can be placed directly on the cornea.

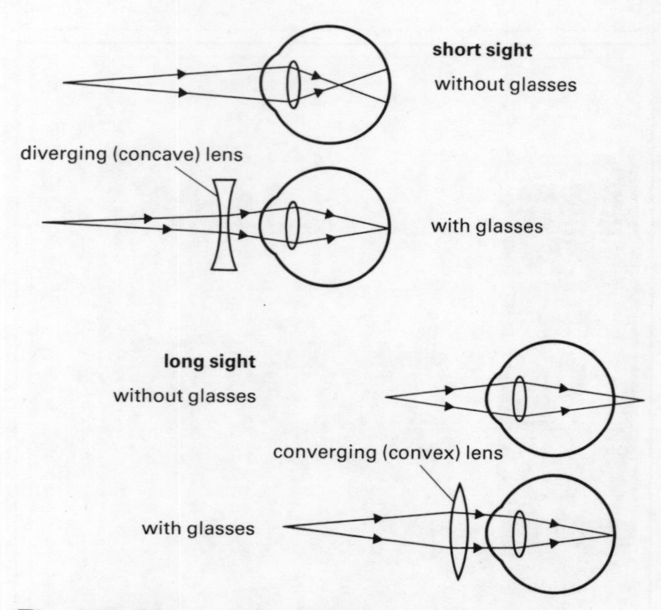

Fig. 3.61 Short and long sight

Assignment – Vision

Find out the answers to the following questions.

1. What is binocular vision? Why is it very useful?
2. What is astigmatism? How can its effects be lessened using lenses?
3. Why do some animals have eyes on the front of their head (like us) but others have them on the side of their heads?
4. The owl is a bird of prey. What part do its eyes play when it is hunting? What is special about the vision of owls?

Colour

If you look at a source of white light through a triangular prism, you will see a number of colours (Fig. 3.62). In fact white light can be split up into seven different colours. You will probably not be able to distinguish all of these using a simple prism. On p.147 it was said that waves are refracted when they pass from one material into another. The different colours in white light are refracted by different amounts and so they become spread out. Red light is refracted the least, violet light the most.

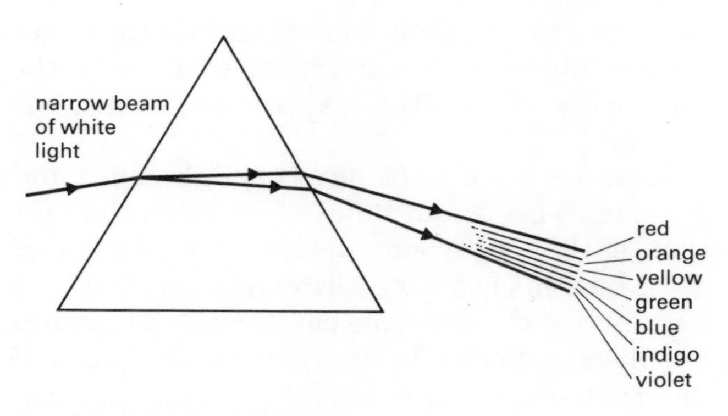

Fig. 3.62 The spectrum

> **Assignment – Over the rainbow**
>
> How is a rainbow formed?

Primary colours

It is possible by combining red, green and blue light to produce other colours, called secondary colours. It is not possible to produce red, green or blue by combining any other colours. If all three primary colours of light are combined together, then white is produced.

Primary colours	Secondary colour
red + green	yellow
red + blue	magenta (purple)
blue + green	cyan (blue–green)

Colour television

A colour television uses the three primary colours. The screen is made up of three different types of phosphorescent material. Phosphorescent means that the material glows when hit by electrons inside the television tube. These three materials emit red, green and blue light. They are arranged in the form of tiny triangles, with each type of material at one corner of each triangle (R = red, B = blue and G = green). If the red and green phosphorescent materials are hit by the electrons then the triangles will appear yellow and so that part of the screen will appear yellow.

The eye and colour

On the retina at the back of the eye, there are two types of cells that detect light. One type responds to the brightness of light and these are called RODS. The other type is sensitive to colour and these cells are called CONES. It is thought there are three different types of cone

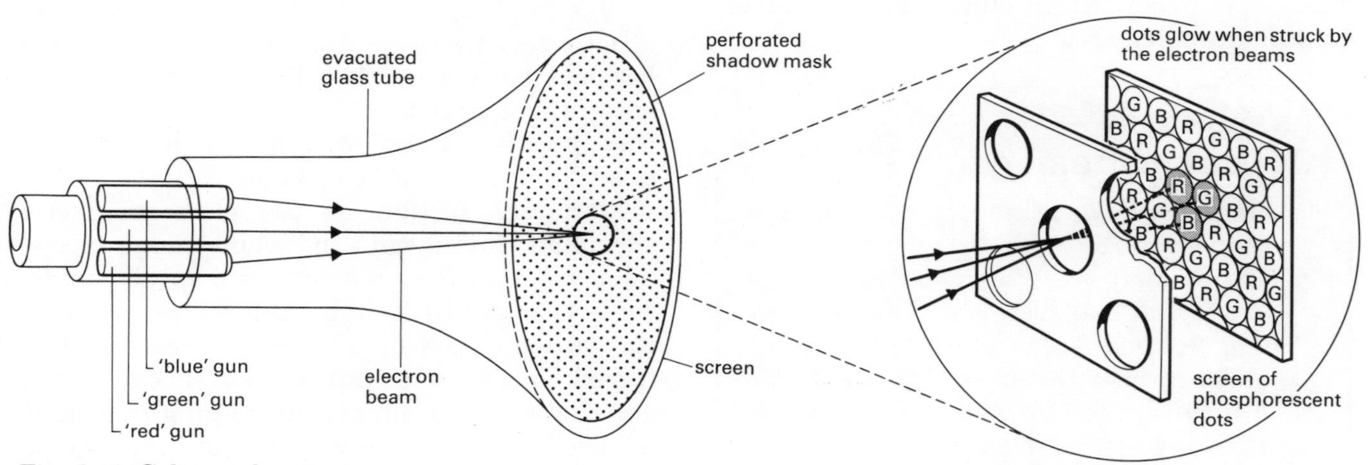

Fig. 3.63 Colour television

and each type responds to one of the three primary colours. Only in daylight do the cones work effectively. So at night, when only the rods operate, everything appears to be a shade of grey.

Some people do not have enough cones and are said to be 'colour blind'. It is often thought that such people can detect only shades of grey, but this is a very rare condition. The vast majority of these people can see all the colours but have difficulty in distinguishing shades of different colours (usually the colours are red and green). Many people do not realise they are 'colour blind' until they take one of the special tests designed to show up any colour defects in their eyes.

Assignment – Colour blindness

Some animals are said to be 'colour blind'. Devise an experiment to determine whether a common laboratory animal, say a gerbil or guinea pig, could distinguish between two different colours. The method must not cause discomfort to the animal involved.

Extending our ability to see

Even with perfect vision, there is a limit to how much the unaided eye can see. The object might be too small to be clearly visible. Or the object might be very faint because it is far away. Optical instruments are usually designed to solve one of these problems. Microscopes make small objects appear much larger. The magnifying glass is the simplest example. Telescopes make distant objects appear nearer, i.e. larger and clearer.

Assignment – Optical instruments

1 Find out how microscopes and telescopes work.
2 The camera is a very useful optical instrument that allows us to make a permanent record of what is seen. Find out how the camera works. Make a list of the similarities and differences between the eye and the camera.

Electromagnetic waves

This is the name of the most important group of waves that there is. We have already met two members of the group: infrared waves (on page 140), and light waves (on page 153). In Fig. 3.64 there are the names (with other information) of the different types of electromagnetic wave. The names of the different types do not appear to have much resemblance to each other. When each was discovered, no one knew that they were all the same type of wave. For example, X-rays were discovered by a scientist called Roentgen. He managed to produce radiation that showed wave properties. As nobody had any idea what these 'waves' were, Roentgen decided to call them X-rays. After a great deal of experimentation it was found that the different types of wave had many properties in common. Here are some of their important properties.

All electromagnetic waves:

(a) travel through a vacuum at the same speed of 300 000 000 m s^{-1} (3×10^8 m s^{-1});
(b) show the wave properties of reflection, refraction, interference and diffraction;
(c) are transverse waves.

The basic difference between the six named types of electromagnetic waves in Fig. 3.64 is their frequency range. Gamma rays have the highest frequency range and radio waves the lowest. You will see that some ranges overlap. This happens because the different types of wave are named after the way they are produced. Since all the waves travel at the same speed in a vacuum, which type of electromagnetic wave will have (a) the longest wavelengths, (b) the shortest wavelengths?

It is important to remember that there are differences between the wave types. This happens because of their different frequencies and wavelengths. Radio waves can pass through the bricks of a house (how do we know?) but light waves cannot.

The only section of the electromagnetic spectrum (this means the whole range of frequencies) that does not require a special detector is that of visible light. Our eyes can detect the various wavelengths in this region. It is incredible how much information we can obtain about our surroundings, considering that light is such a very tiny part of the whole electromagnetic spectrum.

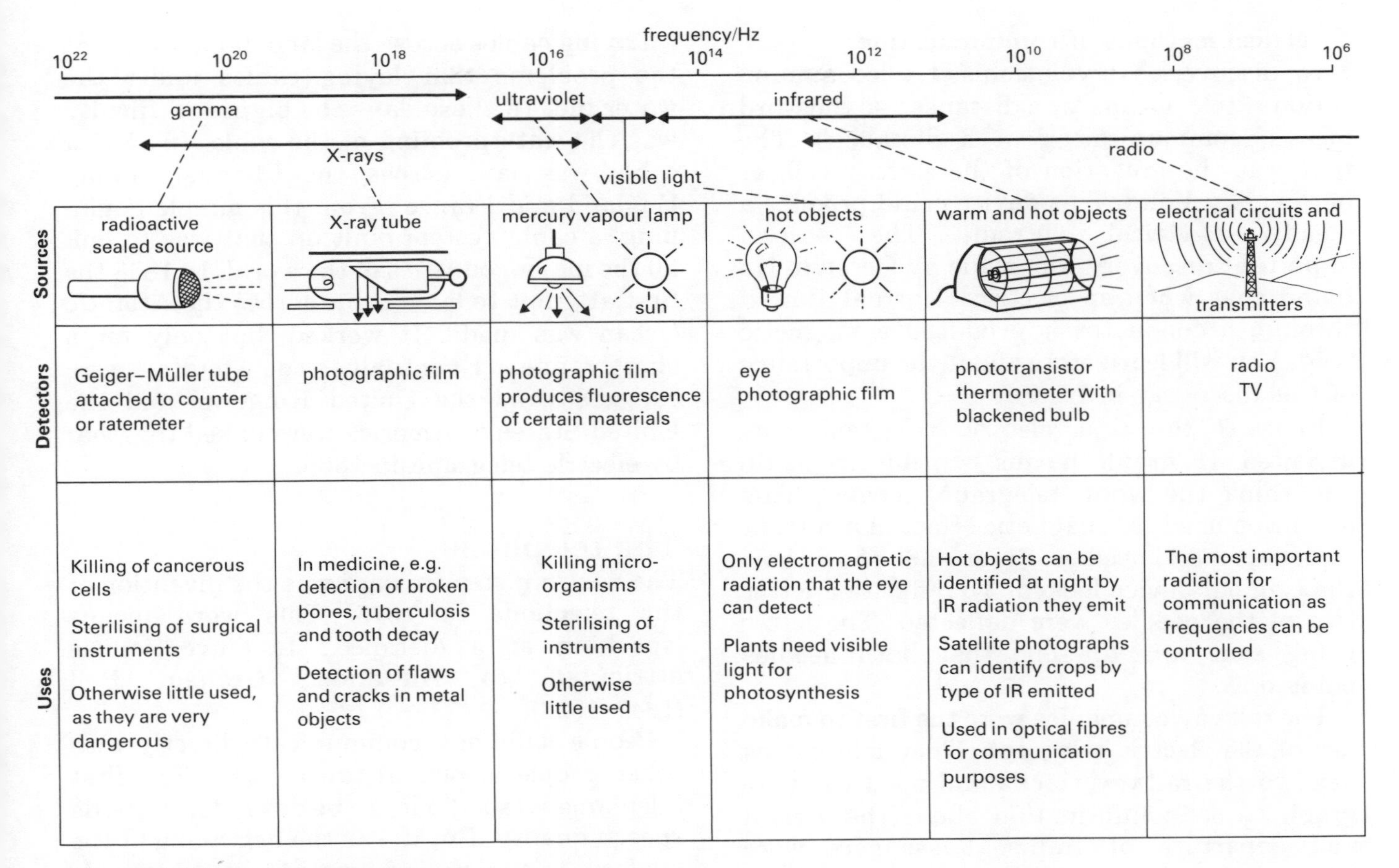

	gamma	X-rays	ultraviolet	visible light	infrared	radio
Sources	radioactive sealed source	X-ray tube	mercury vapour lamp / sun	hot objects	warm and hot objects	electrical circuits and transmitters
Detectors	Geiger–Müller tube attached to counter or ratemeter	photographic film	photographic film produces fluorescence of certain materials	eye photographic film	phototransistor thermometer with blackened bulb	radio TV
Uses	Killing of cancerous cells Sterilising of surgical instruments Otherwise little used, as they are very dangerous	In medicine, e.g. detection of broken bones, tuberculosis and tooth decay Detection of flaws and cracks in metal objects	Killing micro-organisms Sterilising of instruments Otherwise little used	Only electromagnetic radiation that the eye can detect Plants need visible light for photosynthesis	Hot objects can be identified at night by IR radiation they emit Satellite photographs can identify crops by type of IR emitted Used in optical fibres for communication purposes	The most important radiation for communication as frequencies can be controlled

Fig. 3.64 Electromagnetic spectrum

COMMUNICATION

Early methods of communication

What would it have been like 150 years ago if you had wanted to communicate urgently with a friend who was about 50 kilometres away? How would you have done it? You could have employed a messenger to carry the message to your friend. Or you could have jumped on a horse and ridden over to your friend's home. If your friend had been expecting a message, it might have been possible to set up a simple signalling system using fires from the tops of hills to send the information. You might even have had a tame carrier pigeon.

In those days, it was virtually impossible to send information rapidly over a long distance. The only efficient system available then was a semaphore system used by the military authorities (Fig. 3.65). This used mechanical or human arms to send messages. Semaphore is still sometimes used at sea for signalling from one ship to another.

Fig. 3.65 The semaphore system

The military authorities used semaphore signals from the tops of towers placed on convenient hills. It was possible to send a brief message from London to Portsmouth in 15 minutes using this system, relaying the message from hill to hill. But it did not work at night or in misty or foggy conditions.

Electrical methods of communication
Two discoveries revolutionised telecommunications ('tele' means 'at a distance', so the word means 'communicating at a distance'). The first was the invention of the electric cell, or battery, by Volta. This device could produce a steady electrical current. The second important discovery was made by Oersted. He found that when an electrical current flowed through a conductor it produced a magnetic field. You will learn more about the importance of this discovery in Unit 4.

In 1837 the first electric telegraph was invented. If 'graph' means 'writing', what do you think the word 'telegraph' means? This invention used the magnetic effect of a current to deflect metal needles. Figure 3.66 shows how the original device looked. To indicate a letter two of the needles were deflected. The letter being sent was the one that both needles pointed at.

The railway companies were the first to make use of the electric telegraph. They laid cables next to the railway tracks and used the telegraph to send information about the arrival and departure of trains. Passengers were allowed, for a small fee, to send messages to other stations. Once it was realised how useful this device could be, telegraphs were set up to link cities or towns. Unfortunately, it required a skilled operator to send messages quickly - it could not be done by the person wishing to send the message.

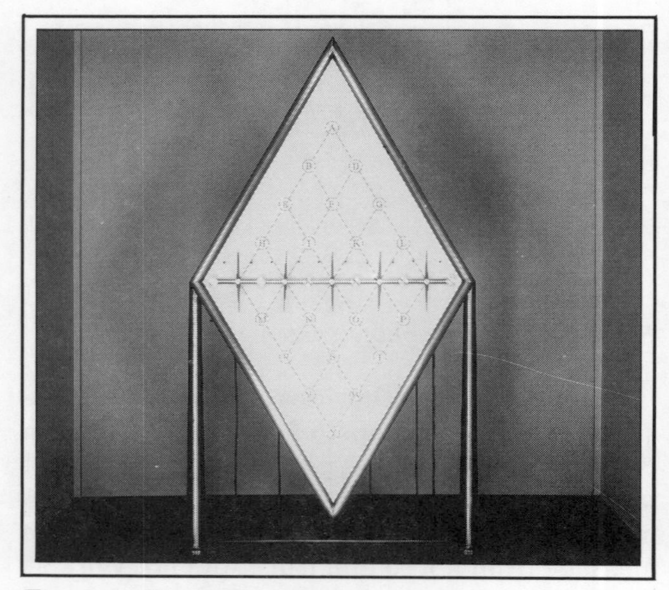

Fig. 3.66 The first telegraph

Laying cables across the land did not present the problems that laying cables under the water did. In those days the biggest difficulty was the waterproofing of the cable. In 1850 a cable was laid across the Channel linking England and France. From this simple beginning, a cable system built up that was to link all the major countries of the world. In 1858 the first attempt to lay a cable across the Atlantic Ocean was made. It worked, but only for a short time. Eventually the faults were corrected and the United Kingdom and the United States of America were linked together by electric telegraph in 1866.

The telephone

The next big step forward was the invention of the telephone in 1875. This word means 'speaking at a distance'. Its invention is attributed to Alexander Graham Bell (1847–1922).

People could now communicate directly with other people a long distance away. The first telephone sets left a lot to be desired as regards speech quality. But it was the beginning of the modern telephone system. As more people wished to use the system, it was essential to set up a central exchange that could link the individual subscribers with each other. The first such telephone exchange opened in London in 1879. However, different companies tried to set up their own exchanges. Eventually the system became so complex that it was all nationalised by the government and the Post Office was asked to run the telephone system.

Fig. 3.67 The first telephone

Radio communication

You will remember that radio waves have the smallest frequencies in the electromagnetic spectrum. In 1867 Heinrich Hertz managed to produce and detect radio waves. In Hertz's original apparatus, the distance between the transmitter (sending out the waves) and the receiver (detecting the waves) was very small. To be useful for commercial purposes, the distance between these had to be greatly increased. The method of sending information was the Morse Code that is still used today. A series of short (dots) and longer (dashes) pulses represented letters and numbers. The operator changed the message into 'dots' and 'dashes' which were transmitted. The operator at the receiving end then changed these back into the original message.

The greatest achievement at this time was the sending (in 1901) of a radio message by Guglielmo Marconi from England to Newfoundland, Canada, a distance of over 2000 miles. In the beginning the greatest use of radio communication was at sea. Most ships were fitted with radio and this allowed them to communicate with other ships and with land based radio stations. It was a radio message that helped to save more than 700 people from the Titanic when it sank after hitting an iceberg in the Atlantic in 1913.

Soon it was possible to transmit speech and music by means of radio waves and then the way was open for large-scale transmissions. The listener had a receiver that would be tuned to the particular frequency of the transmission. In 1923 the British Broadcasting Company (BBC) was established. The next development was the sending of 'pictures' by means of radio waves. The BBC, in 1936, produced the first television broadcast. What does the word 'television' mean?

Modern methods of communication

The need for more and more communications has contributed to the development of the existing systems and the invention of new methods of communication. There are about 60 million calls made on the telephone system each day in the UK. It is possible to dial direct from here to more than 100 countries. With TV, colour is now standard and it is possible to receive TV broadcasts from the other side of the world as they happen. Two new systems have helped in the development of the telephone network and television broadcasting. These are radio microwaves and satellites.

Radio microwaves

This system uses radio waves of short wavelength (hence the name 'microwave'). The first radio microwave system was set up in 1952. Since then a network of transmitters and receivers has been built up all over the UK, with the centre of the system the Post Office tower in London (built in 1965). The network is used for transmitting telephone calls between city centres and for television transmissions. It also provides microwave radio links with the continent.

Satellites

The first satellite used for telecommunication purposes was launched in 1962. It was called Telstar. The early satellites just reflected the radio waves from the transmitter back down to the receiver. The present satellites take the incoming signal, 'clean' it up and amplify it before re-transmitting it to earth.

The ideal position for a satellite is to remain exactly above the same point on the earth's surface. This is called a geo-stationary orbit and means the satellite takes 24 hours to make one complete orbit (why is this?). The first satellites could not be put high enough (the altitude required is 36 000 km) for this and they went round the earth every 90 minutes or so. Can you explain why this would cause problems if the satellite has to reflect signals between a transmitter and receiver? Most inter-continental telephone calls are routed via satellites nowadays. The latest communication satellites can handle 45 000 telephone calls and two TV broadcasts simultaneously.

Optical fibres

The most important recent development in communications has been the OPTICAL FIBRE. This consists of a hair-thin fibre of pure glass. Electromagnetic waves in the infrared region are used to transmit information along the fibres. The infrared waves are produced by special lasers.

Because the frequency of the waves used is so high compared with microwave, the amount of information that can be sent along one of these fibres is very large. A pair of these optical fibres can handle 2000 telephone calls at the same time. In fact optical fibres now form part of the national telephone network. It is hoped that, in the near future, an optical fibre cable will be laid across the Atlantic.

Communications in the near future

When communication systems are used together with computers, the range of innovations is virtually endless. With a home computer and a TV set, you will be able to do such things as ordering goods from shops, booking a holiday or asking the bank how much money is in your account. You will have access to vast amounts of information inside computers at the press of a button. You might even do all your work from home. The uses of computers and communication networks is limited only by our own ingenuity.

Fig. 3.68 Visible light being shone through several metres of optical fibre

Waves

WHAT YOU SHOULD KNOW

1. All waves are produced by oscillators.
2. All waves carry energy. Also, many types of waves can carry information.
3. The speed of a wave $= \text{frequency} \times \text{wavelegnth}$
4. The most important properties of waves are reflection, refraction, diffraction and interference.
5. There are two main types of wave: transverse and longitudinal.
6. Sound travels as a longitudinal wave.
7. The human ear has a limited frequency range.
8. Most sounds have complex waveforms which can be observed using an oscilloscope.
9. Loud sounds can cause temporary or permanent hearing loss.
10. It is often necessary to amplify sounds for them to be heard adequately.
11. Light travels as a transverse wave.
12. Short and long sight are two common defects of the human eye.
13. White light can be split up into seven different colours: red, orange, yellow, green, blue, indigo and violet.
14. The three primary colours are: red, green and blue.
15. It is often necessary to use optical instruments because of the limitations of the human eye.
16. All electromagnetic waves travel at $3 \times 10^8\,\text{m s}^{-1}$ in a vacuum and are transverse.
17. Gamma, X-rays, ultraviolet, visible light, infrared and radio make up the electromagnetic spectrum.

QUESTIONS

1. The diagram shows the appearance of some water waves with a scale superimposed on them. All distances are shown in centimetres.

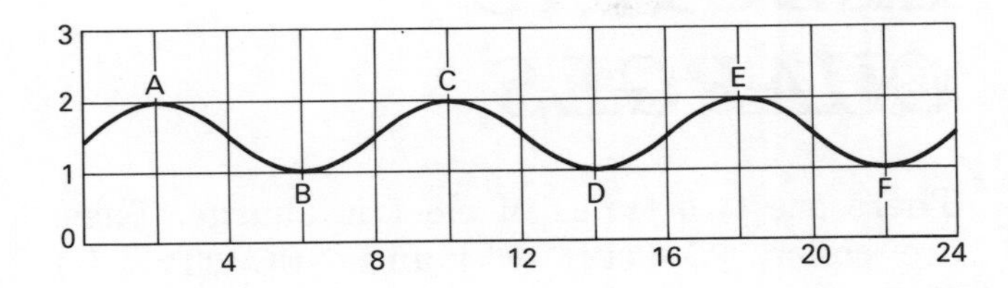

(a) Which letters represent (i) the crests and (ii) the troughs?
(b) What is the amplitude of the waves?
(c) What is the wavelength of the waves?
(d) If the frequency at which the waves are produced is 1.5 Hz, what is the speed of the waves?
(e) If the frequency is now increased to 2.0 Hz, what will be the new wavelength?

2 The diagram shows a transverse wave being sent along a rope by moving the free end up and down. The wave has not yet hit the wall.

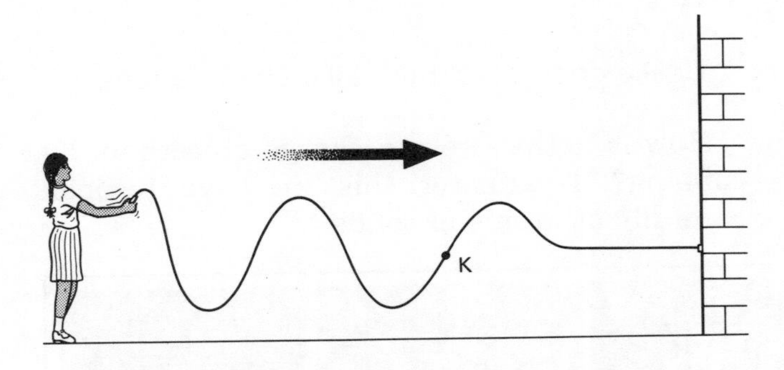

(a) If the girl moves her hand up and down once per second, what is the frequency of the wave produced?
(b) There is a knot in the rope at K. If the distance from her hand to the knot is 4 m, what is the wavelength of the wave? Hence, calculate the wave speed.
(c) How will this knot move as the wave passes it?
(d) What happens to the waves as they hit the wall?
(e) Why does the amplitude of the wave become smaller as it travels along?

3 The diagram below represents a tank used for the testing of model ships before the full-size ships are built. The bar on the left vibrates up and down to produce the waves.
(a) If the bar vibrates up and down 10 times a second and the wavelength is 1.8 cm, calculate a value for the speed of the waves.

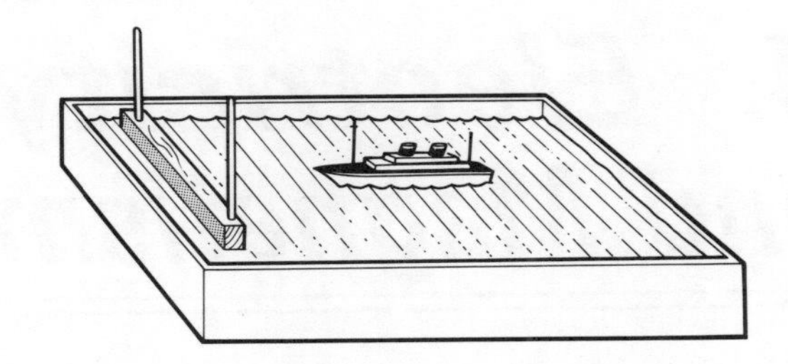

(b) To investigate the effect of longer wavelengths, the wavelength was increased to 3.0 cm. What would be the new frequency of vibration?
(c) What would you do to
(i) increase the size of the waves,
(ii) increase the speed of the waves?
(d) There were found to be unwanted reflections of the waves from the right hand side of the tank. Suggest a way of avoiding this.
(e) The model ship was turned through 90° so as to be parallel to the wave crests and troughs. What wave effect might you notice behind the ship?

4 (a) During a thunderstorm, the sound of the thunder is heard 4 s after the lightning flash is seen. How far away is the thunderstorm? What assumption can you make about the speed at which light travels compared with sound? (Speed of sound in air can be taken as 340 $m s^{-1}$.)
(b) A person stands a distance away from a large wall. If an echo takes 1.6 s to return to the person, how far away is he from the wall?
(c) Sonar is used by ships to detect objects under the surface of the sea. A transmitter sends out pulses of sound waves. These hit the object and are reflected back to the receiver.
(i) If a pulse of sound takes 1.2 s to travel from the transmitter to the receiver, how far down is the seabed?
(ii) A shoal of fish swim under the ship and the pulse time becomes 0.8 s. How far above the seabed is the shoal swimming?

(Speed of sound can be taken as 1500 $m s^{-1}$.)

4. Electricity and Magnetism

ELECTRIC CHARGES

There are two types of electric charge. These are called POSITIVE (+) and NEGATIVE (–). Charges are detected by the electrical forces they produce. The easiest way to place charges on an object, or charge it up, is to rub it. If you rubbed a balloon on your woollen pullover you would charge the balloon negatively. If you charge another balloon the same way and hang it close to the first one, the balloons will repel each other as shown in Fig. 4.2.

If you now rub two balloons together, one balloon will be charged negatively and one positively. If these are hung side by side they will attract. If two positively charged balloons were hung side by side they would also repel each other. We can sum up these results by saying

Unlike charges attract. Like charges repel

How is it that we can charge objects up by rubbing? To answer this we have to look carefully at atoms in solids.

Fig. 4.1

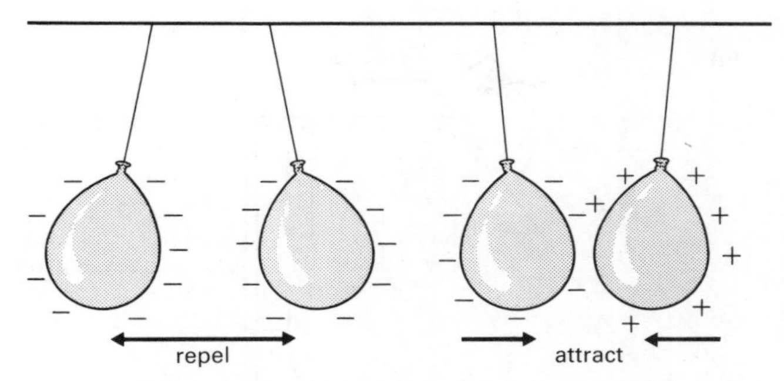

Fig. 4.2 Like charges repel; unlike charges attract

CHARGES AND ATOMS

Matter is made of tiny particles called atoms. You will study atoms in more detail in Unit 5. To understand how charge moves around we need to look at the structure of the atom shown in Fig. 4.3.

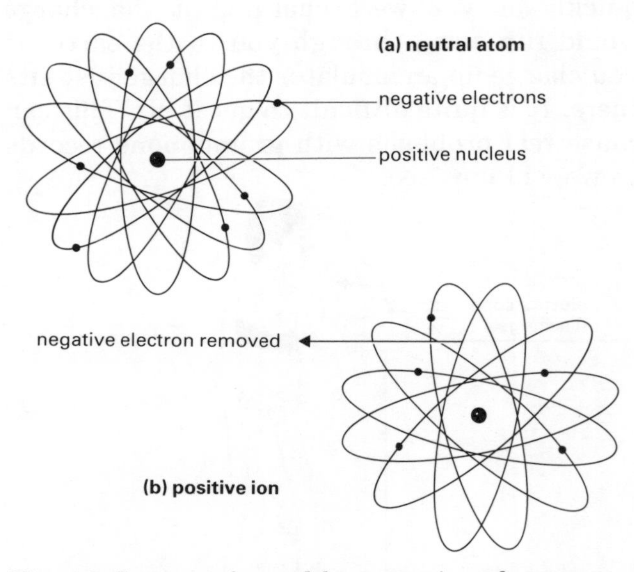

Fig. 4.3 Ions are formed by removing electrons from atoms

The diagram on the left shows an atom. It is made up of a positive nucleus surrounded by a cloud of negative electrons. The amount of positive charge on the nucleus is equal to the amount of negative charge of the electrons. The positive and negative charges cancel each other out and the atom has no total charge. It is *neutral*.

Atoms can lose or gain electrons. Figure 4.3(b) shows what happens. A negative electron is removed. The nucleus is still as positive as before but this is no longer balanced by the negative charge. A positive particle is now left behind. We call this a positive ION.

CHARGING BY FRICTION

The balloon rubbed with wool becomes negative because electrons are rubbed off the wool onto the balloon. When the two balloons are rubbed together electrons go off one balloon, leaving it positive, onto the other balloon, making it negative.

Solids are charged by electrons moving. In solids positive ions cannot move. Look at Fig. 4.4. This diagram shows that the atoms in a solid are like tiny spheres held firmly in place in a neat fixed pattern. Electrons can be pulled away or added to the solid, but the positive ions are firmly held in place and cannot move. In solids charging happens by the flow of electrons.

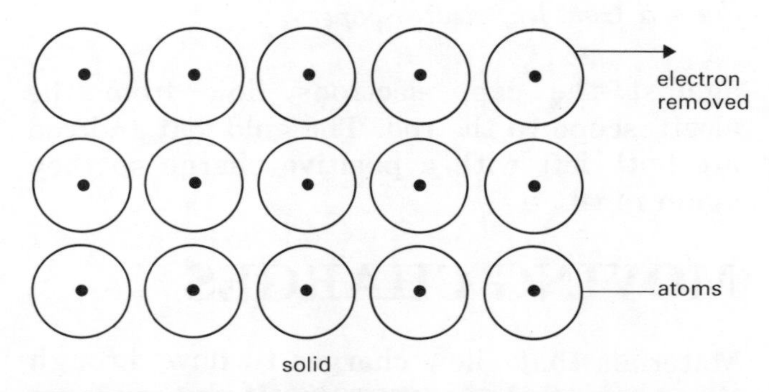

Fig. 4.4 In a solid the atoms are in a fixed pattern - electrons can move but positive ions cannot

DETECTING CHARGE

Figure 4.5 shows a very simple device for detecting charge. It is called a GOLD LEAF ELECTROSCOPE. It is a bit out of date really - your school may have a far better device called an electrometer. The gold leaf electroscope is still useful because it is very easy to use.

The diagram shows how it works. A negatively charged rod is scraped against the cap of the electroscope. Electrons are transferred to the cap and these flow down the metal rod to the gold leaf. Both the metal rod and the gold leaf are charged negatively so the leaf is repelled by the rod. The more charge that is put on the electroscope the further the leaf moves.

The electroscope can also detect positive charge. If a positively charged rod is scraped

Fig. 4.5 *Gold leaf electroscope*

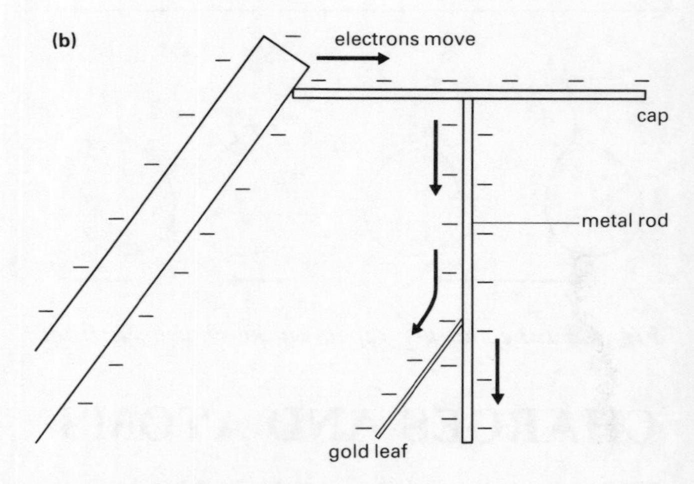

Fig. 4.5(b) Charging a gold leaf electroscope. The gold leaf has the same charge as the rod and so is repelled

against the cap, electrons flow from the electroscope to the rod. The gold leaf and rod are both left with a positive charge so they again repel.

MOVING CHARGES

Materials that allow charges to flow through them are called CONDUCTORS. If you touch an electroscope with a conductor the leaf will fall as the charge flows away through the conductor. If you connect a metal wire to a water or gas pipe and touch the wire to the electroscope cap the leaf will fall completely. All the charge will flow away. We say that we have connected the electroscope to EARTH, or EARTHED the electroscope. By an earth we mean something that will take lots of charge. If you touch the electroscope cap with your finger the leaf will collapse. The charge has travelled through your finger to your body and down to the ground. The electroscope has effectively shared its charge with you and the ground. As you and the ground are much larger, you and the ground take most of the charge!

Materials that do not allow charge to flow through them are called INSULATORS. All the materials that you charge by rubbing are insulators. Can you see why this must be so? If you tried to charge a conductor by rubbing it, as quickly as you were charging it the charge would run away through you to the earth. If you charge up an insulator the charge just sits there. It is quite difficult to get it off. This can cause real problems with gramophone records as we will now see.

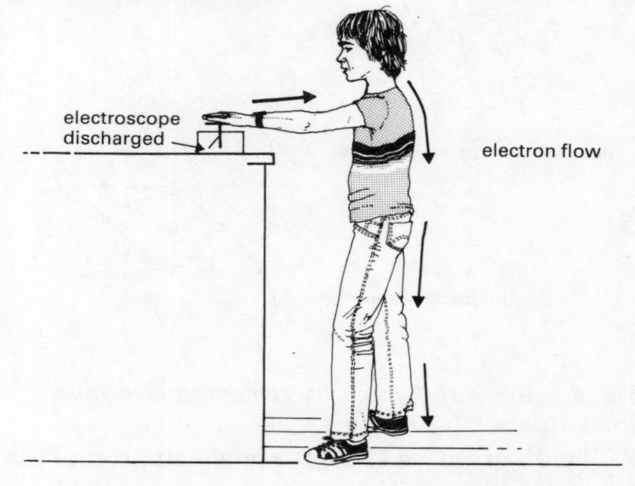

Fig. 4.6 Earthing an electroscope

INDUCED CHARGES

You have probably noticed that a charged object can pick up small pieces of paper. You can do this for yourself. Just rub a plastic biro against the sleeve of your pullover. It is now charged and can pick up small bits of paper - even though these are not charged. Figure 4.7 shows how this happens. As you can see, no

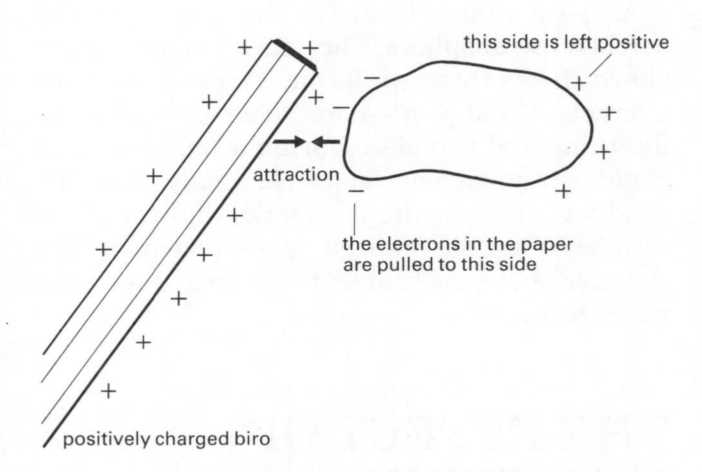

Fig. 4.7 A small piece of paper is attracted to a biro because of induced charges

extra charge is put on the paper. The charges already there separate out, leaving extra negatives on one side and positives on the other. These are called INDUCED charges.

Now think about your record player. When you take the record out of its polythene sleeve you charge it by rubbing. It gets charged more by the turntable as it spins round. You can now see why dust is attracted to it and sticks to the record. Try and work out one way by which this problem can be overcome.

VAN DE GRAAFF GENERATOR

This is a device that can be used for storing up large amounts of electric charges.

Figure 4.8 shows how the Van de Graaff generator works. The rubber belt acts like a conveyor belt. It rubs against a scraper at the bottom which takes electrons off the belt. As the belt rotates it carries positive charges up to the metal dome. The positive charges on the

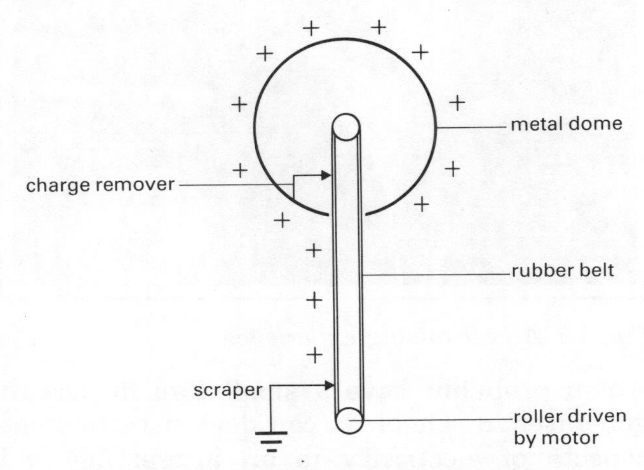

Fig. 4.8 Charging the Van de Graaff generator

Assignment – Electrical smoke catchers

The photograph above shows the chimney of a modern coal fired power station. The station produces about 30 000 kg of smoke and ash per hour. This could of course be a very serious pollution problem. If you look carefully at the photograph you will see very little ash coming from the chimney. This is because 99% of the ash is removed before it reaches the top of the chimney.

The ash is caught by electrostatic precipitation. A precipitator is made up of wires and plates. The wires are negatively charged. The particles of ash are attracted to the wires and given a negative charge. They are then attracted to the positive plates. These are shaken to remove the ash.

1 Why should the particles of smoke be attracted to the negative wire? (There are two possible answers – try to give them both.)
2 Why are the particles of smoke repelled by the wire after they have touched it?
3 Although these smoke catchers remove 99% of the ash, power stations are still a major pollution problem. The power stations in northern England are threatening the forests of Scandinavia by causing 'acid rain'. Explain why this happens.

belt pull electrons off the dome leaving it positive. The charge has lots of energy, which makes the Van de Graaff a useful device for atomic physicists who need to give electrons a good deal of energy so that they can smash into other particles at high speeds.

Fig. 4.9 A hair raising experience!

You probably have a small Van de Graaff generator in school. It can demonstrate some aspects of electricity in an interesting and amusing way.

If a pupil puts her hand on a Van de Graaff generator which is then turned on, the pupil will be charged. This will only happen if the pupil is standing on an insulating material - otherwise the charge will just run through her feet to earth. If the person has fine dry hair it will stand on end. This is because all the hairs get charged negatively, and so they try to get as far away as possible from each other. Once the pupil is charged up, the charge has to be removed carefully. If the person is earthed by touching, say, a gas tap, the charge will be removed quickly and the pupil will feel a shock. This will not be dangerous, but it may be a little uncomfortable.

Discharging the Van de Graaff generator

The charge can be removed by connecting the dome to earth with a conductor. Air is normally an insulator, but if there is enough charge on the dome, air can be made to conduct. You will probably be shown this demonstration.

A spark jumps from the dome to a nearby earthed metal sphere. The spark happens when charge flows through the air. You will also hear a 'crack'. What you see and hear is small-scale lightning and thunder. During a thunderstorm clouds become charged. Suddenly the air conducts and a huge spark (lightning) is formed when the charge flows to earth. The thunder that you hear is the 'crack' that goes with the spark.

THE EFFECT OF MOISTURE

You may have noticed that electrostatic experiments do not work very well on damp days. It is much easier to charge up balloons and your biro by rubbing them against your pullover on a fine dry day. Water conducts electricity. When it is damp all objects, including you, get covered with a thin film of water. This means that instead of staying on your biro, the electrons would flow through the water and away to earth via your body. Moist air is a far better conductor than dry air. The Van de Graaff generator gives a smaller spark on a damp day.

IONS IN GASES

Before we leave the problem of sparks through the air, there is one more important point we should consider. Remember, in solids only the electrons can move. The positive ions are locked in the structure of the solid. In gases the positive ions can move. The spark is actually caused by electrons moving one way and positive ions moving the other way.

Ions are also produced in flames, where the gas molecules get very hot and energetic. Electrons gain enough energy to escape from atoms, leaving positive ions. If you hold a lighted match above a charged gold leaf electroscope the leaf will be discharged by the ions in the flame. You can see the movement of ions by using the apparatus shown in Fig. 4.10.

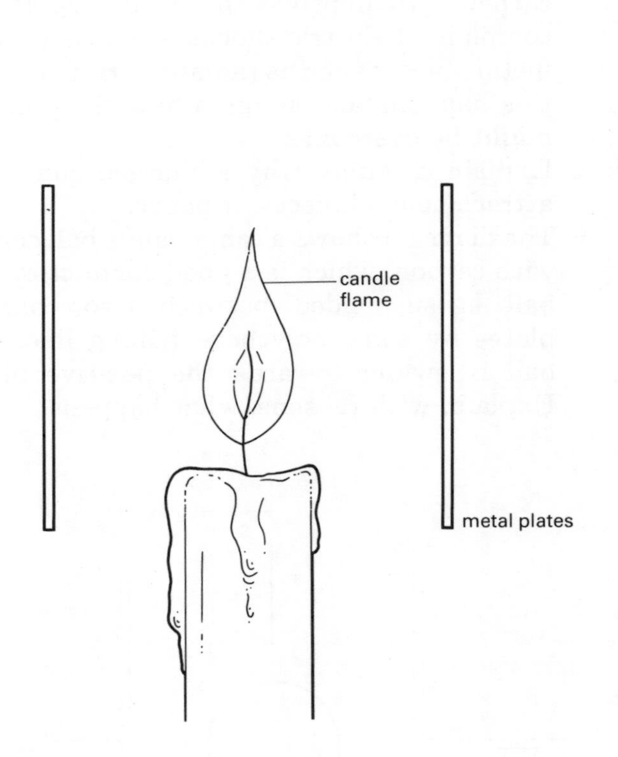

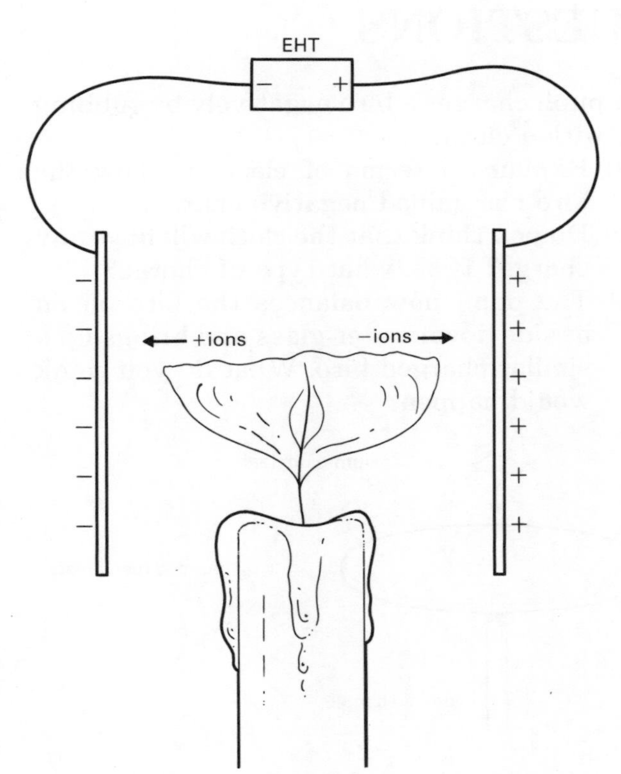

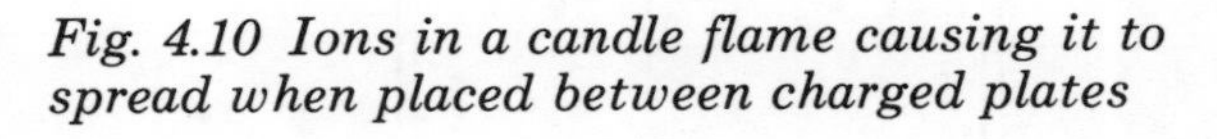

Fig. 4.10 Ions in a candle flame causing it to spread when placed between charged plates

When the metal plates are connected to an electrical supply called E.H.T. (extra high tension), negative charges are put on one plate and positive charges on the other. The positive ions in the flame move towards the negative plate and the negative ions towards the positive plate. You can actually see the top of the flame spread out as this happens.

Assignment – Thunderstorms

You know that lightning happens when charge on a cloud flows through the air to earth. Try to find out more about thunderstorms from the books in your library. Write an essay called 'Thunderstorms'. In the essay say how the charge builds up on clouds. Describe the differences between 'sheet' and 'forked' lightning. Say which sort of objects are most likely to be hit by lightning and explain why. Describe the steps that can be taken to avoid damage by lightning.

Electric charges

WHAT YOU SHOULD KNOW

1. There are two types of electric charge: positive (+) and negative (–).
2. Like charges repel, unlike charges attract.
3. Matter is made of atoms.
4. Atoms have a positive nucleus surrounded by negative electrons.
5. The positive charge on the nucleus is equal to the negative charge of the electrons. The charges cancel out to give zero total charge. The atom is neutral.
6. Atoms can lose or gain electrons to form charged particles called ions.
7. Some materials can be charged by rubbing. Electrons can be removed to leave the material positive or added to leave the material negative.
8. Conductors allow charge to flow through them. Insulators do not.
9. If a charged conductor is connected to earth the charge will flow away to earth.
10. If a charged material is brought near a neutral material, attraction may take place because of induced charges.

QUESTIONS

1 A pupil charges a biro negatively by rubbing it with a cloth.
   (a) Explain in terms of electrons how the biro has gained negative charge.
   (b) Do you think that the cloth will have any charge? If so, what type of charge?
   (c) The pupil now balances the biro on an upside down water glass and brings up a similar charged biro. What do you think would happen?

2

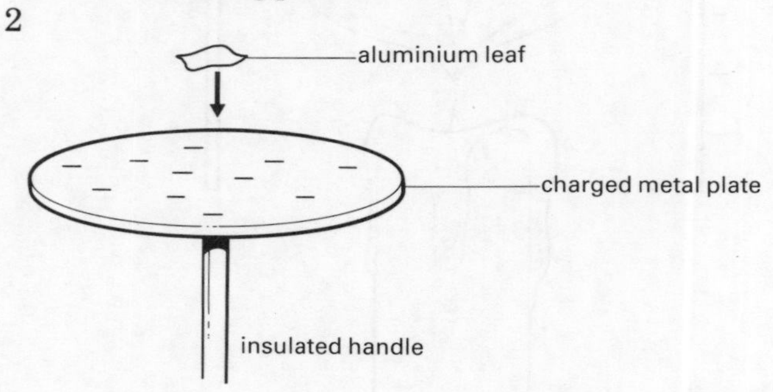

In the above experiment a small piece of aluminium leaf is dropped. It falls slowly until it hits a charged metal plate. It then jumps upwards and remains suspended above the plate.
Explain why this happens.
(This is a delightful demonstration – try to persuade your teacher to show it to you!)

3 (a) A pupil touches a negatively charged electroscope with a metal rod. What would you see happen? Explain your answer.
   (b) The pupil now touches the same charged electroscope with an uncharged biro made of polythene. What would you see happen now? Explain your answer.
   (c) The pupil now touches the same charged electroscope with a biro that has been charged negatively. What would you see happen now? Explain your answer.

4 The Head of Music in a school has her room carpeted to improve the acoustics. Pupils complain of electric shocks when they touch metal objects such as radiators. Explain why this happens and suggest how the problem might be overcome.

5 Explain carefully why a charged comb will attract neutral pieces of paper.

6 The diagram shows a table tennis ball coated with carbon, which is a good conductor. The ball is suspended between two charged plates by some polythene fishing line. The ball is swung towards the positive plate. Explain, with reasons, what happens.

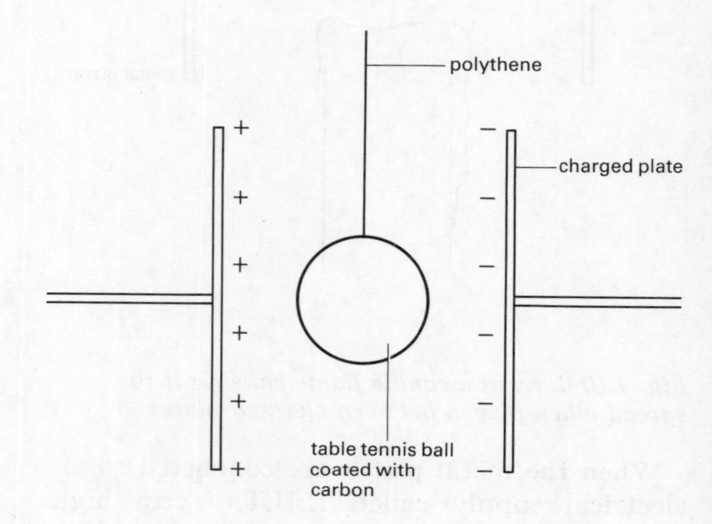

7 (a) A metal rod and a polythene rod are suspended by fishing line. Both are charged negatively. A pupil takes the charge off the metal rod by touching it. She finds that she cannot discharge the polythene rod in the same way. Explain the difference.
   (b) Another pupil suggests that the charge could be removed from the polythene rod by waving it in a Bunsen burner flame. Do you think that this would be successful? Explain your answer.

Electricity and Magnetism

ELECTRIC CURRENTS

When you switch on an electric light it glows. What causes the glow? If you asked most people this question they would give you a quick answer - electricity. If you now asked them what electricity is, the answer would not come so quickly.

You must have asked yourself the question - what is electricity? Now you can answer it. All electricity is a flow of charge. In the case of a lightbulb just a flow of electrons through the lamp. When you switch on, the electrons don't just flow to the lamp and sit there. More and more electrons keep flowing through the lamp transferring energy as they go.

The demonstration illustrated in Fig. 4.11 shows that moving charge is an electric current. The Van de Graaff generator is connected to earth - say a gas-tap - via an ammeter. You know from your work in previous years that an ammeter measures electric current. The ammeter in this demonstration is very sensitive - that means that it can measure very small currents. When you turn on the Van de Graaff you will see the needle in the ammeter move. Now you should see that an electric current is a flow of charge.

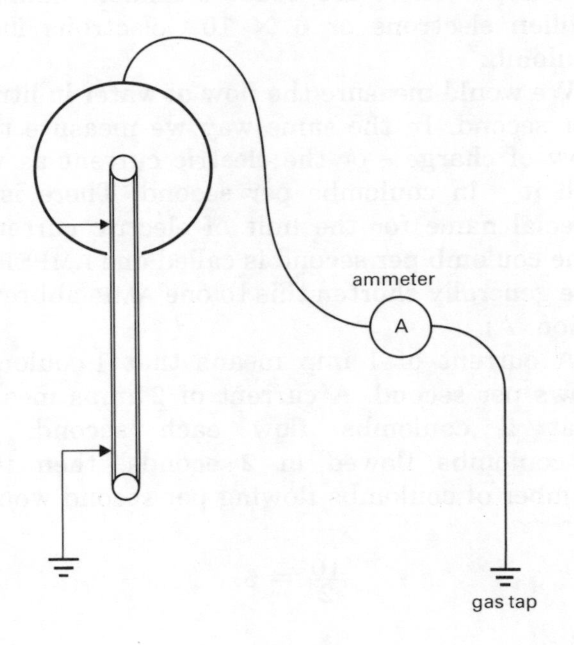

Fig. 4.11 The charge flowing off the Van de Graaff generator shows as an electric current

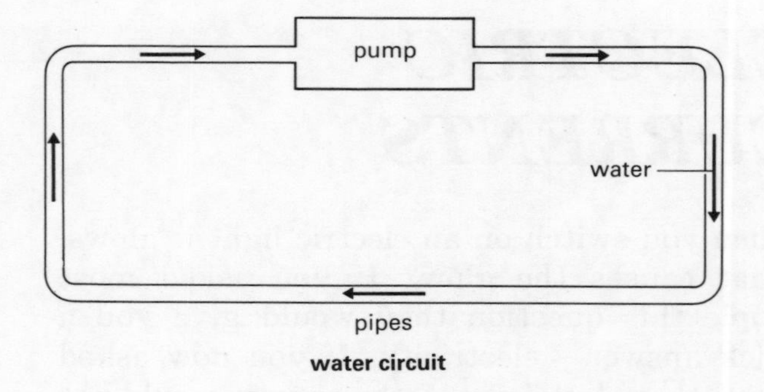

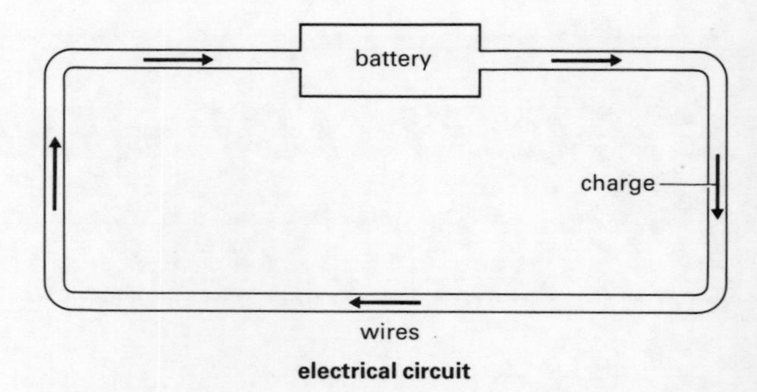

Fig. 4.12 The flow of charge around an electric circuit is like the flow of water around a water circuit

ELECTRIC CIRCUITS

From earlier work you should also know that electric current will flow only if we have a complete circuit. To help us to understand electric currents better, it is helpful to compare the flow of electric charge to the flow of water in pipes. Helpful comparisons like this are called ANALOGIES.

Look at Fig. 4.12. Notice that in the water circuit the water is not used up - it just flows around the circuit. The same is true in an electrical circuit: the charge is not used up. The battery acts like a pump. It gives the charge carriers - in the case shown above, with solid wires, the electrons - some energy and pushes them around the circuit.

We would measure the amount of water in units called litres. In the same way we measure the amount of charge in units called COULOMBS (abbreviation: C). A coulomb is a large amount of charge. There are about 6 million, million, million electrons or 6×10^{18} electrons in a coulomb.

We would measure the flow of water in litres per second. In the same way we measure the flow of charge - or the electric current as we call it - in coulombs per second. There is a special name for the unit of electric current. One coulomb per second is called one AMPERE. We generally shorten this to one AMP (abbreviation: A).

A current of 1 amp means that 1 coulomb flows per second. A current of 2 amps means that 2 coulombs flow each second. If 10 coulombs flowed in 2 seconds, then the number of coulombs flowing per second would be

$$\frac{10}{2} = 5.$$

From this we can see that

$$\text{current} = \frac{\text{charge}}{\text{time}}.$$

We use the symbol Q for charge, I for current and, t for time, so we can write

$$I = \frac{Q}{t}.$$

Now think of the problem the other way round. If we know the current and the time, can we work out the charge? Suppose a current of 3 amps flows for 4 seconds, how much charge flows?

Remember 1 amp means 1 coulomb per second, so 3 amps means 3 coulombs per second. If 3 coulombs flow each second, then in 4 seconds, 12 coulombs will flow. All we did to find the charge was multiply the current by the time.

So we write

$$\text{charge} = \text{current} \times \text{time}$$

or, in shorthand,

$$Q = It.$$

You may like to use the triangle method (see Unit 3).

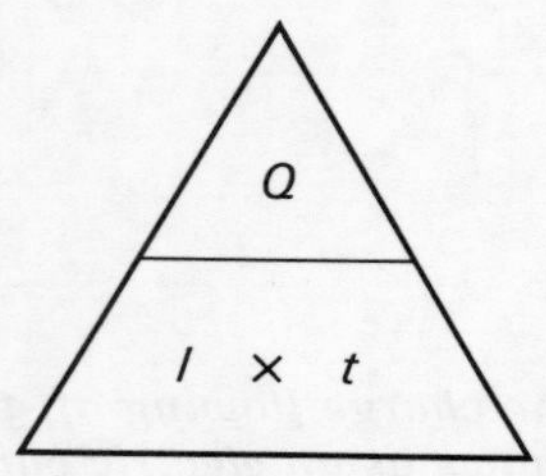

Fig. 4.13

Charge flow in circuits – the scientists get it wrong!

We now know that only the negatively charged electrons can move in solids. The positive ions left when electrons are pulled away from atoms are fixed in the structure of the solid. In a circuit, the battery energises the electrons and pushes them around the circuit. The battery has two ends. One is marked + and the other –. The electrons are repelled by the – and attracted by the + end of the battery. Figure 4.14 shows how the electrons flow.

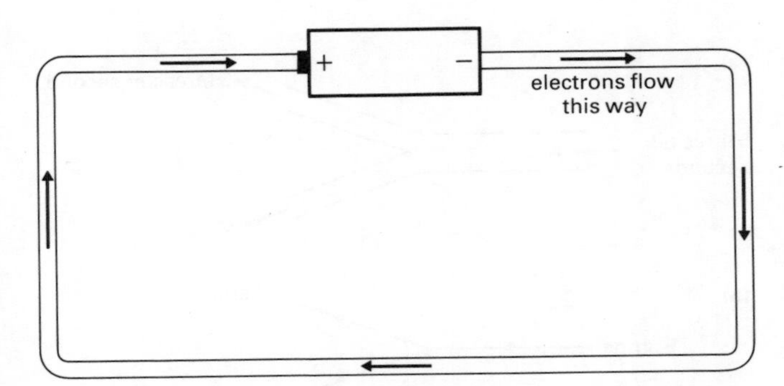

Fig. 4.14 Electrons actually flow away from the negative side of the battery towards the positive

Unfortunately for us, the scientists who worked out important laws of electricity thought that electric currents in wires were caused by the flow of positive charges. The scientists built all the laws of electricity and magnetism around the idea that, in an electrical circuit, positive charge flows away from the positive end of the battery towards the negative. In electrical circuits we still show current going from positive to negative. This is called the 'conventional current' (Fig. 4.15).

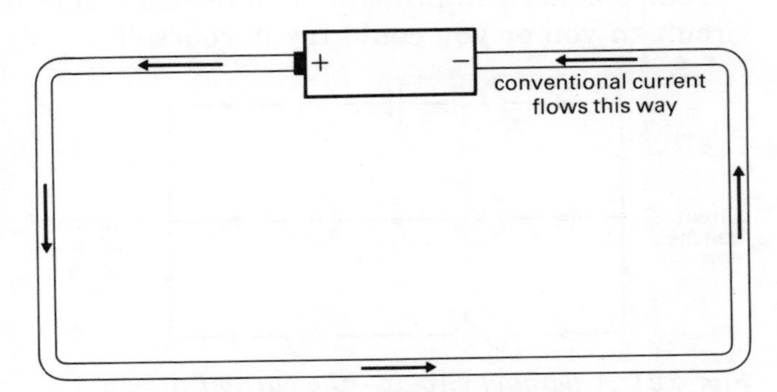

Fig. 4.15 The 'conventional current' goes away from the positive towards the negative

Circuit diagrams

During your studies of electricity in previous years you will probably have drawn circuit diagrams. Just to remind you, the symbols are shown in Fig. 4.16. Together in a circuit diagram they look like Fig. 4.17.

Always use a ruler for circuit diagrams. The lines must be straight and they should join at right angles.

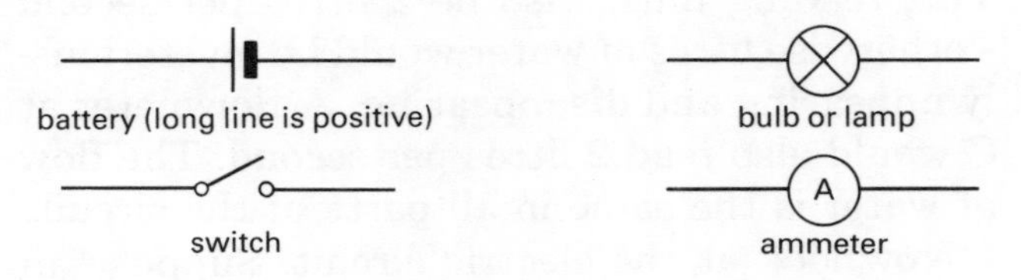

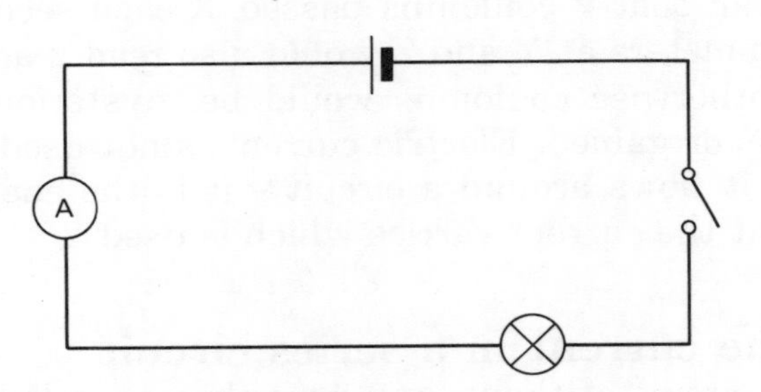

Fig. 4.17 A circuit diagram

The current in electric circuits

To help understand what happens to the current it is useful to compare it to a water circuit. Look at Fig. 4.18.

First look at the water circuit. The pump is pushing water around the circuit. It is not producing water – the water is already there in the pipes. The pump is giving energy to the water. This energy is then transferred to the paddle wheel.

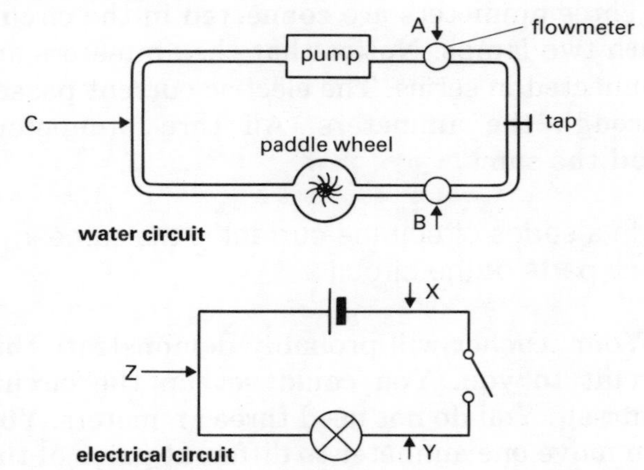

Fig. 4.18 A useful comparison: a water circuit and an electrical circuit

It is the same with the electric circuit. The battery does not produce electrons. The electrons are already in the wires. The battery energises the electrons and pushes them around the circuit. The energy is then transformed to heat and light in the bulb.

Now look back to the water circuit. The flowmeter at A measured 2 litres per second. What would be the reading on the flowmeter at B? This reading must also be 2 litres per second - otherwise litres of water would be mysteriously appearing and disappearing. A flowmeter at C would also read 2 litres per second. The flow of water is the same in all parts of the circuit.

Now look at the electric circuit. Suppose an ammeter put in at X read 2 amps. This would mean that 2 coulombs passed X each second. Ammeters at Y and Z would also read 2 amps - otherwise coulombs would be mysteriously lost or gained. Electric current is not used up as it flows around a circuit - it is the energy that the current carries which is used.

The current in a series circuit

A circuit without any branches is called a SERIES circuit. Figure 4.19 shows a series circuit.

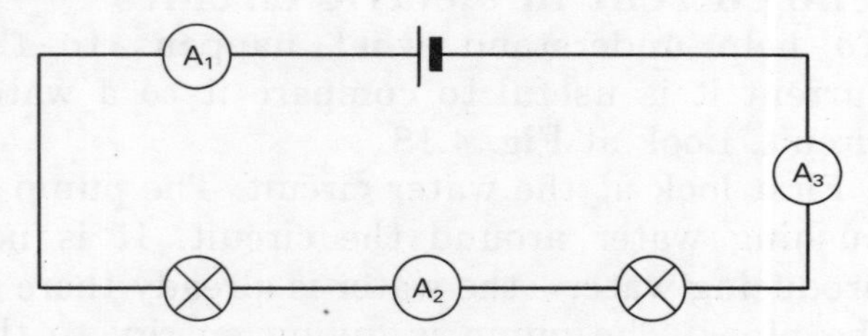

Fig. 4.19 The current is the same in all parts of a series circuit: A_1, A_2, A_3 all read the same

Three ammeters are connected in the circuit with two lamps. Notice that the ammeters are connected in series. The electric current passes through the ammeters. All three ammeters read the same.

In a series circuit the current is the same in all parts of the circuit.

Your teacher will probably demonstrate this circuit to you. You could set up the circuit yourself. You do not need three ammeters. You can move one ammeter to different parts of the circuit and check that the current is always the same.

The current in branching circuits

Water circuits again can help us to understand what happens to the current in branching circuits. Look at Fig. 4.20(a). Five litres per second flow along a pipe towards a branch. Three litres per second flow one way. How much goes the other? It must be 2 litres per second. Otherwise water would be mysteriously appearing or disappearing. Now look at Fig. 4.20(b). Five amps of electric current flow into a branch in the wires. Three amps go one way, how much goes the other? It must be 2 amps!

The current flowing out from the branches must equal the current flowing in.

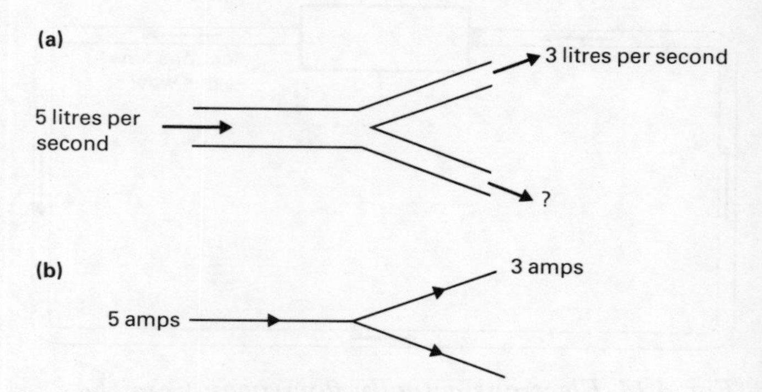

Fig. 4.20 The current in branching circuits

Figure 4.21 shows a circuit for demonstrating this. The ammeter A_1 measures the current drawn from the battery. This divides and some goes through A_2, some through A_3.

$$A_1 = A_2 + A_3 = A_4$$

The two lamps shown in Fig. 4.21 are connected in PARALLEL.

In parallel circuits the total current drawn is equal to the current in the branches added together.

Your teacher will probably demonstrate this circuit to you or you could try it yourself.

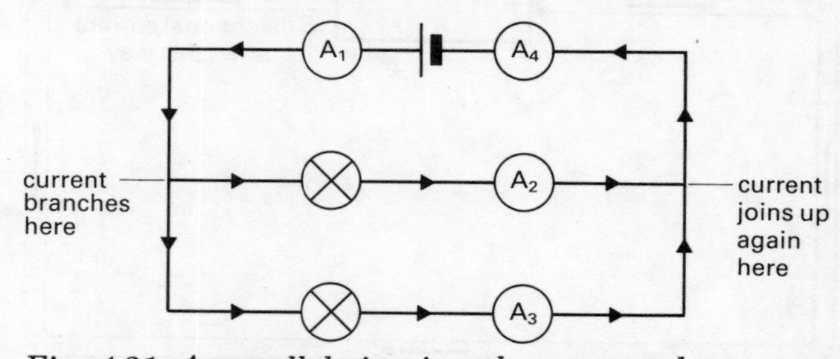

Fig. 4.21 A parallel circuit - the current drawn from the supply is equal to the current in the branches added together

PASSING ELECTRIC CURRENTS THROUGH LIQUIDS

Some materials when in molten form or when dissolved in water will conduct electricity. The fact that they conduct gives us important information about the structure of the material.

Although it is harder to do the experiment, it is easier to understand what is happening when molten solids are used. Figure 4.22 shows what happens when electricity is passed through molten lead bromide. Your teacher may show you this experiment. This process is called ELECTROLYSIS.

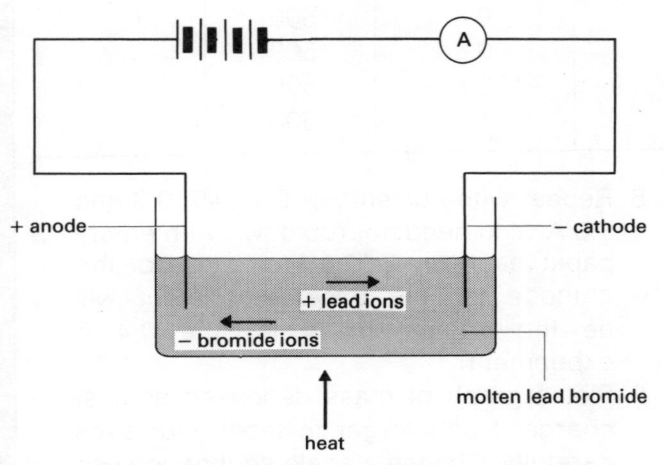

Fig. 4.22 Passing electricity through molten lead bromide

The lead bromide is melted in a heatproof container. The current is passed through the liquid by dipping into it two carbon rods. These are called ELECTRODES. The electrode connected to the positive side of the battery is called the ANODE. The electrode connected to the negative side is called the CATHODE. The liquid conducting the electricity is called an ELECTROLYTE.

In the liquid there are *mobile* ions ('mobile' just means 'free to move'). Lead ions are positive and bromide ions are negative. The lead ions are attracted towards the negative cathode and the bromide ions are attracted towards the positive anode. There is a flow of positive ions one way and negative ions the other way. The ammeter shows that an electric current in flowing.

Notice that a current in a liquid is different from a current in a solid because in a liquid both positive and negative charges move.

When the lead ions reach the cathode they pick up spare electrons to become lead atoms. This lead sinks to the bottom of the heatproof container.

At the anode, the bromide ions lose their extra negative charge to become bromide atoms. The bromine bubbles off as a gas.

Great care is needed with this experiment. Lead bromide is a toxic material. High temperatures are needed to melt the lead bromide. The bromine gas that bubbles off is extremely hazardous. It can cause serious damage to the lungs if inhaled. This is why your teacher will do this experiment in a fume cupboard.

Current flow in solutions

If a material that is made up of ions is dissolved in water, the ions become mobile and the solution will conduct electricity. This fact can be very useful. Look at Fig. 4.23.

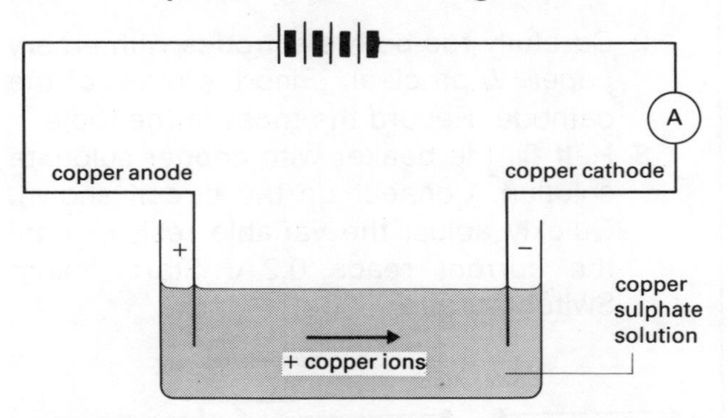

Fig. 4.23 Passing electricity through copper sulphate solution

Copper sulphate is a substance made up of ions. In solution the positive copper ions are attracted to the negative cathode. If you set up the apparatus shown you will see copper deposited on the cathode. If you look carefully at the anode you will see that the copper is dissolving away. Can you guess what is happening?

This process can be used to purify metals. An impure metal is used as the anode and pure metal is deposited at the cathode. It can also be used for putting one metal on top of another. This is called PLATING. The wheels on your bicycle may be plated with a metal called chromium to stop them from rusting.

Investigation 4.1 How does the mass of copper deposited vary with charge?

Collect

2 copper electrodes
emery paper
power supply – low voltage
connecting leads
variable resistor
copper sulphate solution
distilled water
ethanol
propanone
stopclock

What to do

1 Draw this table in your book.

Original mass of cathode/g	Final mass of cathode/g	Gain in mass of cathode/g	Current/A	Time/s	Charge/C
			0.2	500	
			0.4	500	
			0.6	500	
			0.8	500	
			1.0	500	

2 Carefully rub both electrodes with emery paper. Wipe clean. Find the mass of the cathode. Record the mass in the table.

3 Half fill the beaker with copper sulphate solution. Connect up the circuit shown. Quickly adjust the variable resistor until the current reads 0.2 A. Start timing. Switch off after 500 s.

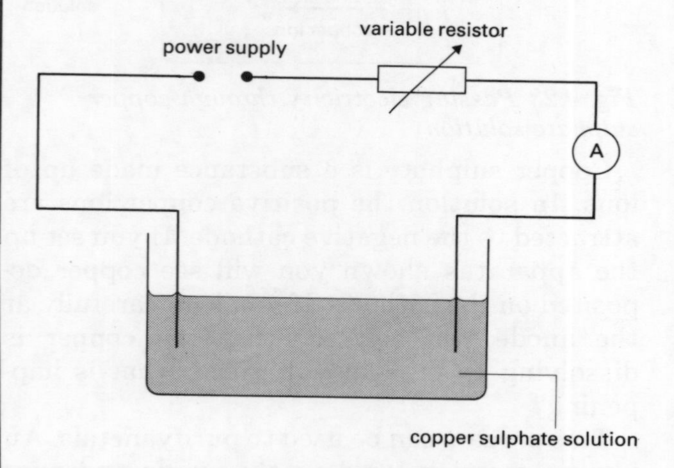

4 Very carefully rinse the cathode with distilled water, then ethanol, then propanone. Leave to dry. Find the mass and record the result in the table.

5 Repeat with currents of 0.4, 0.6, 0.8 and 1.0 A. You need not rub down with emery paper each time. The final mass of the cathode for the 0.2 A experiment will be the original mass for the 0.4 A experiment.

6 Plot a graph of mass deposited against charge. Don't forget to label your axes carefully. Choose a scale so that you use as much of the graph paper as possible.

7 Describe how the mass of copper deposited on the cathode varies with time.

Questions

A pupil carries out this experiment slightly differently. She keeps the current fixed at 1 A and runs the experiment for different times. Her results are shown on page 175.

1 Copy and complete this table in your book.

2 Plot a graph of gain in mass of cathode against charge. Plot charge along the horizontal axis and gain in mass up the vertical axis. Don't forget to label your axes carefully.

3 Describe how the mass of copper deposited on the cathode varies with charge. ▶

Original mass of cathode/g	Final mass of cathode/g	Gain in mass of cathode/g	Current/A	Time/s	Charge/C
123.3	124.3		1	3000	
124.3	126.2		1	6000	
126.2	129.3		1	9000	
129.3	133.4		1	12000	
133.4	138.4		1	15000	

4 Does this pupil's experiment fit in with your results from the investigation?
5 How much mass would 1000 C deposit?
6 How many coulombs would be needed to deposit 1 g of copper?
7 Where do you think that the copper deposited on the cathode has come from?
8 How could you test your answer to 7?

Electric currents

WHAT YOU SHOULD KNOW

1 Electric charge is measured in coulombs (C). There are 6×10^{18} electrons in a coulomb.
2 An electric current is a flow of charge. Electric current is measured in amps (A).

1 amp = 1 coulomb per second.

3 $\text{Current (in amps)} = \dfrac{\text{charge (in coulombs)}}{\text{time (in seconds)}}$

or $I = \dfrac{Q}{t}$.

4 A complete circuit is required before a current will flow.
5 Electric current is measured with an ammeter.
6 In a series circuit the current is the same in all parts of the circuit.
7 In a parallel circuit the current in the branches adds up to give the total current drawn from the supply.
8 In solutions or in molten ionic substances both positive and negative ions are free to move and carry the current.
9 When an electric current passes through a liquid, positive ions move to the cathode. Negative ions move to the anode.
10 The mass of a substance deposited during electrolysis is proportional to the charge flowing.

QUESTIONS

1 (a) What is measured in coulombs?
(b) What is measured in amps?
(c) Two of the following statements are correct, two are false. Write out the correct statements.

1 coulomb = 1 amp per second
1 coulomb = 1 amp for 1 second
1 amp = 1 coulomb per second
1 amp = 1 coulomb for 1 second

2 Work out the current in each of these cases.
(a) If 2 coulombs flow past a point in 1 second.
(b) If 4 coulombs flow past a point in 2 seconds.
(c) If 10 coulombs flow past a point in 2 seconds.
(d) If 4 coulombs flow past a point in 2 seconds.
(e) If 60 coulombs flow past a point in 1 minute.

3 Work out the charge passing a point if:
(a) 1 amp flows for 2 seconds.
(b) 4 amps flow for 3 seconds.
(c) 10 amps flow for 0.5 seconds.
(d) 0.1 amps flow for 1 minute.
(e) 0.1 amps flow for 1 hour.

4

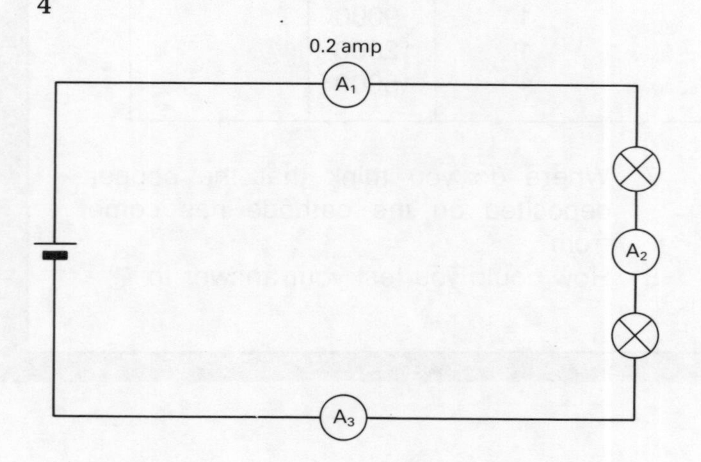

(a) Draw the above diagram into your book. Name the way in which the lamps are connected.
(b) If A_1 reads 0.2 A, show on your diagram what the other two ammeters read.
(c) If one of the lamps was removed, what would happen to the other lamp?
(d) What would the ammeters read?

5

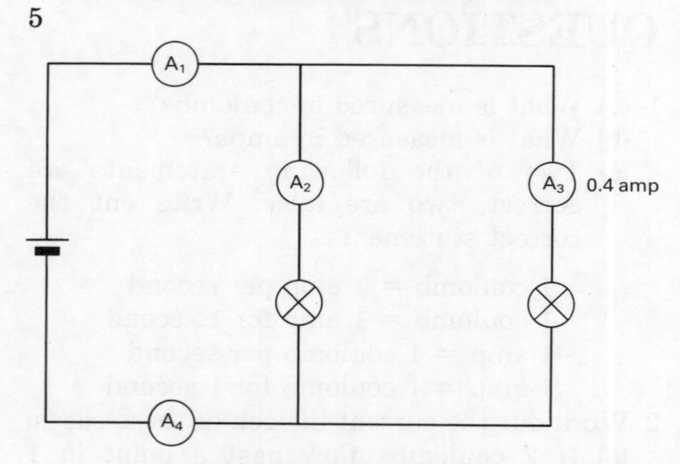

(a) Draw the above diagram into your book. Name the way in which the lamps are connected.
(b) If the lamps are identical and A_3 reads 0.4 A, show on your diagram what the other ammeters read.
(c) If the lamp next to A_3 is removed, what would happen to the other lamp?
(d) What would the ammeter now read?

6

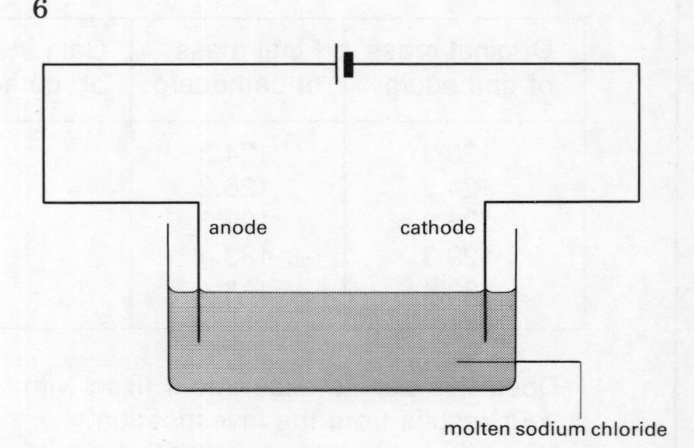

The diagram shows an electric current being passed through molten sodium chloride (sodium chloride is difficult to melt – the method of heating is not shown).
(a) Which battery terminal is the cathode connected to – positive or negative?
(b) What sort of particles carry the current through the molten sodium chloride?
(c) Sodium is seen forming at the cathode. What can you say about sodium ions?
(d) What would you expect to see at the anode? Explain your answer.

7

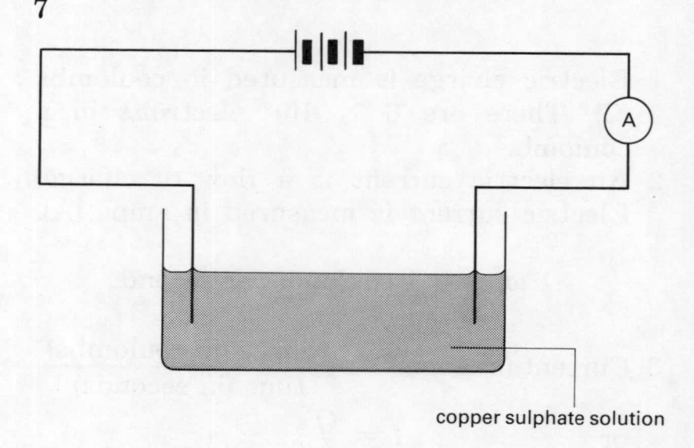

This diagram shows the apparatus required to coat a metal with copper.
(a) Copy the diagram, marking in the anode and cathode.
(b) Show carefully where the copper will be deposited.
(c) What should the anode be made of?
(d) What will happen to it?
(e) 3000 C deposit 1 g of copper. If the ammeter reads 2 A and the current flows for 50 min, how much copper will be deposited?

Electricity and Magnetism

ENERGY IN ELECTRIC CIRCUITS

In your home you are surrounded by electrical devices. They are useful because they transfer electrical energy into other forms (Fig. 4.24).

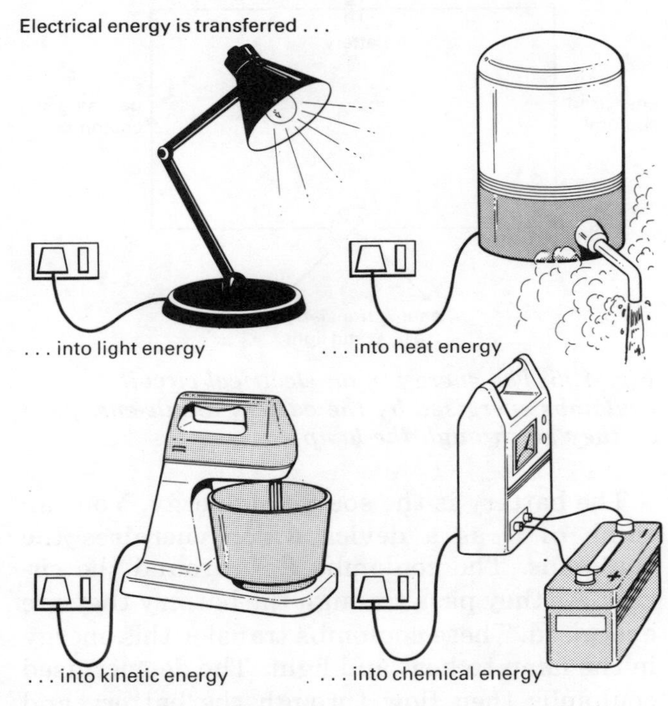

Fig. 4.24 Electrical energy is transferred into more useful forms in the home

These devices have transformed our lives. We have become very dependent upon them. If, during power cuts, electricity is limited, it causes great inconvenience.

Assignment – Energy transfer devices

Draw up a table like the one below.

Device	Heat energy	Light energy	Kinetic energy	Chemical energy	Sound energy	Before electricity
light bulb	√	<u>√</u>				candle

Think of as many devices as you can. Write the name in the column on the left. Put a tick in the column corresponding to the energy transfer that the device brings about. Many devices transfer energy to more than one form. Underline the most important energy transfer. The first one is done for you.

In the final column, write down how the job would have been done before the days of electricity.

ENERGY TRANSFER IN CIRCUITS

In order to understand energy transfer in a circuit more fully we need to go back and consider again a very simple circuit (Fig. 4.25).

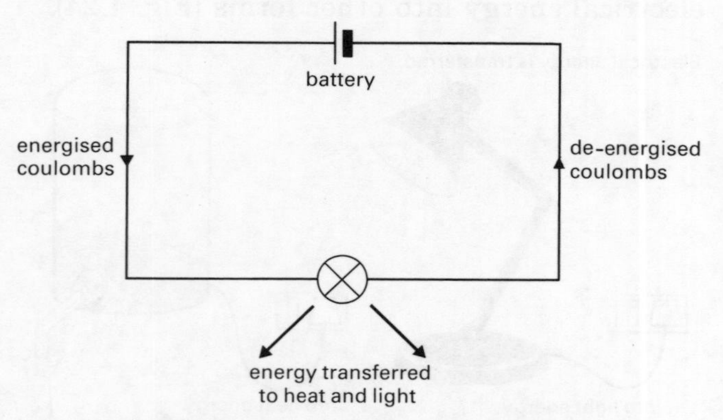

Fig. 4.25 The energy in an electrical circuit - coulombs energised by the battery are de-energised as they go through the lamp

The battery is the source of energy. You can think of it as a device which energises the coulombs. The coulombs flow around the circuit. As they pass through the battery they are energised. These coulombs transfer this energy in the lamp to heat and light. The de-energised coulombs then flow through the battery and are re-energised. This is a very simplified view but it should help you to grasp this very difficult idea.

Now concentrate on the coulombs flowing through the lamp. Before they get to the lamp they have electrical energy. After they leave the lamp they have transferred most of this energy. We have a special name for the energy transfer that each coulomb makes as it moves between two points. It is called the POTENTIAL DIFFERENCE between the points. It is measured in VOLTS (abbreviation: V) and is often called VOLTAGE.

Look at Fig. 4.26. If each coulomb passing through the lamp transfers 1 joule, then the potential difference across the lamp will be 1 volt.

potential difference or voltage between two points = energy transfer per coulomb moving between those points

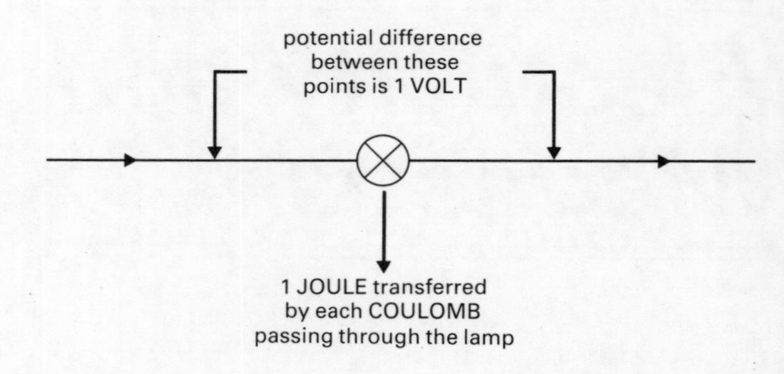

Fig. 4.26 There is a potential difference of 1 volt across the lamp if each coulomb transfers 1 joule

Another way of writing this is to say

$$\text{voltage} = \frac{\text{energy transfer}}{\text{charge}}$$

or $$V = \frac{E}{Q}$$ for short.

If you prefer the triangle method, see Fig. 4.27.

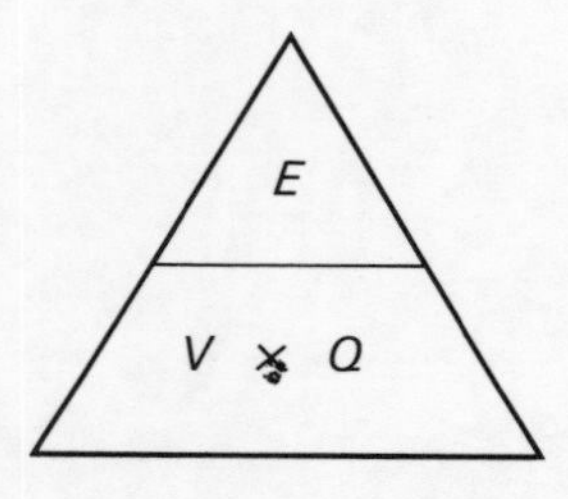

Fig. 4.27

Measuring voltage

To measure potential difference a VOLTMETER is used. It is most important that you know how to connect a voltmeter up. If we wish to measure the voltage *across* the bulb we have to connect a voltmeter *across* the bulb.

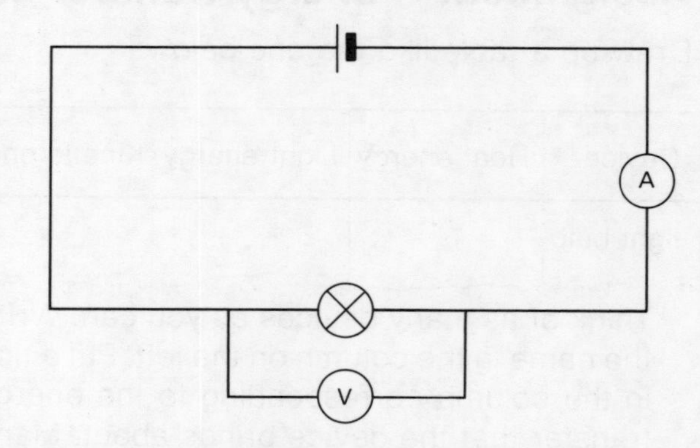

Fig. 4.28 An ammeter is connected in series; a voltmeter is connected in parallel

In Fig. 4.28 the voltmeter is shown by the symbol V . Notice that it is connected across the bulb. There is also an ammeter in the circuit. The ammeter measures current. The current actually flows through the ammeter - it is connected in series.

The current does not flow through the voltmeter. It is connected in parallel. You can actually think of the voltmeter as diverting a very small amount of current past the lamp so that it can check on how much energy the coulombs transfer. Most of the coulombs just go straight past the voltmeter and through the lamp (Fig. 4.29).

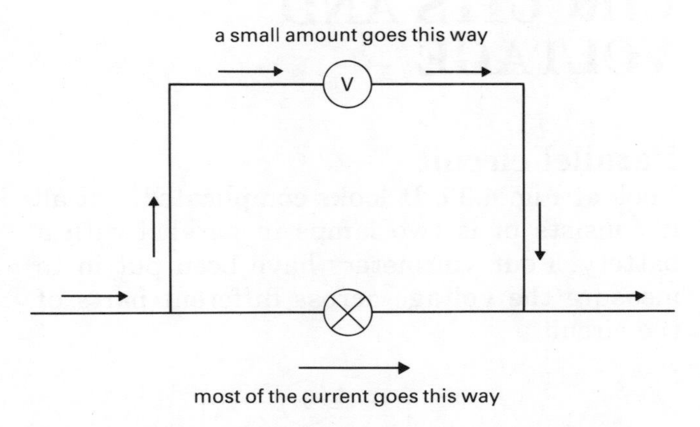

Fig. 4.29 A voltmeter draws very little current

Assignment – A problem of charge

You are a manager in the Loamshire Electricity Board. One day you receive this letter:

Dear Sir.

My teacher taught me today that electricity is not used up as it flows around a circuit. She showed us this diagram:

A A

The two ammeters read the same. I told this to my mum and dad. I said that all the electricity that flows into our house flows out again. We don't see therefore why we have to pay our electricity bill. My mum and dad have decided not to pay, and are going to give me extra pocket money instead.

Write a polite reply on behalf of the L.E.B.

CIRCUITS AND VOLTAGE

Parallel circuit

Look at Fig. 4.30. It looks complicated, but all it consists of is two lamps in parallel with a battery. Four voltmeters have been put in to measure the voltage across different parts of the circuit.

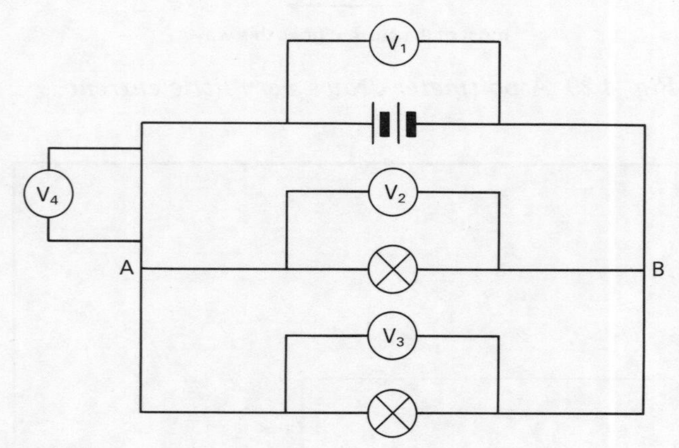

Fig. 4.30

Suppose that V_1 reads 4 volts. This means that each coulomb passing through the battery is given 4 joules of electrical energy. Now look at V_4. What do you think this will read? It is just connected across a piece of connecting wire. If the connecting wires are good conductors there is very little electrical energy transferred to heat - so little that we can ignore it. V_4 will therefore read 0 volts.

Now think of the coulombs reaching point A. They divide up. Some go through one lamp, some through the other. They have still got 4 joules each at point A, but by the time they get to B they must have transferred this energy. By the time they get back to the battery they are completely de-energised.

So V_2 and V_3 must both read 4 volts.

In a parallel circuit the potential difference (p.d.) across each device is equal and is the same as the p.d. across the supply.

The p.d. across a piece of connecting wire is zero.

Series circuit

Now look at the circuit in Fig. 4.31, showing lamps connected in series. V_1 again reads

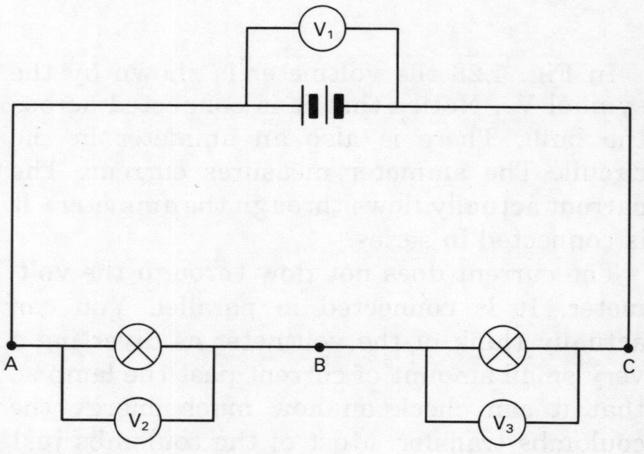

Fig. 4.31

4 volts. As the energised coulombs move round the circuit they transfer some energy in the first lamp and some in the second. Each coulomb transfers 4 joules altogether. If the lamps are identical, each coulomb will transfer 2 joules in each lamp. So V_2 and V_3 will both read 2 volts.If different lamps are used the p.d. across each lamp will not be the same *but* each coulomb must transfer 4 joules to the lamps. So the readings on V_2 and V_3 will add up to 4 volts.

In a series circuit the voltage across the devices will add up to give the voltage across the supply.

THE VOLTAGE AND CURRENT IN A CIRCUIT

Does the voltage across a device affect the current that flows through it? It is useful to think of an analogy using flowing water (Fig. 4.32).

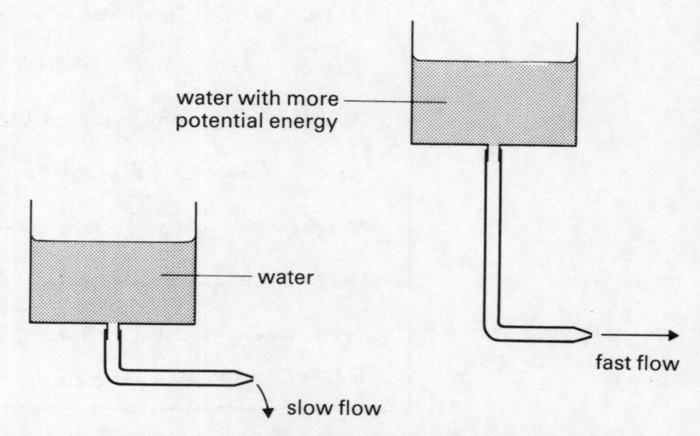

Fig. 4.32 Water flowing from a tank. The greater the energy of water the greater the current

Investigation 4.2 Using a voltmeter

In this investigation you will practise using a voltmeter. You will also check the voltages in parallel and series circuits.

Collect

2 batteries
2 lamps
connecting leads
voltmeter

What to do

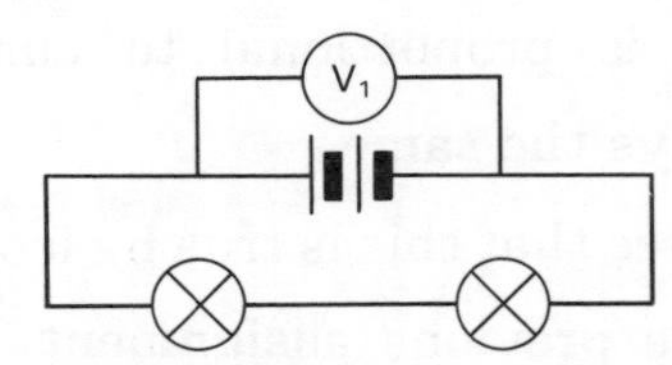

1 Connect up the lamps and batteries as in this circuit. Make sure the lamps light.
2 Connect the voltmeter as shown. Make sure that you connect it the right way round. Make a note of the reading on the voltmeter.
3 Now connect the voltmeter in the three places shown in Fig. 4.32 on p. 180. Draw the circuit diagram and write down the voltmeter readings.
4 Repeat the above experiment with the lamps connected in parallel as shown on Fig. 4.31 on p. 180. Draw the diagram and write down the four voltmeter readings.

Questions

Copy and complete this table.

Series circuit	Parallel circuit
Current the ____ in all places.	Current in branches ______ ______ to give current drawn from supply.
Voltage across each device ______ ______ to give voltage of supply.	Voltage across each device is the ______ as voltage of the supply.

Investigation 4.3 How does the current flow through a wire depend on the voltage across it?

In this experiment you will use a variable resistor to change the voltage across a coil of wire. You will see how the current depends on the voltage.

Collect

voltmeter
ammeter
coil of wire
power supply
variable resistor
6 connecting leads

What to do

1 Copy down this results table.

voltage/V	current/A	$\frac{\text{voltage}}{\text{current}}$

2 Connect up this circuit. Adjust the variable resistor to make sure that you get a good range of readings on the ammeter.

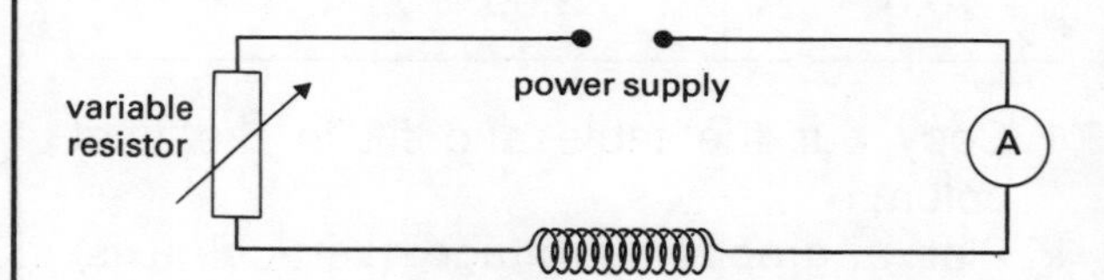

3 Now put the voltmeter into the circuit. It goes across the coil. Check that the needle moves the correct way.
4 Use the variable resistor to change the voltages across the coil of wire. Take a set of readings of voltage and current and fill in the results table. You need about six pairs of results. Plot a graph of voltage (vertical axis) against current (horizontal axis).

Questions

1 Look at your results. If you increase the voltage, what happens to the current?
2 If the voltage is doubled, what happens to the current?
3 Work out the values of $\frac{\text{voltage}}{\text{current}}$. Fill in the table. What can you say about these values?
4 Describe as fully as you can how the current depends on voltage.

The diagram on the left shows water flowing from a can through rubber tubing and out through a thin nozzle. The diagram on the right shows the same equipment but with the water in the can lifted to a higher level - it is given more potential energy. Does this change the water flow from the nozzle? Yes, the water current increases. This shows that more energy leads to more current. Is the same true of electricity?

Assignment – Voltage and current

A pupil carries out an investigation. The results are shown below.

voltage/V	current/A	$\frac{\text{voltage}}{\text{current}}$
2.0	0.40	
4.0	0.80	
6.0	1.20	
8.0	1.60	
10.0	2.00	

1 Copy out the table and fill in the final column.
2 Plot a graph of voltage (vertical axis) against current (horizontal axis).
3 What happens to the current if the voltage doubles?
4 As the voltage changes, what can you say about $\frac{\text{voltage}}{\text{current}}$?
5 How many volts are needed to drive 1 A through the wire?
6 How many volts would be needed to drive 5 A through the wire?

OHM'S LAW

From the results of Investigation 4.3 you can see that if you double the voltage across a piece of wire you double the current. If you triple the voltage you triple the current and so on. When two things are related to each other like this they are said to be *proportional*. If they are plotted on a graph they give a straight line.

voltage is proportional to current

or, in shorthand,

$$V \propto I$$

This relationship was first discovered by a German schoolteacher called Georg Ohm. It was a very useful discovery and is called OHM'S LAW.

Resistance

If voltage is proportional to current, then $\frac{\text{voltage}}{\text{current}}$ stays the same.

You can see that this is true by looking at the table in the previous assignment. $\frac{\text{Voltage}}{\text{current}}$ is constant and equal to 5. Investigation 4.3 also showed this to be true, but the value of voltage/current was different - it depends on the piece of wire that you use.

If you think about $\frac{\text{voltage}}{\text{current}}$ it is telling us the number of volts needed to drive 1 A of current. We give this a special name: RESISTANCE.

$$\frac{\text{voltage}}{\text{current}} = \text{resistance}$$

or

$$\frac{V}{I} = R$$

The greater the resistance of a piece of wire, the more volts are needed to push 1 A of current through it. The unit of resistance is the OHM (symbol: Ω).

If a piece of wire has a resistance of 1 ohm then 1 volt is needed to push 1 amp through it. If you like the triangle method you should remember Fig. 4.33.

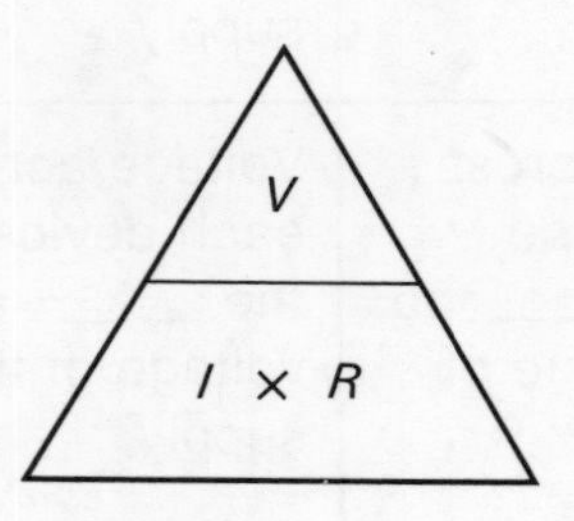

Fig. 4.33

For example: If a lamp draws 0.5 A from a 2 V supply, what is the resistance of the lamp?

$$R = \frac{V}{I} \qquad V = 2\ \text{V} \quad I = 0.5\ \text{A}$$

$$R = \frac{2}{0.5} = 4\ \Omega$$

What current flows when a 10 Ω resistor is connected to a 2 V supply?

$$I = \frac{V}{R} \qquad V = 2\ \text{V} \quad R = 10\ \Omega$$

$$I = \frac{2}{10} = 0.2\ \text{A}$$

Always make sure your answer has the right unit.

RESISTORS

Devices that are specially made to resist an electric current are called RESISTORS. There are many resistors in your home, particularly in something like a television or radio. Figure 4.34 shows some of the different types of resistors that you might find.

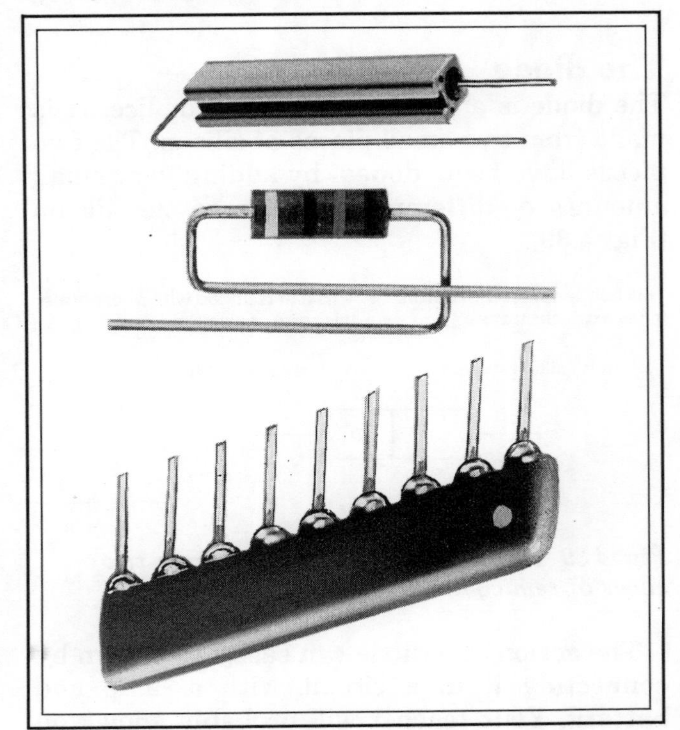

Fig. 4.34 Different types of resistor

Assignment – What is the volt–amp relationship for a solution?

You know that for a coil of wire $V \propto I$. Design and carry out an experiment to investigate the volt-amp relationship for copper sulphate solution. You should think clearly about how you record your information. You should write the experiment up clearly and carefully showing what you did, the results you expected and the conclusion that you can draw from your results.

Most resistors are made either from lengths of wire or from layers of carbon. How does the resistance of a wire vary with length and thickness? Think about people moving along a corridor. The flow of people is rather like the flow of charge along a wire (Fig. 4.35). A long thin

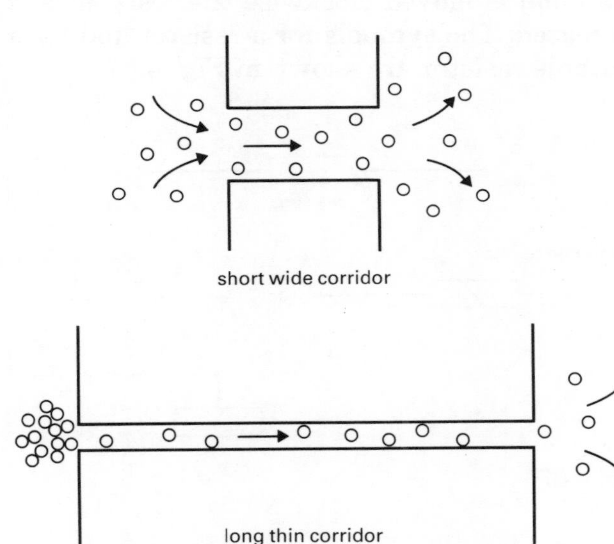

Fig. 4.35 A short, wide corridor has a low resistance; a long, narrow corridor has a high resistance

corridor resists the flow of people much more than a short wide corridor. The same is true of wires.

The longer the wire the greater the resistance.
The thicker the wire the less the resistance.

Variable resistors

These are very useful devices. Your television and cassette player both have them – it should not be too difficult to work out where.

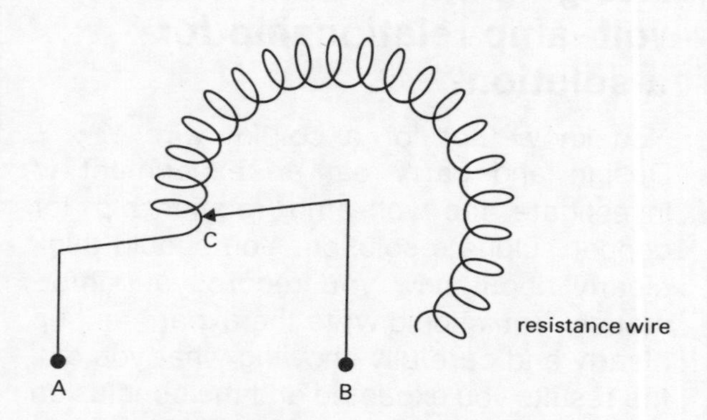

Fig. 4.36 A variable resistor - the resistance is changed by moving contact C to different places on the resistance wire

Look carefully at Fig. 4.36. The variable resistor is connected up at points A and B. The sliding contact C can be moved round to include different lengths of the resistance wire in the circuit and so increase the resistance. As the knob is moved clockwise the resistance is increased. The symbols for a resistor and for a variable resistor are shown in Fig. 4.37.

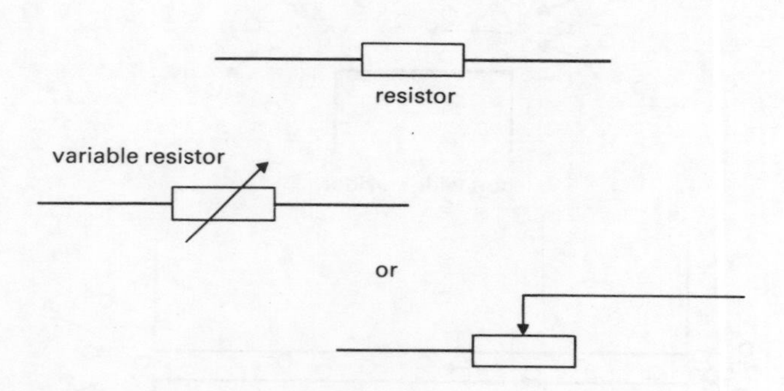

Fig. 4.37

SEMICONDUCTORS

Although many materials obey Ohm's Law, there is a group of materials called SEMICONDUCTORS that do not. These materials are neither good insulators nor good conductors but lie somewhere in between. Their resistance can be dramatically changed by adding small amounts of other substances.

The most important semiconducting material is silicon. It is used to make devices such as diodes and transistors. It is also used to make silicon 'chips' (Fig. 4.38). These are whole circuits which would once have needed thousands of different components, but are now made of a tiny slice of silicon. They are also called INTEGRATED CIRCUITS.

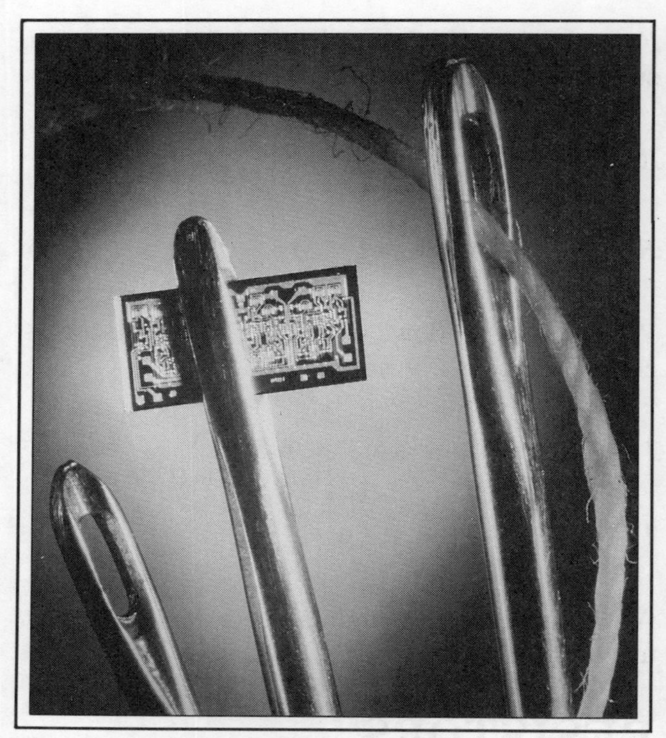

Fig. 4.38 A silicon chip: a complex integrated circuit which is so small it will go through the eye of an ordinary sewing needle

The diode

The diode is a particularly useful device. It is made from two small pieces of silicon. The two pieces have been 'doped' by adding very small amounts of different materials to the silicon (Fig. 4.39).

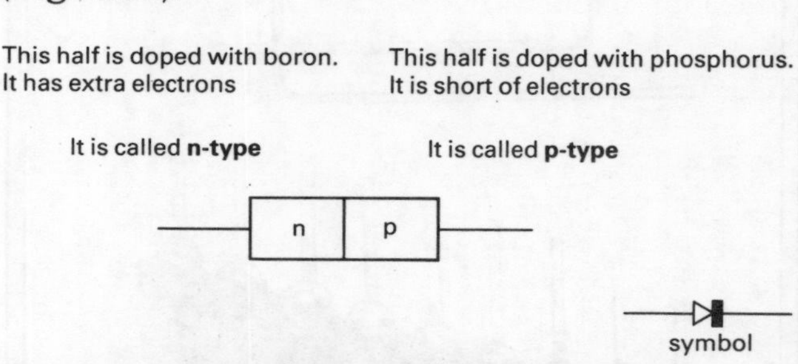

Fig. 4.39 The diode - a device made from two slices of semiconductors joined together

The action of a diode can easily be shown by connecting it in a circuit with a lamp and battery. Your teacher will probably show you this demonstration (Fig. 4.40).

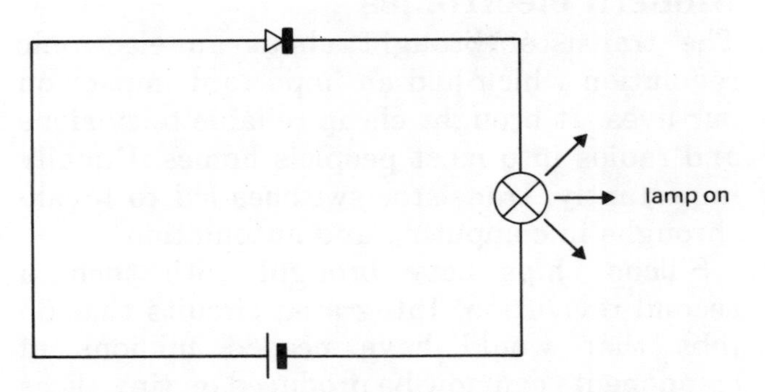

Fig. 4.40 The diode conducts a current in only one direction

If the p-type half of the diode is connected to the positive side of the battery it has low resistance. The lamp lights.

If the battery is reversed, then the diode has a high resistance. The lamp does not light. The diode conducts electricity in only one direction. A full graph of current against p.d. looks like Fig. 4.41. You can clearly see that the diode does not obey Ohm's Law.

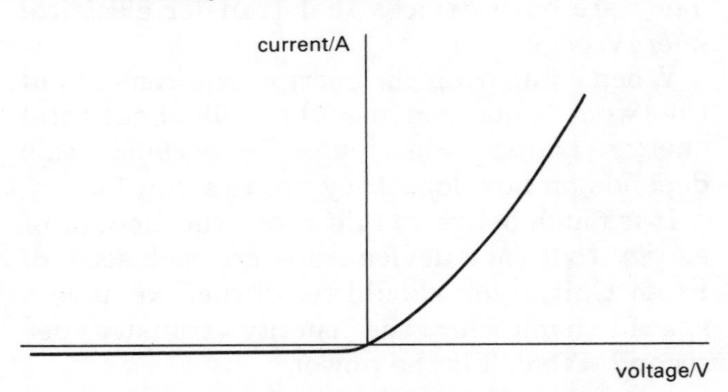

Fig. 4.41 Voltage/current graph for a diode

Transistors

The TRANSISTOR is a device that has two n-p junctions back to back. The three connections are called the BASE, the EMITTER, and the COLLECTOR (Fig. 4.42).

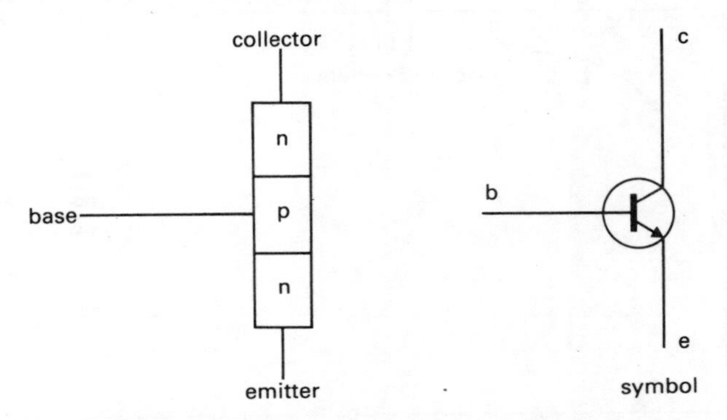

Fig. 4.42 The transistor – a device made from three slices of semiconductors

A transistor can work rather like a gate (Fig. 4.43). A single person can control a gate that lets thousands of people through: a small current to the base of a transistor can control a larger current through the collector and emitter.

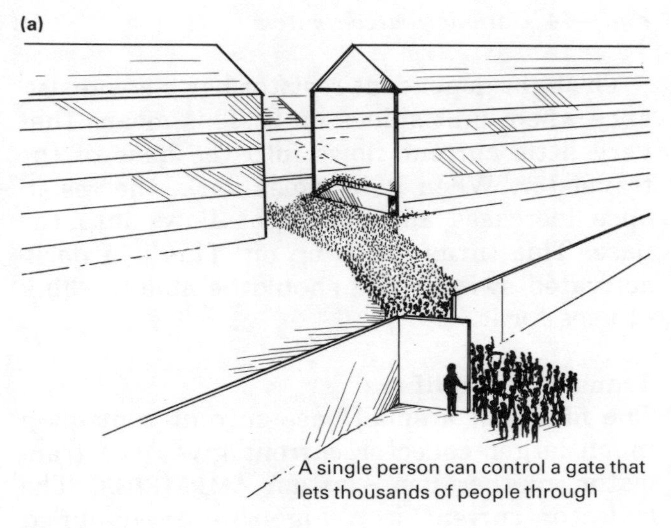

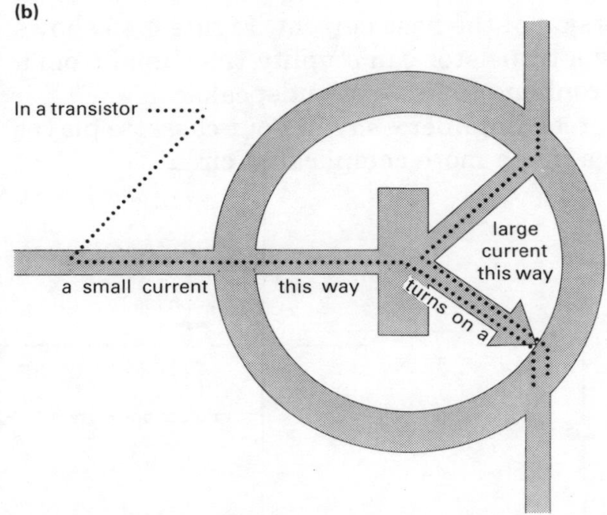

Fig. 4.43 The transistor can act like a gate where a single person controls the flow of many people

Transistor switches
The fact that a small base current turns on a much larger collector current can be used in automatic switches. The circuit in Fig. 4.44 is an example that your teacher could demonstrate to you.

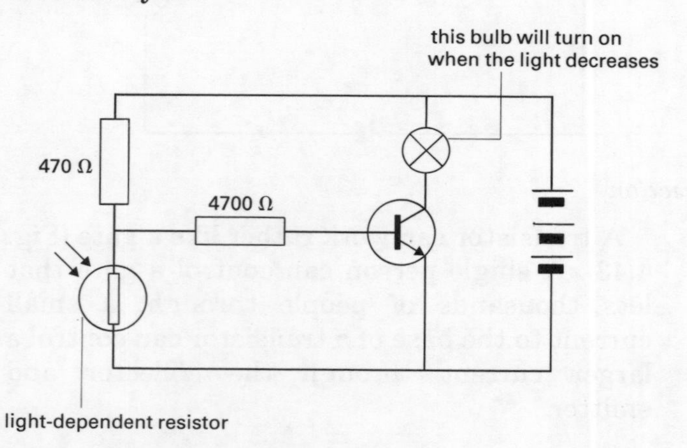

Fig. 4.44 Dark-activated switch

The light-dependent resistor has a low resistance when light shines on it. This means that very little current flows into the base of the transistor. When it becomes dark, the resistance increases and a current flows into the base. This turns the lamp on. This is a dark-activated switch. You should be able to think of uses for it.

Transistor amplifiers
The fact that a small base current controls a much larger collector current gives the transistor another use - as an AMPLIFIER. The collector current is a bigger - or amplified -version of the base current. Figure 4.45 shows how a transistor can amplify the signal from a microphone to drive a loudspeaker.

A real amplifier - say in your cassette player - is a much more complicated circuit.

Modern electronics

The transistor brought about an electronic revolution which had an important impact on our lives. It brought cheap reliable televisions and radios into most people's homes. Equally importantly, transistor switches led to breakthroughs in computing and automation.

Silicon chips have brought with them a second revolution. Integrated circuits that do jobs that would have needed millions of components can now be produced on tiny slices of silicon. This great breakthrough in microelectronics will have a tremendous influence on our future.

THE POWER IN ELECTRIC CIRCUITS

Think about an electric torch and a floodlight. They are both devices that transfer electrical energy.

When comparing the energy requirements of the two it is not very useful to talk about total energy transfer in joules - because this depends on how long they are running for.

It is much better to talk about the amount of energy that each device transfers each second. From Unit 3 you should recall that we have a special name for the energy transfer per second. We call it the power.

$$\text{power} = \frac{\text{energy transfer}}{\text{time taken}}$$

or

$$P = \frac{E}{t}$$

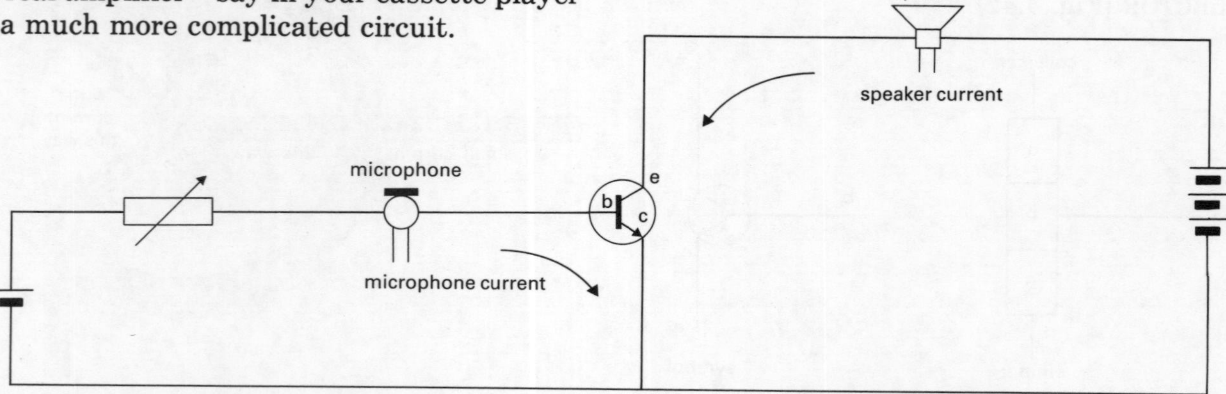

Fig. 4.45 A transistor amplifier - a small microphone current produces a large speaker current

Assignment - Microelectronics

The photographs show important applications of microelectronics. Write an essay about the impact of microelectronics. Describe a number of applications. Say what the benefits will be - and what the disadvantages will be. Say whether you are looking forward to the new electronic age and which things in particular you think you will benefit from. Mention also which things particularly concern you.

The power of the floodlight is much greater than the power of the torch.

How does the power in an electric circuit depend on the p.d. and current? Look at Fig. 4.46. A 3 volt battery is connected to a lamp. The current in the circuit is 2 amps. This means that 2 coulombs per second flow through the lamp. Now since the p.d. is 3 volts, this means that each coulomb transfers 3 joules as it flows through the lamp.

If there are 2 coulombs per second and each coulomb transfers 3 joules, how many joules are transferred each second?

The answer is 6 joules each second - so the power of this lamp is 6 watts (see page 111).

You can see that the power can be calculated by multiplying the current by the p.d.

$$\text{power} = \text{current} \times \text{p.d.}$$

or

$$P = V \times I$$

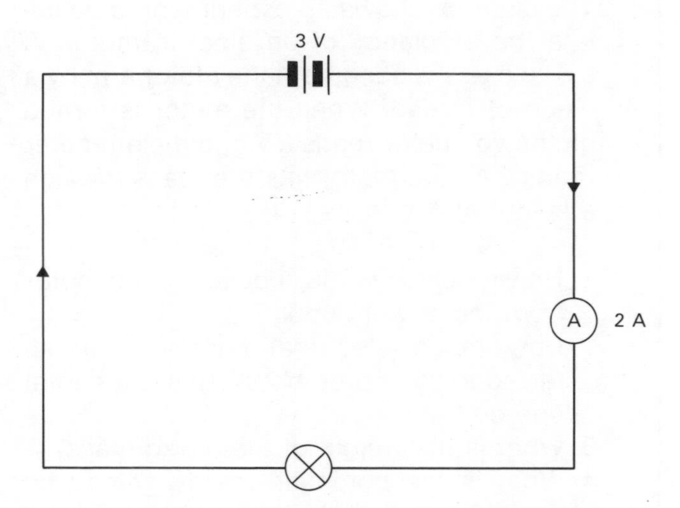

Fig. 4.46 Power in an electric circuit - 3 joules per coulomb and 2 coulombs per second means 6 joules per second

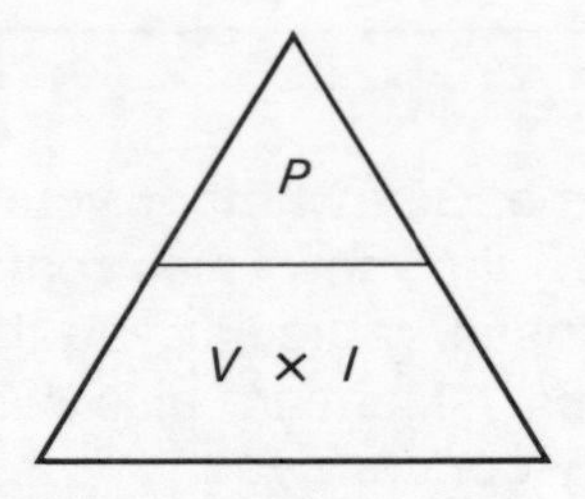

Fig. 4.47

If you use the triangle method you should remember Fig. 4.47.
This means that if you know any two of these you can calculate the third.

For example: What is the power of a lamp that draws 3 A from a 12 V car battery?

$$P = VI \qquad V = 12\text{ V}, \quad I = 3\text{ A}$$

$$P = 12 \times 3 = 36\text{ W}$$

In the case of a lamp the energy is mostly turned into heat which spreads out. We often talk about power being *dissipated* - this means spread out. So we would say that 36 watts of power are dissipated in the lamp.

Large amounts of power are measured in KILOWATTS (abbreviation kW).

1 kilowatt = 1000 watts

Assignment – Energy and efficiency in electric motors

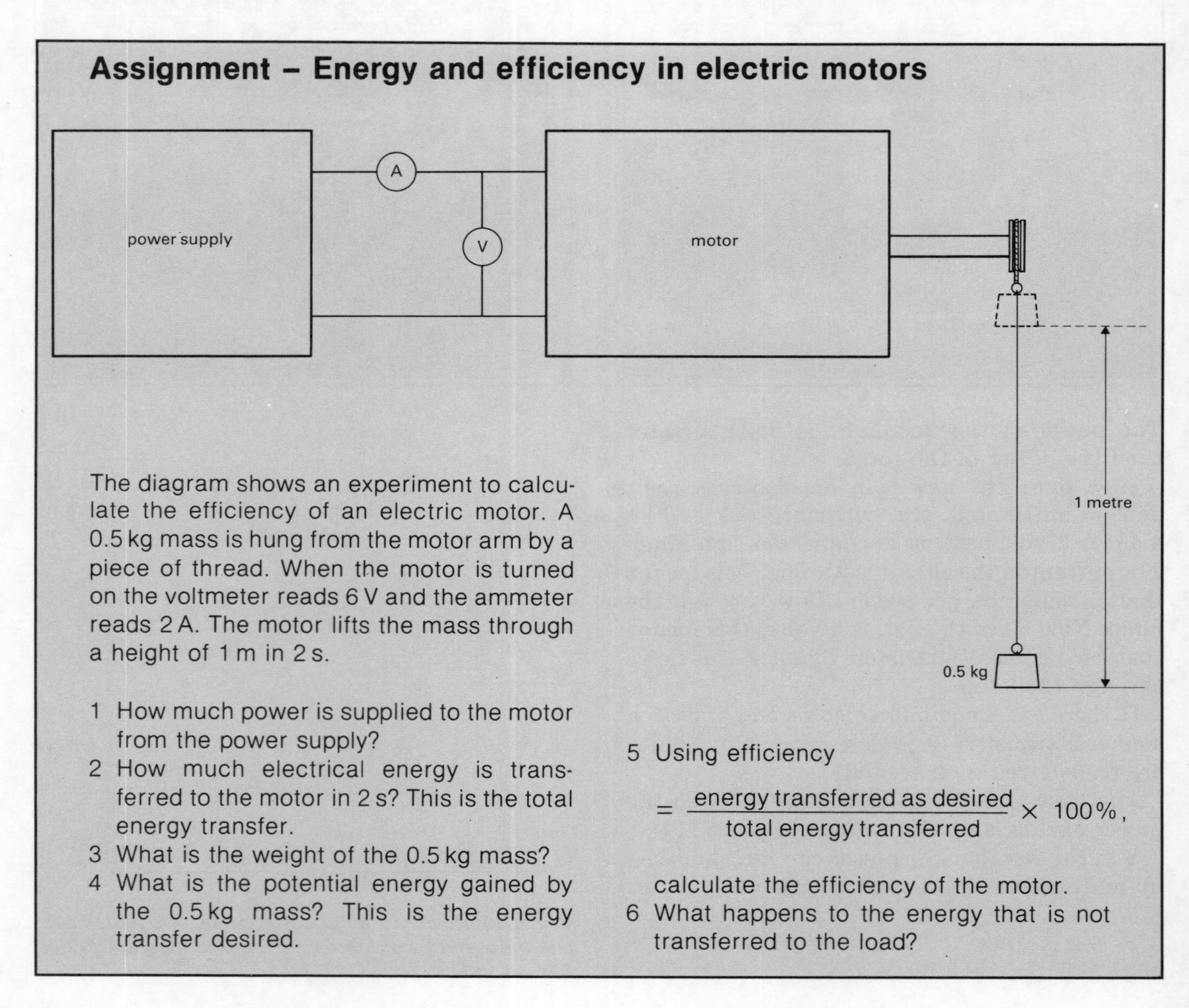

The diagram shows an experiment to calculate the efficiency of an electric motor. A 0.5 kg mass is hung from the motor arm by a piece of thread. When the motor is turned on the voltmeter reads 6 V and the ammeter reads 2 A. The motor lifts the mass through a height of 1 m in 2 s.

1 How much power is supplied to the motor from the power supply?
2 How much electrical energy is transferred to the motor in 2 s? This is the total energy transfer.
3 What is the weight of the 0.5 kg mass?
4 What is the potential energy gained by the 0.5 kg mass? This is the energy transfer desired.
5 Using efficiency

$$= \frac{\text{energy transferred as desired}}{\text{total energy transferred}} \times 100\%,$$

calculate the efficiency of the motor.
6 What happens to the energy that is not transferred to the load?

WHAT YOU SHOULD KNOW

1 The potential difference between two points in an electric circuit is the energy transfer per coulomb as the coulombs pass between these points.
2 Potential difference is measurement in volts (V) and is often called voltage.

1 volt = 1 joule per coulomb

3 In an electric circuit ammeters are connected in series, voltmeters in parallel.
4 In a series circuit the current is the same everywhere. The voltages across each device add up to give the voltage of the supply.
5 In a parallel circuit the voltage across each device is the same as the voltage of the supply. The currents in the branches add up to give the current drawn from the supply.
6 For a piece of wire at a fixed temperature the current is proportional to the voltage. This is called Ohm's Law.
7 Resistance = $\frac{\text{voltage}}{\text{current}}$ ($R = \frac{V}{I}$).
8 The unit of resistance is the ohm (Ω).

$$1 \text{ ohm} = \frac{1 \text{ volt}}{1 \text{ amp}}$$

9 The resistance of a wire increases with length and decreases with thickness.
10 Semiconductors do not obey Ohm's Law.
11 A diode allows a current to flow in only one direction.
12 A transistor uses a small current into the base to turn on a large current between the collector and emitter.
13 Power dissipated = voltage × current ($P = VI$).
14 Power is measured in watts (1000 watts = 1 kilowatt).

QUESTIONS

1 A lamp is connected up to a 2 V battery.
(a) What type of energy transfer takes place in the lamp?
(b) What type of energy transfer takes place in the battery?
(c) How many joules of energy does each coulomb transfer as it passes through the lamp?
(d) How much energy would 100 coulombs transfer?
(e) Copy and complete this:

There is a potential difference of 1 volt between two points if each ______ that passes between them transfers 1 ______ of energy.

(f) Draw the circuit diagram and include a voltmeter to measure the voltage across the lamp.

2 (a) What is meant by electrical resistance?
(b) What can you say about a resistor if it obeys Ohm's Law?
(c) What is meant by saying that a wire has a resistance of 1 ohm?

3 In an experiment to measure the resistance of a wire the following apparatus is used: a battery, an ammeter, a voltmeter, a switch, the wire, connecting leads.
(a) Draw a circuit diagram to show how the apparatus would be set up.
(b) When the switch is closed the voltmeter reads 6 V and the ammeter 0.5 A. What is the resistance of the wire?
(c) If a wire of the same material but twice as long was used, what do you think its resistance would be?
(d) What would the ammeter read now?

4 (a) What current does a 6 V battery supply when joined to a 2 Ω resistance?
(b) When a 200 Ω resistance is joined to a mains supply, the current is 1.15 A. What is the mains voltage?
(c) A 2 V battery is joined to a resistance and the ammeter reads 0.25 A. What is the resistance?

5 A mains lamp is marked '100 W'. The lamp is connected to the mains.

(a) How many joules of electrical energy are converted each second in the lamp?

(b) How many joules are converted in (i) 10 seconds (ii) 1 minute (iii) 1 hour?

(c) If the mains voltage is 240 V, how much current flows through the lamp?

6

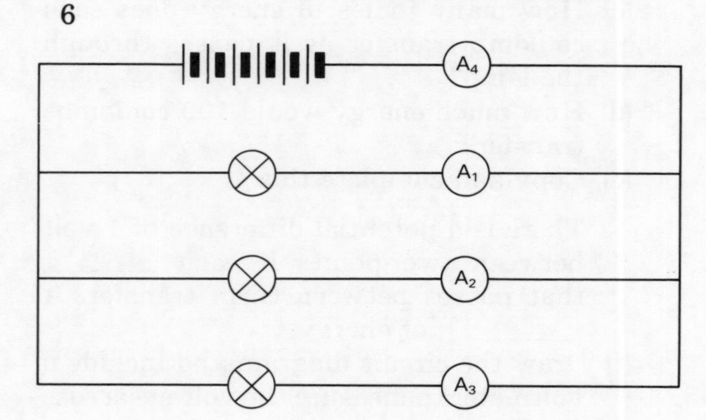

In the above diagram, each of the lamps is marked 6 W and is glowing at normal brightness. A_1 reads 0.5 A.

(a) How are the lamps connected?

(b) What do A_2 and A_3 read?

(c) What would A_4 read?

(d) What is the voltage across each lamp?

(e) What is the voltage of the supply?

7

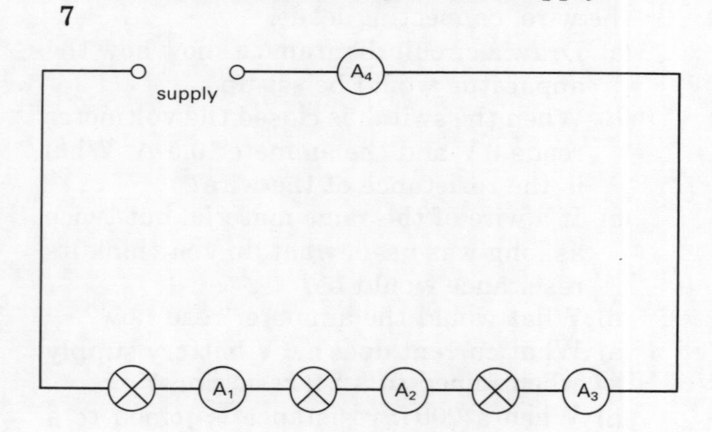

The circuit above shows three lamps, each marked 80 W. They are glowing with normal brightness. A_1 reads 1 A.

(a) What do A_2 and A_3 read?

(b) What does A_4 read?

(c) What is the voltage across each lamp?

(d) What is the voltage of the supply?

8 You are given three boxes. They are identical and have two leads coming out of them. The boxes are sealed. You are told that one contains a light bulb, one contains a diode and the third contains a wire whose resistance does not change with temperature.

Describe how you could find out which was which. Say what apparatus you would use, what results you would expect and how you would make your decision. You cannot open the boxes.

9

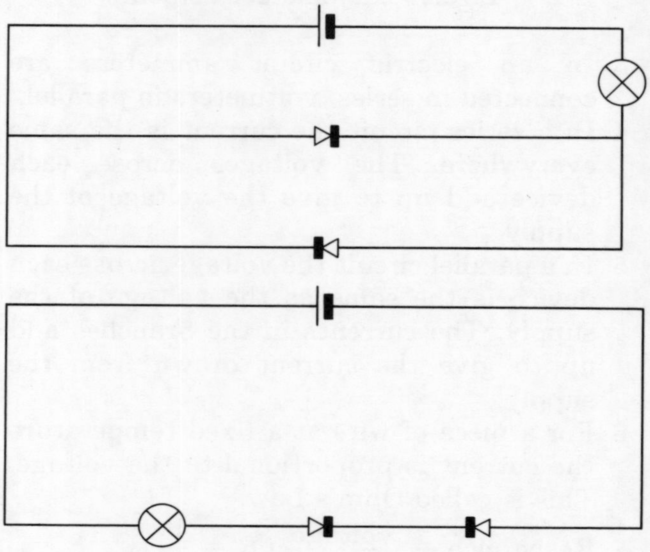

(a) Would the lamps in these circuits be lit? Explain your answer.

(b) What would happen in each case if the battery connections were reversed?

10

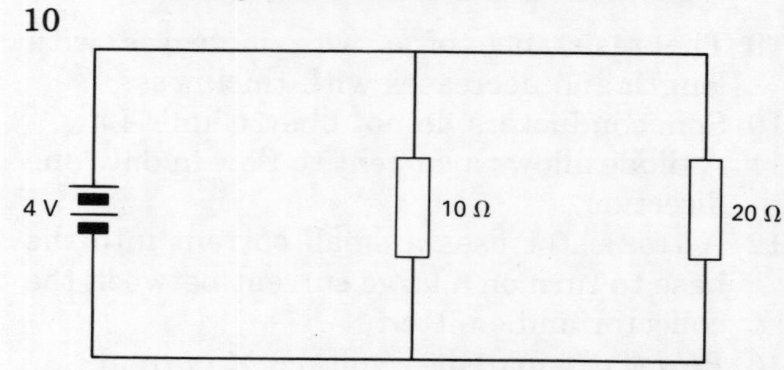

(a) How much current flows through the 10 Ω resistor?

(b) How much current flows through the 20 Ω resistor?

(c) What is the total current drawn from the supply?

Electricity and Magnetism

ELECTRICITY IN THE HOME

The electrical supply in our houses differs from the supply we get from a battery in two ways. Firstly it has a much higher voltage. Mains electricity is supplied at 240 V in Britain (the voltage varies from country to country). It is this high voltage that makes it dangerous, because the high voltage can drive large currents. The other difference is that the mains supplies an ALTERNATING CURRENT.

If you connect a resistor up to a battery, a steady current flows one way. If you connect a resistor up to the mains, the current flows backwards and forwards, 50 times a second. This changing current is called alternating current or a.c. for short.

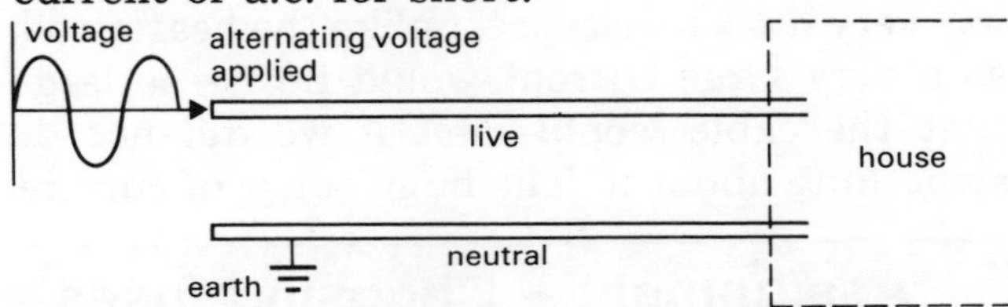

Fig. 4.48 The mains supply - a high-voltage live cable and an earthed neutral cable

The electricity supply to your home comes from the power station along two cables. (Fig. 4.48). The *live wire* has an alternating voltage applied to it at the power station. The *neutral wire* completes the circuit. The neutral is earthed by the Electricity Board. These wires are carefully wrapped in insulation and buried under the ground.

When you plug your television into the socket at home you will have noticed that there are three pins in the plug. The bottom two are the live and neutral connectors, the third is a safety device called the earth (see page 194).

The circuit diagram in Fig. 4.49 also shows another safety device called the FUSE.

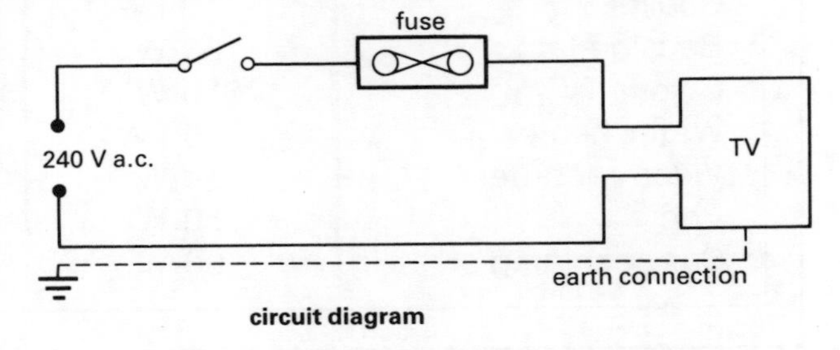

Fig. 4.49 Plugging into the mains

THE FUSE

This is a simple device which melts and breaks the circuit if a fault develops. There are two reasons for a fuse melting or 'blowing'.

Overloading

A fault may develop which causes too much current to flow. In the case of your hairdryer, the motor may jam. This larger current would cause your hairdryer to overheat and could damage it.

Short circuit

This is a serious problem. It usually results from a faulty connection. Figure 4.50 shows the wires leading to the heating coil of a hairdryer. Suppose that a strand of wire came loose and touched across to the neutral. The strand has very little resistance, unlike the heater coil, so a very large current would flow - so large that the cable would melt if we did not do something about it. The huge surge of current melts the fuse in the plug, cuts off the current and so protects the hairdryer.

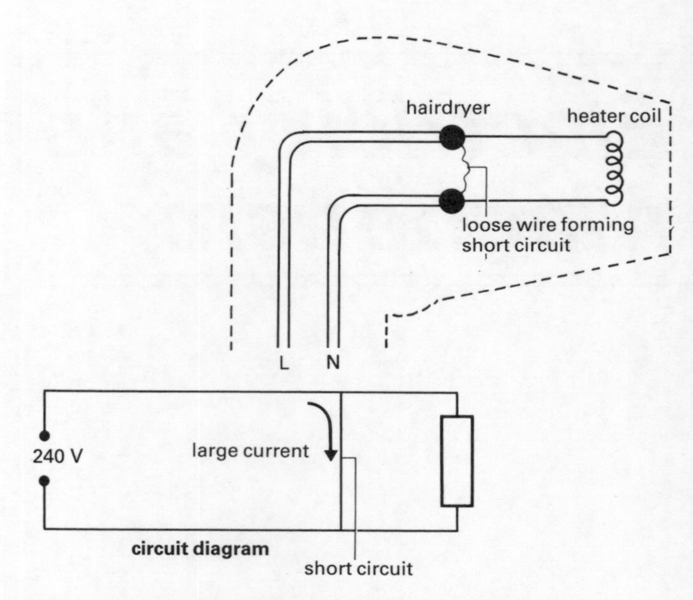

Fig. 4.50 A short circuit - a low resistance connection between live and neutral

If the fuse in a plug 'blows' it means there is a fault. It is no use just replacing the fuse - the fault has got to be corrected first.

Assignment – Choosing fuses

Plugs are normally fitted with a 3 A or a 13 A fuse. When choosing a fuse, the new fuse must have a current rating slightly higher than the normal current through the device. So if the device is overloaded, too much current flows and the fuse melts.

To work out the correct fuse you need to know the current drawn by the device. If you buy an electric kettle you will not be told the current, but you will be told the power. A typical electric kettle is marked 2.4 kW. Remember this means 2400 W.

Using $P = VI$ $\quad V = 240$

or $I = \dfrac{P}{V}$ $\quad P = 2400$

$$I = \frac{2400}{240} = 10\text{ A.}$$

So the current is 10 A. Therefore a 13 A fuse would be used.

Now copy the table below into your notebook and complete the end two columns.

Device	Power	Voltage	Current	Fuse
Cooker	10 kW	240 V		
Bedside lamp	60 W	240 V		
Colour TV	350 W	240 V		
Water heater	5 kW	240 V		
Video recorder	50 W	240 V		
Iron	750 W	240 V		
Car headlamp	36 W	12 V		

ELECTRIC SHOCKS

An electric shock is caused by an electric current flowing through your body. The greater the current that flows, the more painful and serious the shock. The table below shows how the current relates to the shock.

Table 4.1 Electric shocks

Current/mA	Effect on the body
0–1	just detectable
1–15	cramp in muscle, unable to let go let go of wire if held, painful
15–30	severe muscle cramps, difficulty with breathing, limit of tolerance
30–50	loss of consciousness if sustained
above 50	burns, probable death if sustained

1 milliamp (mA) is a thousandth of an amp

The amount of current depends on the voltage and the resistance.

$$\text{resistance} = \frac{\text{voltage}}{\text{current}}$$

$$\text{current} = \frac{\text{voltage}}{\text{resistance}}$$

You can see that if there is a large voltage and a small resistance then the current will be large. The mains is always 240 V, but the resistance of the human body varies.

Figure 4.51 shows two situations that would probably prove fatal. If you touched the live wire with one hand and the neutral with the other there would be 240 V across you. Because your hands make good contact you would have a low resistance and so a current would flow that would probably be lethal.

In the second diagram the person touches the live wire with one hand and a good earth such as a metal pipe with the other. Again,

Fig. 4.51 Electric shocks

there would be 240 V across the person and a current would flow that could be lethal.

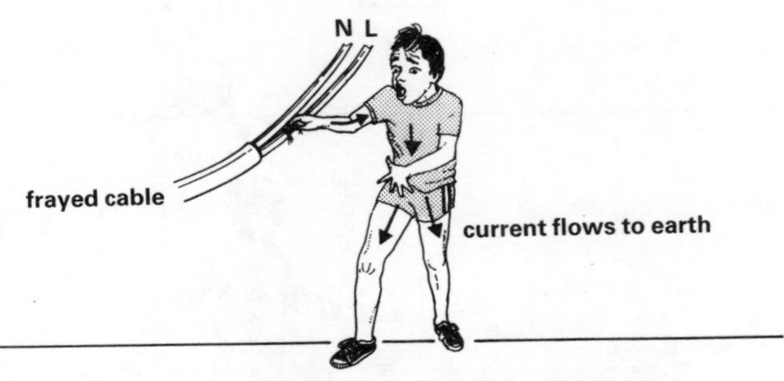

Fig. 4.52 Common type of electric shock from a frayed cable

People normally get shocks when they touch just the live wire (Fig. 4.52). In this case the current flows through the person to earth. Again the voltage is 240 V, but the resistance depends on a number of factors. If the person was wearing rubber shoes or standing on a plastic floor the resistance would be high. The experience would be very painful but not fatal. If the person was barefoot and standing on a stone floor the resistance would be low and a fatal current could flow. The presence of water reduces resistance and so increases the danger of serious shocks.

Assignment – Safe wiring

To make sure that houses are safely wired the Institute of Electrical Engineers (ICE) produces a set of rules, the ICE Wiring Regulations. Try and find out what special rules there are for wiring bathrooms in houses and explain why these rules are needed.

EARTHING

The presence of the earth connection on a device greatly reduces the likelihood of a serious shock.

Look at Fig. 4.53(a). This shows an electric kettle with no earth connection. A fault could develop and the kettle could become live. A person touching the kettle could then receive a fatal shock.

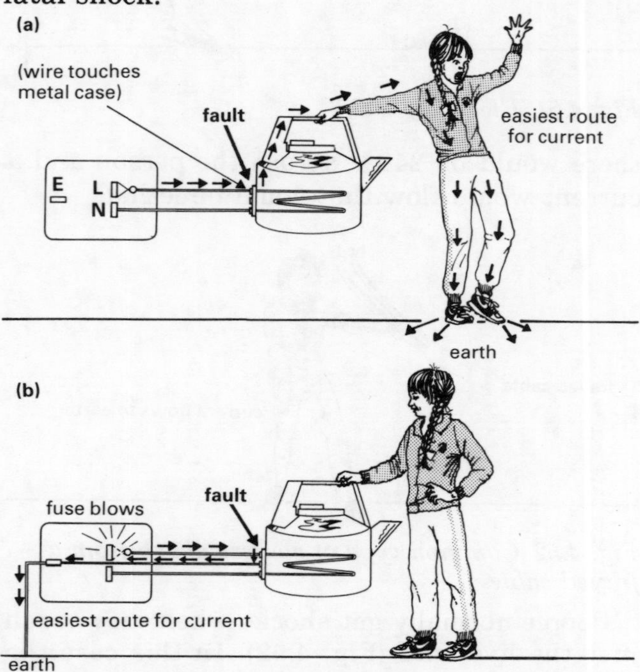

Fig. 4.53 (a) Faulty kettle without earth leads to serious electric shock. (b) With earth - a large current flows to earth and blows the fuse

Now look at Fig. 4.53(b). This time the outside of the kettle is connected to earth. If the live wire touches the kettle, instead of the person getting a shock a large current will flow to earth and the fuse will blow - isolating the kettle from the live wire.

All metal electrical appliances must be earthed. Many plastic devices such as hairdryers do not have an earth lead on the cable. This is because the casing is itself an insulator and therefore cannot become live. Devices like this do not need an earth.

Three pin plugs

Most electrical devices that you buy just have a length of cable on them. A plug has to be fitted. The most common is the three pin plug (Fig. 4.54).

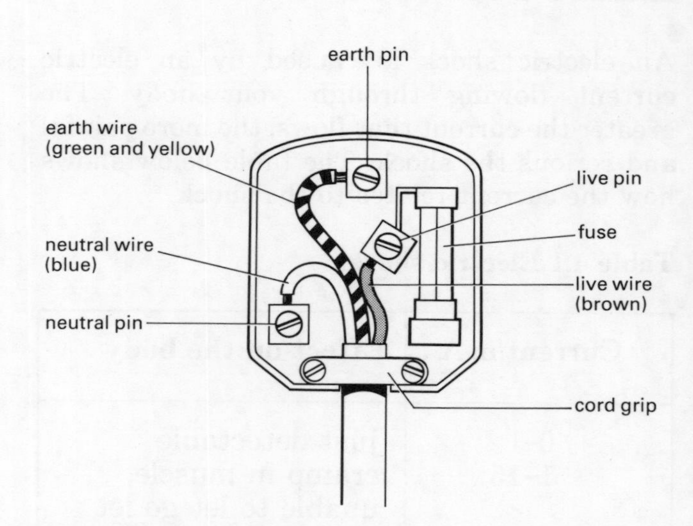

Fig. 4.54 Wiring a three pin plug

When connecting up a plug you must make sure that the three wires in the cable are connected up to the correct terminals. If you connect the live and neutral up the wrong way your hairdryer would still work, but it would mean that the switch was in the neutral rather than the live part of the circuit. So you could switch the hairdryer off but it would still be connected to live.

HOUSEHOLD CIRCUITS

A diagram of the electrical circuits in your house looks like Fig. 4.55 (if you have a ring main). It appears very complicated, but if you follow the supply it is not too difficult.

The current comes into your house or flat via the *main supply cable*. There are two wires, the live and the neutral. The live wire has a *main fuse*.

The current then flows through a *meter* so that the Electricity Board can measure how much electricity you have used, and so decide how much you should pay. Next comes the *consumer unit*. This is where the cables branch off into a number of parallel circuits. Notice that there is a *fuse* in the live cable of each of the circuits. The consumer unit is also called the *fuse box*. Notice also that the different circuits have different sized fuses.

Assignment – Circuit breakers

The illustration shows an advertisement for a special kind of circuit breaker. It is called a RESIDUAL CURRENT DEVICE (rcd) or EARTH LEAKAGE CIRCUIT BREAKER (elcb).

Don't confuse it with the sort of circuit breaker that replaces a fuse – the type that trips out when too much current flows through a device. The elcb allows a large current to flow to a device but prevents a small current flowing through you. The electrical power supply in your lab may have an elcb. It soon may be a legal requirement for electrical devices used outside, like lawnmowers.

Find out more about the elcb. Your local electricity board should be able to help you. Describe how they work. Say why they are particularly important with lawnmowers, hedge trimmers, extension leads, power tools, etc.

One of the circuits is shown connecting the mains sockets. This is a RING MAIN. The cables begin and end at the consumer unit, forming a long loop or ring around the house.

The fuses in the consumer unit are needed to protect the circuits against overload. If they were not there, the cables under the floor of your house could overheat and cause a fire. There are different sized fuses because the circuits are designed to carry different currents. For the mains sockets, a fairly thick cable is needed to carry 30 A. For a lighting circuit, smaller currents flow so thinner cable is used – because it is cheaper. Thinner cable, however, heats up more easily so it needs a lower-rated fuse – 5 A.

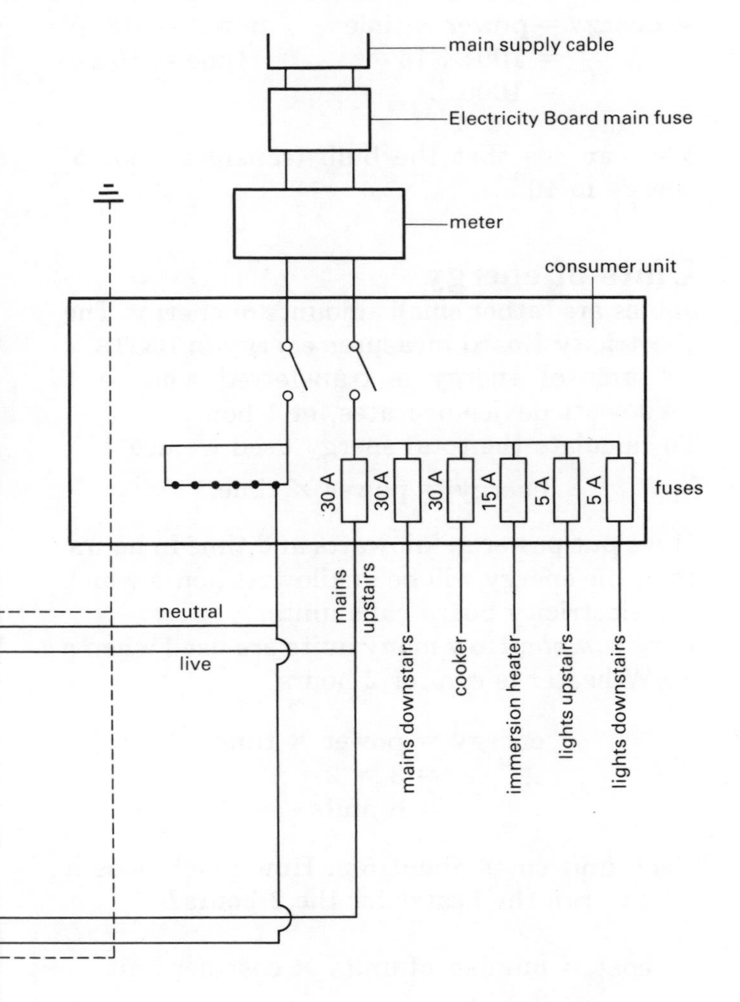

Fig. 4.55 A domestic supply

PAYING FOR ELECTRICITY

When you pay your electricity bill each quarter of the year, what are you actually paying for? Not electric current - we have seen that the current is not used up as it flows around the circuits of your home. Not the voltage either. This is always the same no matter what appliance you use.

You are paying for the *energy* that you use up. Think about an electric light bulb marked 100 W. This means that the *power* of the bulb is 100 W. Each second it is transferring 100 joules of energy. Say the bulb runs for 10 seconds, how much energy is transferred?

Remember, $$\text{power} = \frac{\text{energy}}{\text{time}}$$

so $$\begin{aligned}\text{energy} &= \text{power} \times \text{time} \\ &= 100 \times 10 \\ &= 1000\,\text{J}.\end{aligned}$$

power = 100 W
time = 10 s

You can see that the bulb transfers a lot of energy in 10 s!

Units of energy

Joules are rather small amounts of energy. The Electricity Board measures energy in UNITS.

1 unit of energy is transferred when a 1 kilowatt device operates for 1 hour.

To calculate the total energy used we use

$$\text{energy} = \text{power} \times \text{time}.$$

If we put power in kilowatts and time in hours, then the energy will be in kilowatt hours which the electricity board calls units.

For example: How many units are used when a 3 kW heater is run for 2 hours?

$$\begin{aligned}\text{energy} &= \text{power} \times \text{time} \\ &= 3 \times 2 \\ &= 6\ \text{units}\end{aligned}$$

Each unit costs about 5 p. How much does it cost to run the heater for the 2 hours?

$$\begin{aligned}\text{cost} &= \text{number of units} \times \text{cost per unit} \\ &= 6 \times 5 \\ &= 30\,\text{p}\end{aligned}$$

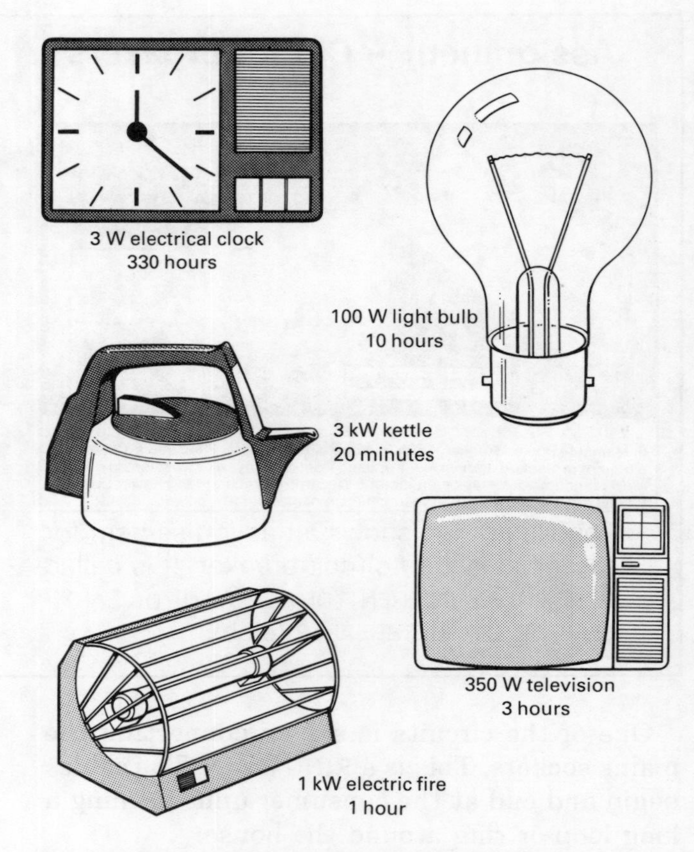

Fig. 4.56 1 unit or kilowatt hour of energy

Figure 4.56 shows what you get for 1 unit.

Assignment – Batteries or mains electricity?

Batteries are fairly cheap but how expensive is the electricity they supply compared to the mains electricity?

An electronic keyboard uses five batteries. The batteries will supply 1 W of power when they are all connected together and they last for about 10 hours. The batteries cost 20 p each.

1. Use this information to compare the cost of electrical energy from the mains and batteries. A unit (1 kW h) of mains energy costs 5 p. How much would a unit of battery energy cost?
2. A mains adaptor costs about £5. For how long would you have to play your keyboard before this paid for itself?

Electricity in the home

WHAT YOU SHOULD KNOW

1 Mains electricity is an alternating current flowing backwards and forwards 50 times each second.
2 Fuses are safety devices. They prevent overloading by melting and breaking the circuit when the current reaches a certain level.
3 Electric shocks are caused by an electric current flowing through people's bodies. The lower the resistance of the body the greater the current and the more serious the shock.
4 The presence of water reduces the body's resistance.
5 An earth is a safety device. All metal electrical appliances must be earthed.
6 Cables for electrical appliances are colour coded: brown = live, blue = neutral, green and yellow = earth.
7 The fuse always goes in the live part of the circuit.
8 Electrical energy = power × time ($E = P \times t$).
9 1 kilowatt running for 1 hour is 1 kilowatt hour. This is called a unit of electricity.

QUESTIONS

1 (a) Explain what a fuse is and what it is used for.
(b) Explain what 'overloading' means.
(c) Explain what a 'short circuit' is.
(d) If a fuse blows, suggest how you could tell by looking at the fuse whether the fault was a short circuit or an overload.
(e) What must you always do when a fuse blows?

2 The diagram below shows a plug that has been incorrectly wired.

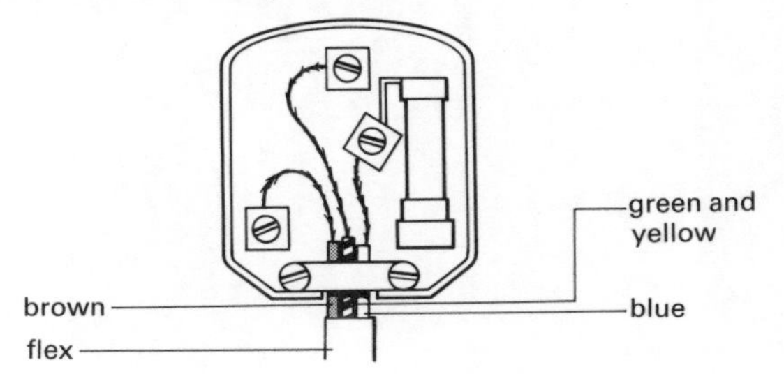

(a) What connections have been incorrectly made?
(b) State two other faults shown in the diagram.

3 (a) Explain what the earth connection on a plug is used for.
(b) Explain why some devices, such as a plastic hairdryer, do not have an earth wire in the cable.

4 Look at Fig. 4.52 (page 193), which shows a person receiving an electric shock.
(a) Copy the diagram and show how a current can flow through the person even though they only touch one wire.
(b) If the person has wet hands how will this affect the resistance?
(c) How will the current be affected?

5 The diagram below shows a kettle with a fault: the live wire touches against the case, causing it to be live.

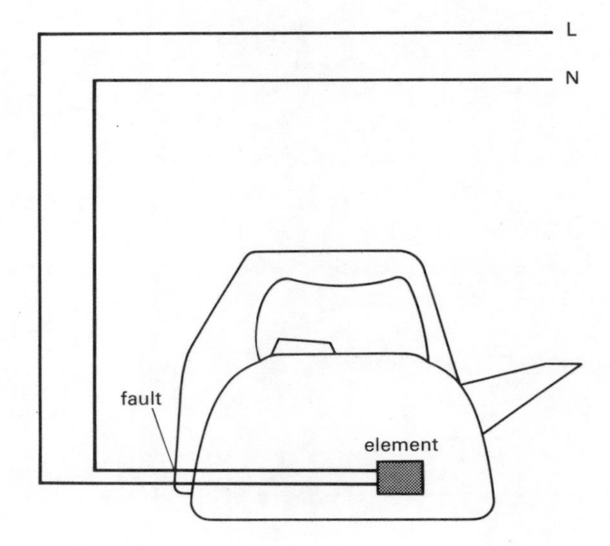

(a) Draw the diagram and include a fuse that would protect someone using the kettle.
(b) Explain how the earth wire and fuse make the circuit safe.
(c) What would happen to this fuse when the fault developed?

6 If electricity costs 5 p for 1 kW h, work out how much these would cost:
(a) a 2 kW fire left on for 3 hours;
(b) a 2 kW kettle used for 2 hours;
(c) a 100 W lightbulb left on for 10 hours;
(d) a 350 W television running for 10 hours.

7 Look at the diagram of the household supply (Fig. 4.55). The immersion heater is rated at about 3000 W.

(a) Explain why it has a circuit all to itself.

(b) The fuse for the immersion heater in the consumer unit is 15 A. Show that the fuse size is correct.

(c) How much would it cost to run the heater for 30 hours if electricity is 5 p per unit?

8 The diagram (right) shows the live and neutral wires for a lighting circuit coming out of a consumer unit.

(a) Copy and complete the diagram, showing how three lamps and switches can be connected so that the lamps can be turned on and off independently.

(b) Now mark on your diagram the position of a single switch that could turn all the lights on and off together.

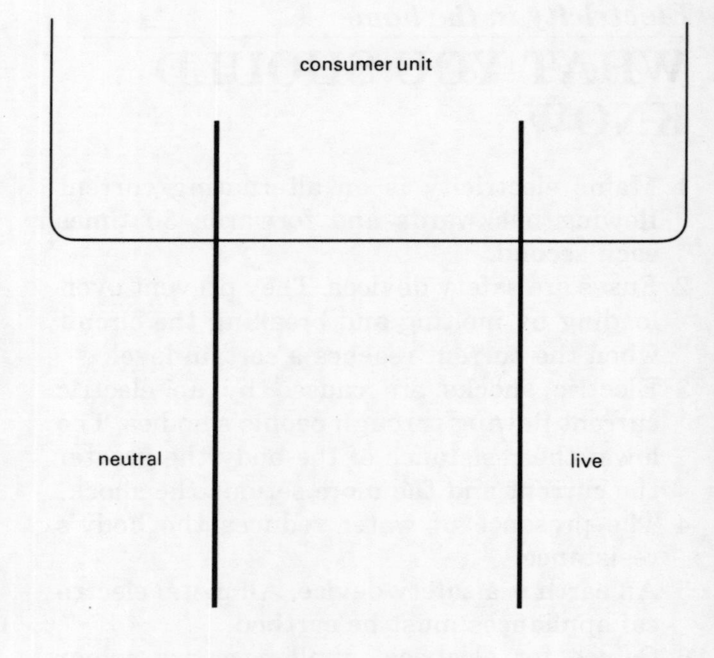

MAGNETIC FIELDS

PERMANENT MAGNETS

Magnets are mysterious and fascinating things. The fact that a magnet could produce 'force at a distance' - in other words pull a piece or iron that was some distance away - puzzled people for hundreds of years. It still remains one of the great problems of modern physics.

Magnetic poles

If a magnet is dipped into iron filings and then shaken lightly, the filings stick mainly to the ends of the magnet. The force produced by the magnet is concentrated at the ends. These are called MAGNETIC POLES. If the magnet is suspended from a thin thread it always points in the same direction - towards the north.

The end of the magnet that points north is called the north-seeking pole or, for short, N pole. The other end is called the south-seeking pole or S pole. In electricity we have + charge and − charge, in magnetism we have N pole and S pole. When we bring magnets together we find the situation is similar to that for electric charges - unlike means attraction.

MAGNETIC FIELDS

Look at Fig. 4.57. It shows what happens when iron filings are sprinkled on a card that is on top of a magnet. The filings line up in a neat pattern. It is as though the magnet has done something to the space around it so that when bits of iron are put there they are pushed into a pattern. We say that the magnet has created a MAGNETIC FIELD which pushes the iron around.

Fig. 4.57 Field pattern produced by a bar magnet

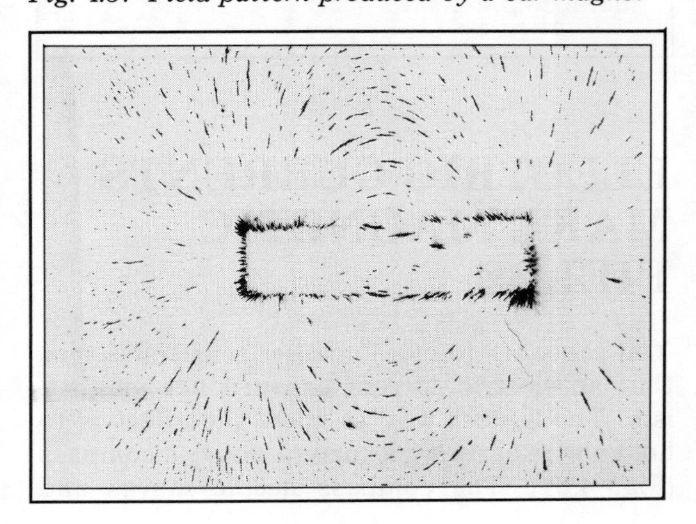

The idea of the magnetic field does not explain the mysterious 'force at a distance', but it does help us to understand how magnets behave when they are put near to one another or near to wires carrying electric currents.

There are two important things we need to know about magnetic forces: how strong they are and in what direction they act.

To show this we use MAGNETIC FIELD LINES. You can see from Fig.4.57 that the iron filings have been lined up to show the field lines. Another way to track down the field lines is to use a plotting compass. This is just a tiny magnet balanced on a pivot.

In Fig. 4.58 you can see a magnet surrounded by plotting compasses. The plotting compasses are pointing along the magnetic field lines. Notice that the field lines begin at the N pole and end at the S pole. You can see that the further away from the poles, the further apart the field lines are. This is because the magnetic fields get weaker further away from the poles.

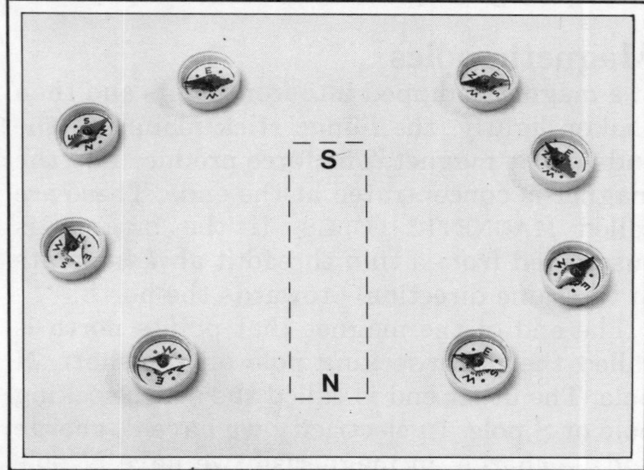

Fig. 4.58 Plotting compasses show the direction of the magnetic field lines

Magnetic field lines are a useful way of showing the strength and direction of magnetic force around a magnet.

The direction of the field line shows the direction of force.
The closeness of the field lines shows the strength of the force.

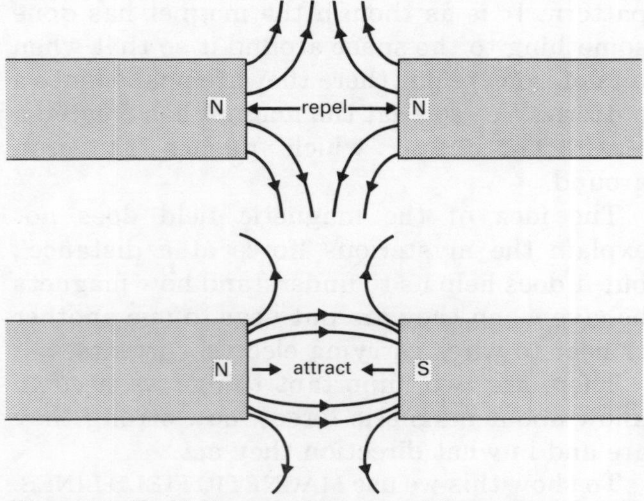

Fig. 4.59 Field lines and forces with pairs of magnets

Fields and forces

You can see how magnetic field lines help us to understand forces by looking at Fig. 4.59. In (a) the field lines are pushing each other aside - the magnets repel. In (b) the field lines run from the N pole of one magnet to the S pole of the other, pulling the poles together.

A magnet puts a force on another magnet. It will also attract materials that are not magnets. Iron and nickel are the only common materials that are attracted to magnets.

If a piece of iron is put near a magnet the field lines look like Fig. 4.60. You can see what has happened: the iron has been turned into a magnet and is now attracted to the permanent magnet. In the same way that bits of dust are attracted to gramophone records because of induced charges, so iron is attracted to a magnet because of induced magnetism.

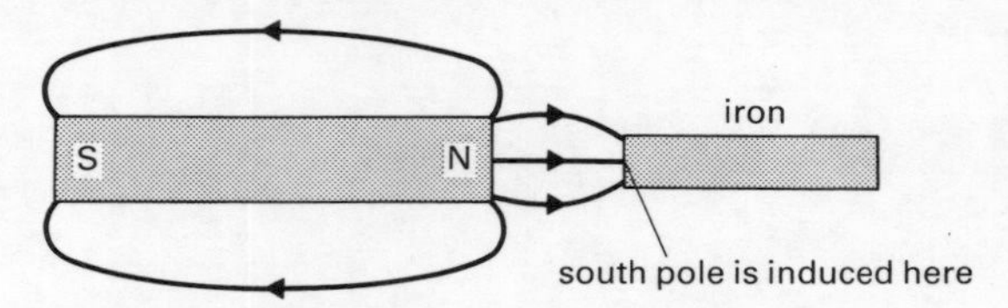

Fig. 4.60 Induced magnetism in a bar of iron

Assignment – The earth's magnetism

The earth is a giant magnet and has a magnetic field. Find out about the earth's magnetism. Draw a diagram showing the earth's magnetic field.

Find out what compasses are and how they work. Describe a simple compass.

Explain how a compass can work on a metal ship.

ELECTRIC CURRENTS MAKE MAGNETIC FIELDS

You probably found, in earlier science lessons, that an electric current can produce magnetism. The easiest way to show the effect is to hold a wire carrying a current above a compass (Fig. 4.61). The compass needle moves and points at right angles to the direction of the current. If the current direction is reversed, so too is the direction of the compass needle.

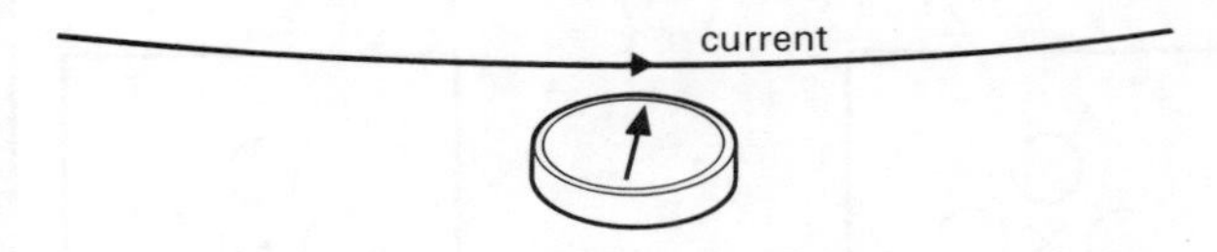

Fig. 4.61 A compass needle is affected by the current in the wire

This experiment was first carried out by a Danish science teacher called Hans Christian Oersted early in the 19th century. He actually first discovered the effect by accident. He was connecting a wire to a battery when he noticed a nearby compass needle move. Oersted's work was continued by a French scientist called Ampère. He realised that the field lines around a wire carrying current form circles.

Figure 4.62 (a) shows the wire carrying the current. Diagrams (b) and (c) show what the field looks like. In (b) the current is flowing down into the paper. That is what the symbol ⊗ means. Think of an arrow going along the wire. If it was going into the paper you would see the flight (the opposite end to the point).

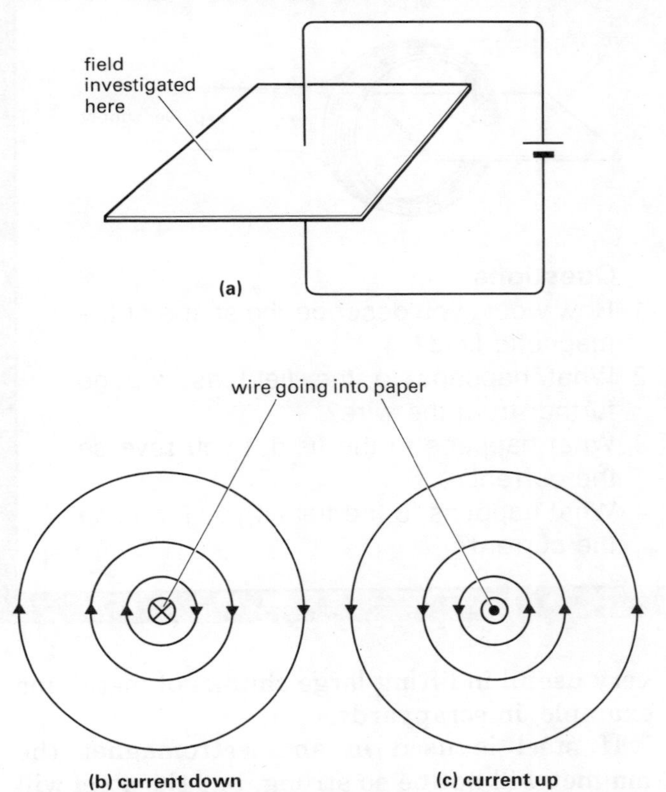

Fig. 4.62 The magnetic field lines around a wire carrying an electric current

Diagram (c) shows the field with the current direction reversed. The current is coming up out of the paper. Again think of the arrow. This time you see the point coming towards you, hence the symbol ⊙ .

From the diagrams you can see these results of Ampère's experiments:

(a) The field lines are concentric circles around the wire.
(b) The further away from the wire, the weaker the field.
(c) Reversing the current direction reverses the field direction.

Ampère also found that increasing the current increases the field strength.

FIELD LINES AROUND COILS

Coils of wire (called SOLENOIDS) are very important. Not just to scientists - you probably have a number in your home. Look at Fig. 4.63.

You can see from the diagram of the loose coil that the field lines are complete loops. The lines go down the centre of the coil, out, around the sides and back in through the other end. The coil is like a magnet. It has an N pole at one end and an S pole at the other. You can see this even more clearly if you wind a tight coil. This has a magnetic field that looks very much like the field around a permanet magnet.

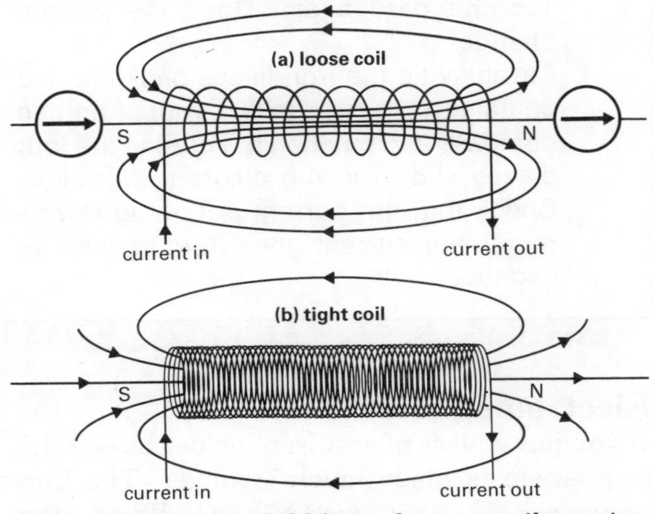

Fig. 4.63 Magnetic field lines due to a coil - notice they form loops. The field is the same as for a bar magnet

Investigation 4.4 Magnetic fields around wires

Collect

power pack
connecting wire
card
iron filings
plotting compass

What to do

1 Set up this circuit.

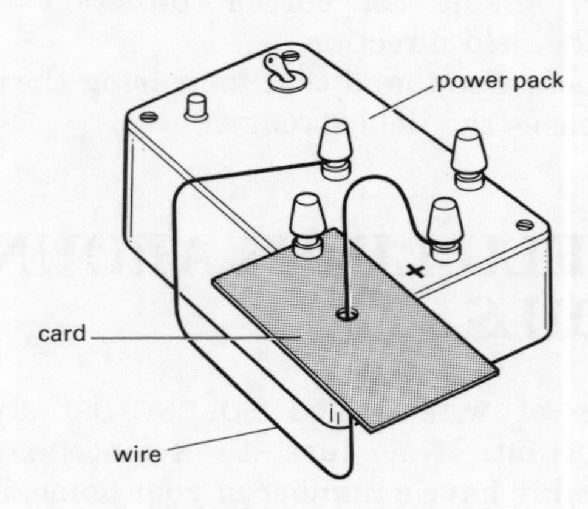

2 Set the power pack at 2 V d.c. Make sure the card is at right angles to the wire. Sprinkle the iron filings on the card. Tap the card with your pencil.
3 In your book, draw a top view of the card. On this, draw the pattern formed by the iron filings.
4 Now reverse the current by changing over the connections to the power pack. Tap the card again. Does the pattern change?
5 Carefully tip the iron filings back into the container. Now take your plotting compass and put it on the card in the four places shown in the diagram (a) below. Check that the current is flowing *down*- remember current goes from positive to negative.

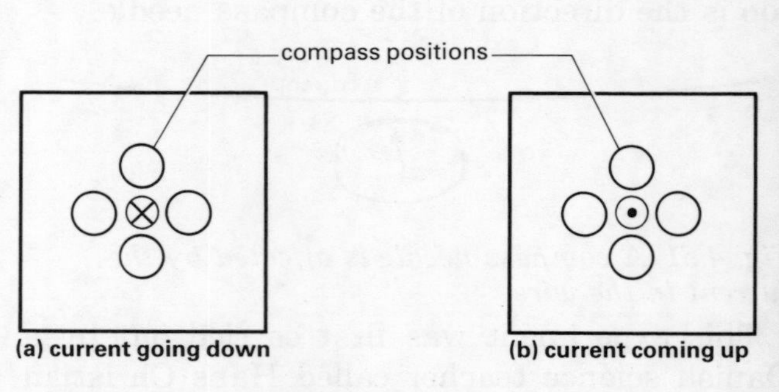

6 Copy the diagram and draw in the compass needles on (a).
7 Now reverse the current. Copy diagram (b). Place the compass at the four places shown and fill in the diagram.
8 Increasing the current may prove difficult because you may not be able to adjust your power pack. There is, however, another way to do this.
Thread the wire through the hole in the card five times (you may have to make the hole larger). Connect up to the power pack. Re-test with iron filings and plotting compasses. Write down any differences.

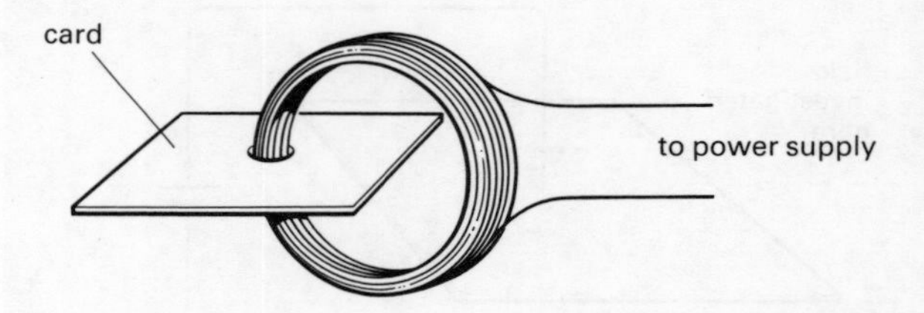

Questions

1 How would you describe the shape of the magnetic field?
2 What happens to the field as you go further from the wire?
3 What happens to the field if you reverse the current?
4 What happens to the field if you increase the current?

Electromagnets

If you put a piece of soft iron inside the coil the magnetism is made much stronger. The iron becomes an ELECTROMAGNET. When the current in the coil is turned off, the iron loses its magnetism. Electromagnets like this are very useful in lifting large chunks of metal, for example, in scrapyards.

If steel is used in an electromagnet the magnet will not be so strong, but the steel will keep some magnetism when the current is turned off.

Assignment – The relay

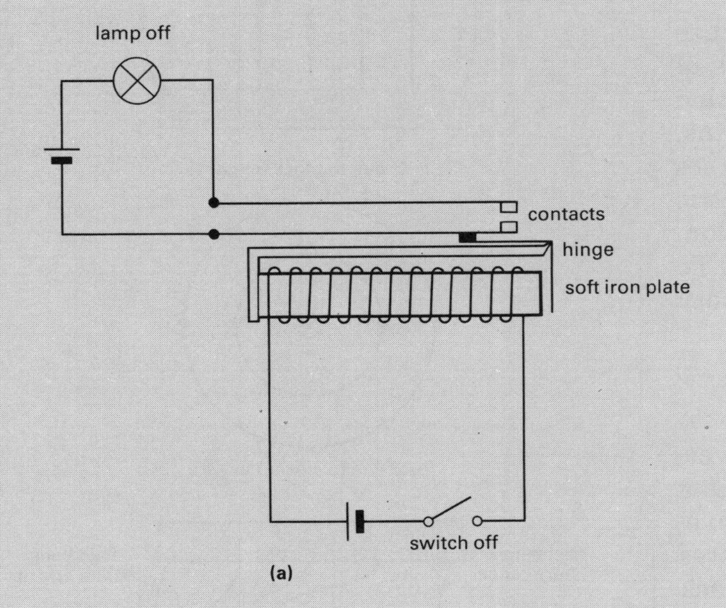

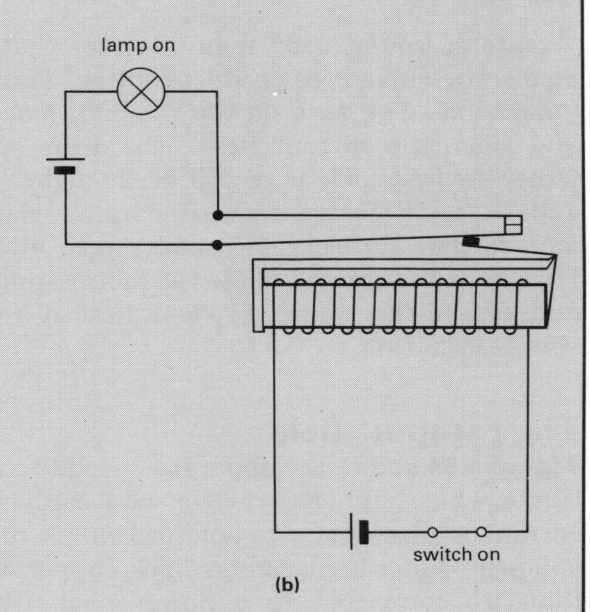

A RELAY is an electromagnetic switch. Relays are used in washing machines, cars and in many other places. The relay consists of an electromagnet, a hinged lever and a pair of contacts.

When no current flows to the electromagnet as in diagram (a), the contacts are apart and the lamp is off.

When a current flows around the electromagnet as in diagram (b), the iron plate is attracted and the lever pushes the contacts together. This completes the circuit and the lamp is turned on.

But this relay circuit does not seem very useful. Why bother to use a relay to switch the light on, when the relay itself has to be switched?

The relay is particularly useful when a small current is used to turn on a very large current. The diagram below shows a relay used to turn on a car. The car is started using a starter motor. The starter motor takes a very large current (up to 200 A) from the battery. This means that very thick wires have to be used. Rather than thread thick wires all the way from the ignition to the starter motor, a relay is used.

Try to find out more about relays. Explain, using diagrams, *two* other uses of electromagnetic relays.

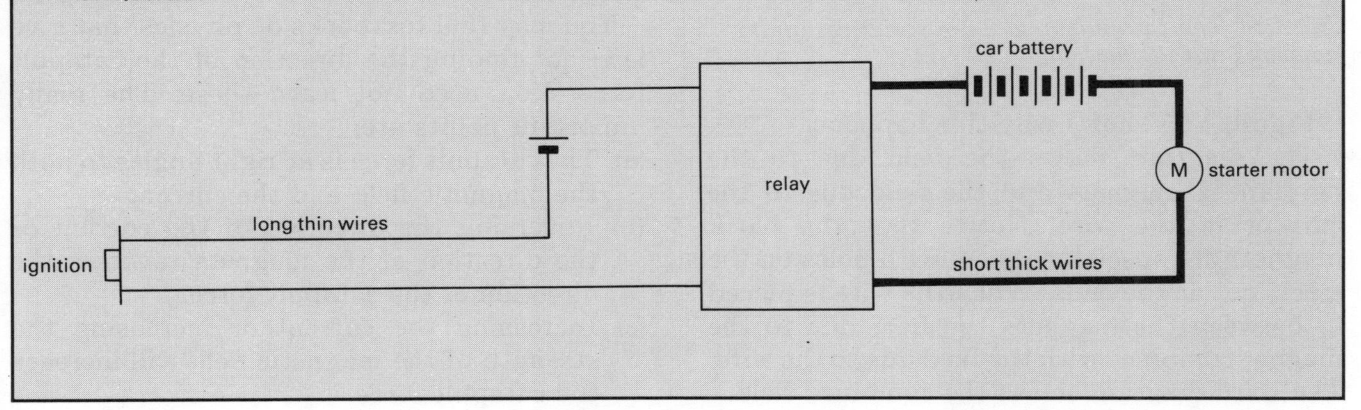

FIELDS, CURRENTS AND FORCES

We are surrounded by many devices that use an electric current to produce motion. Think of your home. You turn on the washing machine and when the current flows, the drum spins. Other devices such as record decks, hairdryers and cassette recorders also contain electric motors that turn electric energy into motion. How does the current make the motor spin? To understand this we need to look first at a very simple situation.

The catapult field

Figure 4.64 shows the apparatus used to investigate what happens when a wire carrying a current is placed in a magnetic field. A moveable wire slider is placed on thick copper wires that are connected to a power pack. Block magnets with opposite poles facing are held on either side of the wire. When the current is turned on the slider jumps sideways out of the magnetic field.

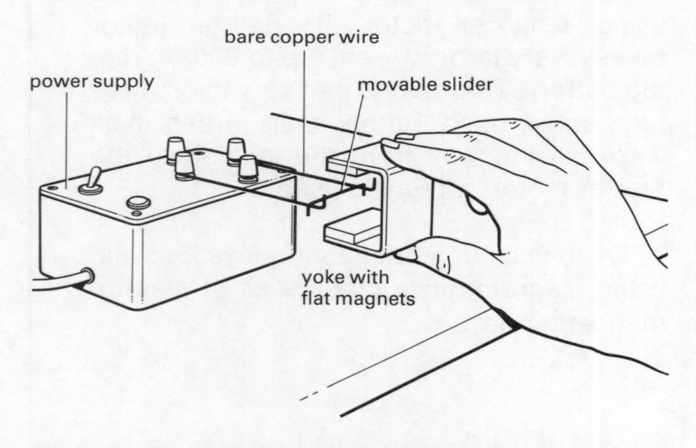

Fig. 4.64 The catapult field - the slider catapults sideways out of the field

Figure 4.65 shows why this happens.

The diagram shows the field due to the permanent magnets and the field due to the current in the wire. Notice that the block magnets are special magnets with poles on the sides, not at the ends. When the wire is placed in between the magnets the field due to the magnet combines with the field due to the wire. The third diagram shows the combined field.

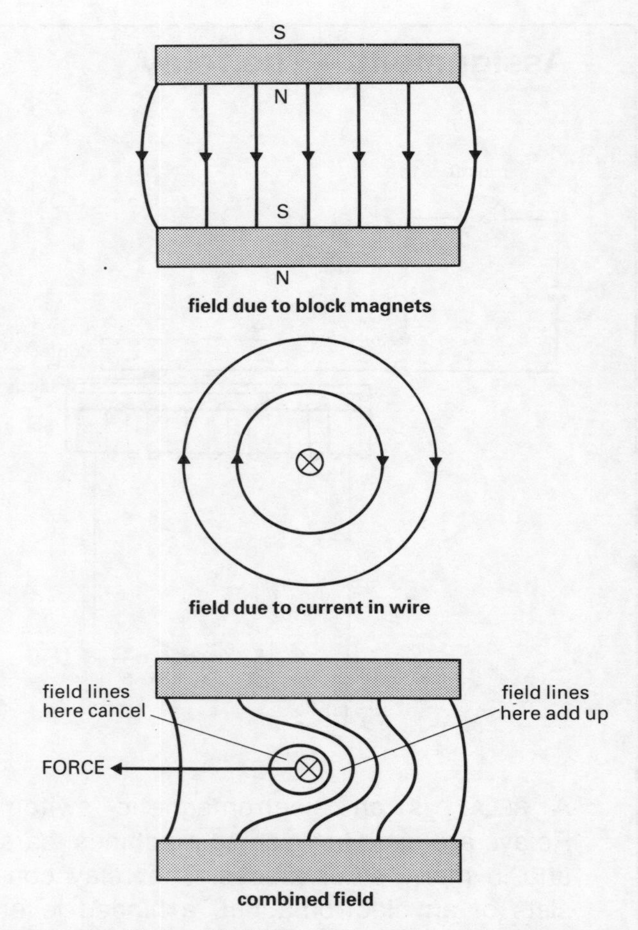

Fig. 4.65 Using field lines to explain the catapult force

On the left hand side of the wire, the two sets of field lines go in opposite directions and so cancel.

On the right hand side, the field lines go in the same direction and so add to give a greater field.

Imagine the field lines are like elastic bands. You can see that the wire would be fired out sideways. That is what happens. A force is produced on the wire by the current and the magnetic field. It is called the CATAPULT FIELD.

You may find textbooks on physics that give laws for finding the direction of the catapult force. You need not learn these. The really important points are:

(a) The catapult force is at right angles to both the magnetic field and the current.

(b) Reversing the direction of the current *or* the direction of the magnets reverses the direction of the catapult force.

(c) Increasing the current or increasing the strength of the magnetic field will increase the catapult force.

AMMETERS

It is the catapult force that makes motors spin. The catapult force can also be used to measure electric current.

Figure 4.66 shows what happens if a coil of wire with a current flowing through it is put between magnets. The diagram on the right shows the combined field. The two wires shown are those running along the long sides of the coil. Notice that because the current in these wires runs in opposite directions the catapult forces will be in opposite directions. The left hand side of the coil is pulled down; the right hand side is pulled up. The coil will be twisted or rotated in an anticlockwise direction. We can use this idea now to make a model ammeter, in Investigation 4.5 on the next page.

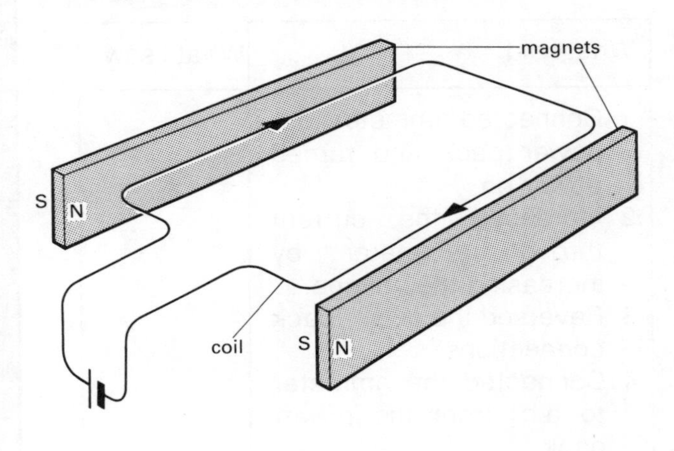

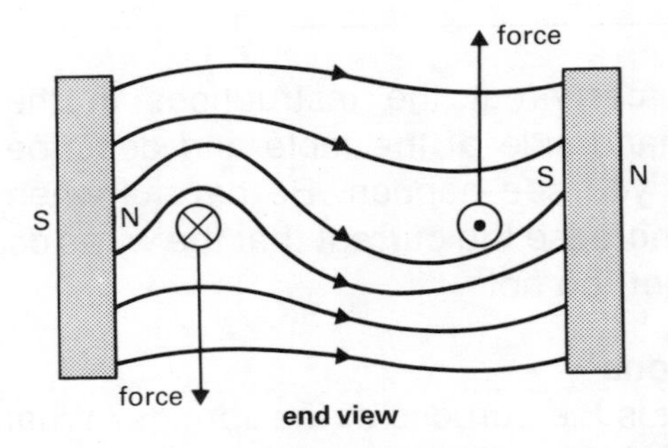

Fig. 4.66 A coil in a magnetic field - the catapult forces rotate the coil in an anticlockwise direction

THE ELECTRIC MOTOR

If you did not have the springs on the model ammeter how far would it rotate?

Look at Fig. 4.67. The catapult forces cause the coil to rotate. By the time it reaches position 3 it will no longer experience a twisting force. The top wire is pulled up and the bottom wire is pulled down. The coil stops rotating.

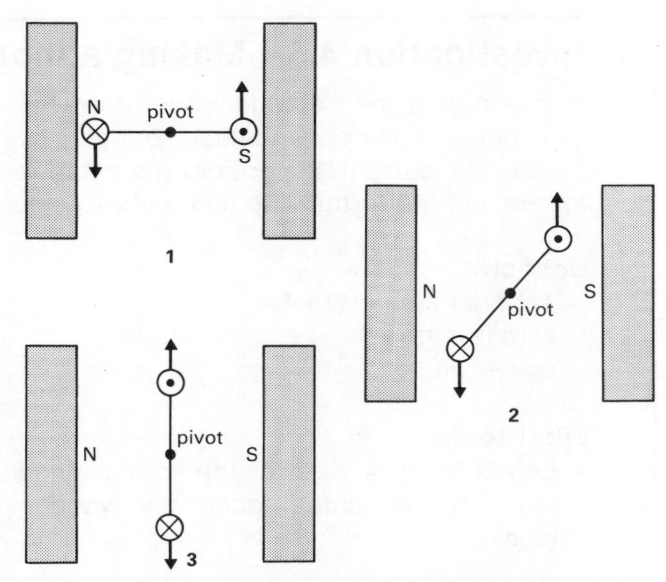

Fig. 4.67 The coil in the magnetic field will only rotate by a quarter turn then stop

In an electric motor we want the coil to spin. Look at the right hand diagram. In order to make the coil continue to turn we have got to change the direction of the force. The top must be pulled down and the bottom pushed up.

This can be done by reversing the current. Figure 4.68 shows how this is done. The wire's to the coil end on a split ring. The current is supplied to the split ring using a brushing contact. This allows a current to flow to the coil without tangling the wires. It also allows the current to change direction each half turn. A split ring is called a COMMUTATOR and the contacts that brush against it are called *brushes*.

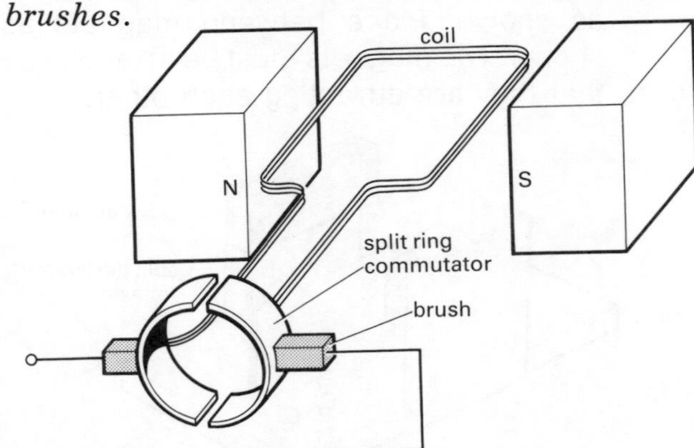

Fig. 4.68 Using a split ring commutator and brushes to keep the coil spinning

Investigation 4.5 Making a model ammeter

In this model the catapult forces turn the coil against a pair of handmade springs. The greater the current, the greater the catapult forces and the further the coil turns round.

Collect

1 model ammeter kit
wire (3 metres)
power pack

What to do

1 Leave ½ m of wire for the spring. Wind the wire ten times round the wooden block.

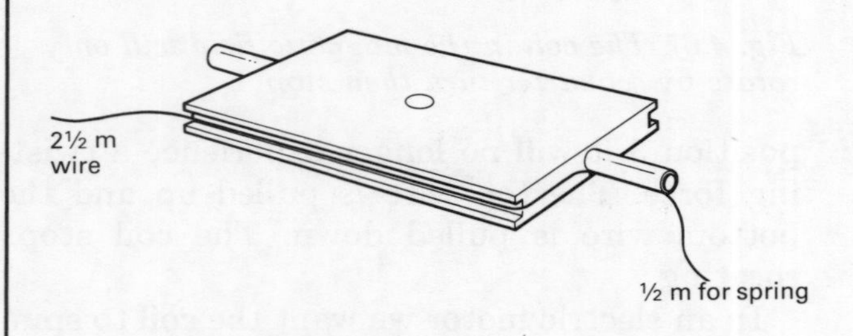

2 Wind a couple of tight turns of wire round the metal tube at each end of the block. Make opposing springs at each end of about four or five loose turns of wire.

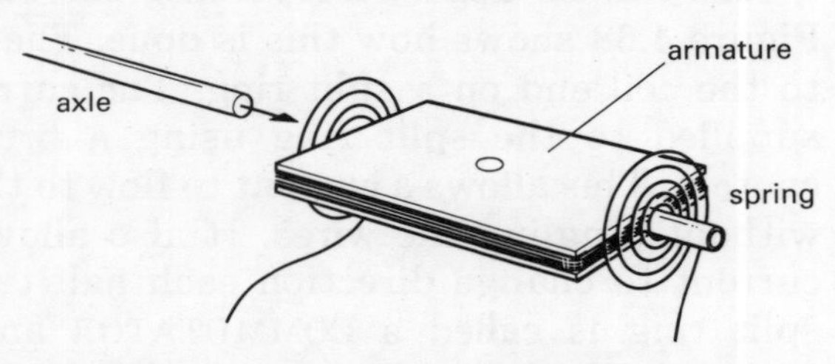

3 Assemble base with rivets and split pins as shown. Place between magnets as shown. The magnets must be arranged so that they are attracting each other.

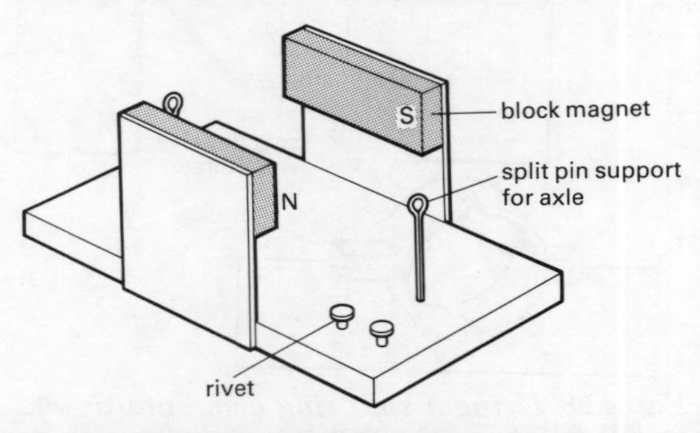

4 Hold the armature in place. Slide the axle through the split pins and the armature. Anchor ends of coil with rivets and insert a straw 'pointer' as shown.

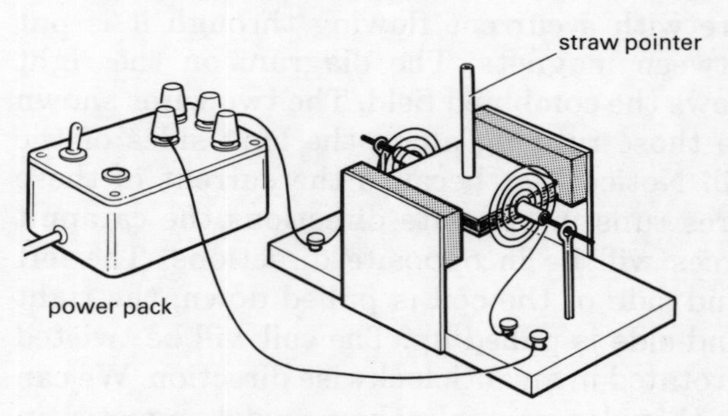

5 Copy this table.

What I did	What I saw
1 Connected ammeter to power pack and turned power on. 2 Increased the current through ammeter by increasing the voltage. 3 Reversed the power pack connections. 4 Connected the ammeter to a.c. from the power pack.	

6 Now carry out the instructions in the left hand side of the table and describe what you see happen. Be careful when you increase the current that the wires do not get too hot.

Questions

1 What is the purpose of the springs? What would happen if they were not there?
2 What would you have to do to your model ammeter to use it actually to *measure* electric current?
3 Give three reasons why your model would never be as good at measuring current as a commercial ammeter.
4 Explain why your ammeter could not be used to measure an alternating current.

Investigation 4.6 Making a model motor

You will now make a model motor using bare wire as the brushes and doubling back the ends of the coil to form commutators.

Collect

motor kit
wire ($2\frac{1}{2}$ metres)
power supply

What to do

1 Insulate the metal tube on one end of the wooden block by wrapping Sellotape round it.

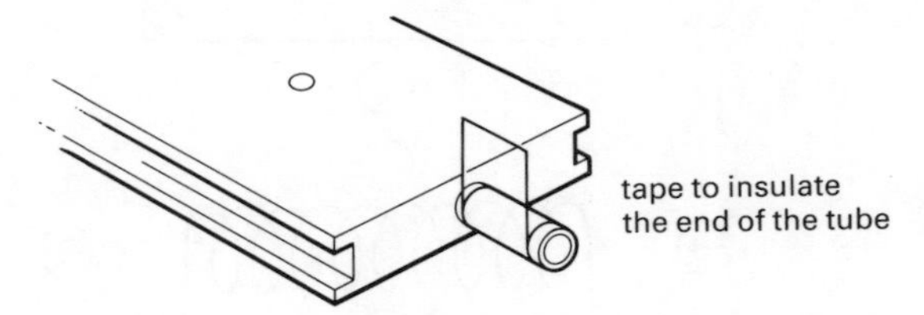

2 Wind 10 turns of wire onto the coil. Start and finish at the same end with about 10 centimetres of wire left over at the beginning and end. Wind these tightly round the metal tube to hold the coil in place.

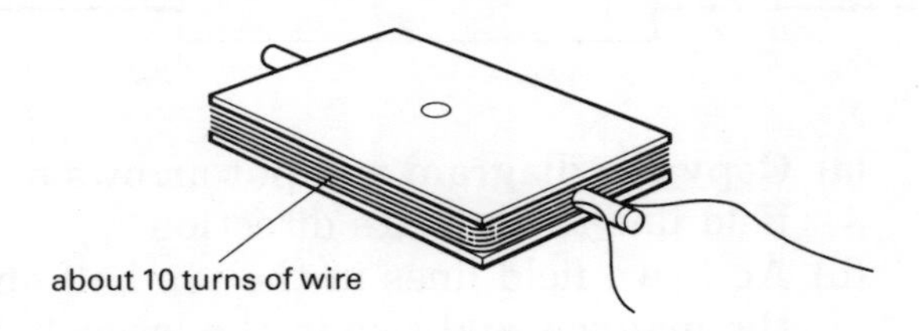

3 Make the commutator as shown above by stripping insulation off the ends of the wire and folding the wire back double. Hold the wires in place with the rubber bands, as shown.

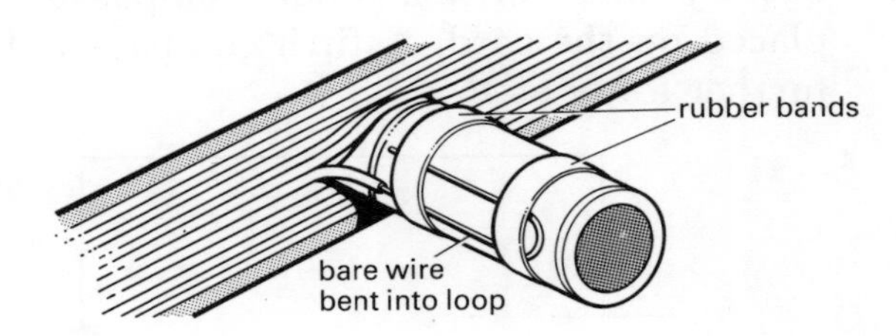

4 Place the wooden block on the base using the split pins and axle. Put the magnets in place. Make sure the coil spins freely.

5 Connect up the brushes as shown. Wind the wire round the rivets and make sure the bare ends touch the commutator. Turn on the d.c. power and give the coil a nudge to set it spinning.

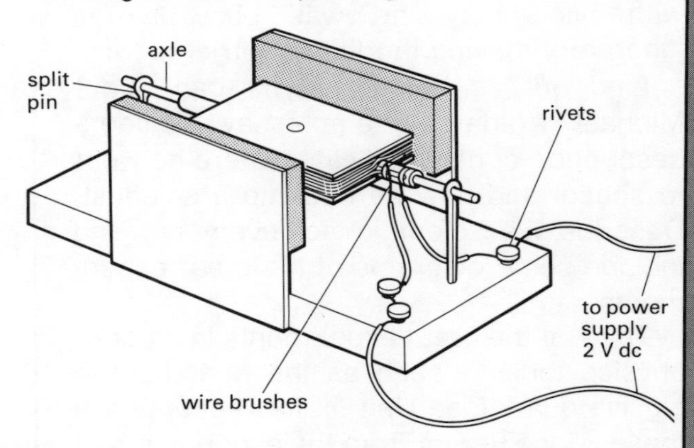

6 If nothing happens, don't be disappointed - check out these points.
(a) Have you connected up the power supply properly? You must use d.c.
(b) Have you insulated the tube on the block?
(c) Have you bared the wire for commutator and brushes?
(d) Are the brushes making contact with the commutator? (It may help to hold them on at first.)
(e) Is there anything preventing the coil from spinning?
(f) Are your magnets arranged so that they attract each other?

Questions

1 You will probably be very pleased with your motor but will notice that its motion is a bit jumpy. Suggest reasons for this. Look carefully at your commutators and brushes. Is there always a current going through the coil?
2 Can you think of a way of making a motor without permanent magnets?
3 If possible, look at the inside of a commercial electric motor. On your motor only two contacts formed the commutator. How many segments are there in the commercial motor? Why is this? How is the magnetic field produced?

Assignment – Michael Faraday

Michael Faraday was probably the greatest experimental scientist of modern time. So dedicated was he to research that in 1821 on Christmas Day, instead of enjoying his Christmas pudding at home with his family, he was busy in his laboratory making the first electric motor!

Find out as much as you can about Michael Faraday. Write an essay. Include a description of his family life, where he went to school and how he became a scientist. Describe his scientific achievements and include some diagrams of his great experiments.

Think of the great experiments in modern physics today - such as the hunt for the elusive quark (see Unit 5). In what way are these very different from the experiments of Michael Faraday?

Magnetic fields

WHAT YOU SHOULD KNOW

1 All permanent magnets have two poles, called north-seeking or N pole and south-seeking or S pole.
2 Like poles repel, unlike poles attract.
3 Magnets create magnetic fields around them. A magnetic material such as iron, cobalt or nickel will feel a force if placed in a magnetic field.
4 When an electric current flows along a wire a magnetic field is produced. The field lines form circles around the wire.
5 A coil carrying an electric current produces a field identical to that of a bar magnet.
6 The presence of soft iron in a coil greatly increases the magnetism.
7 When a wire carrying a current is placed in a magnetic field a force is produced. This force is used in the ammeter and motor.
8 The direction of the force is at right angles to both the current and the magnetic field. The force is increased if the current, the magnetic field or the length of wire is increased.

QUESTIONS

1 These scrap metals were dumped on a scrap yard.

old lead piping	a steel car body
copper car radiators	old nickel wire
an iron bath	a zinc bucket
aluminium pans	brass pans

Which of them could be lifted with an electromagnet?

2 (a) Copy these diagrams and show how the compass needles point.

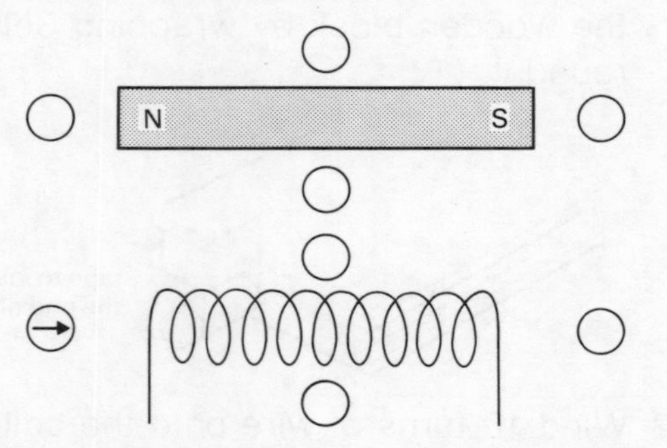

3 The diagram below shows a permanent magnet. One field line is shown.

(a) Copy the diagram and put arrows on the field line to show its direction.
(b) Add two field lines in the top half above the magnet, and two in the lower half.
(c) Where are the field lines closest together?
(d) What does this tell you about the strength of the magnetic field?

4 A piece of wire goes up through a piece of card as shown. The wire is connected to a battery and switch. Four compasses are placed on the card. A fifth compass is held further away at P.

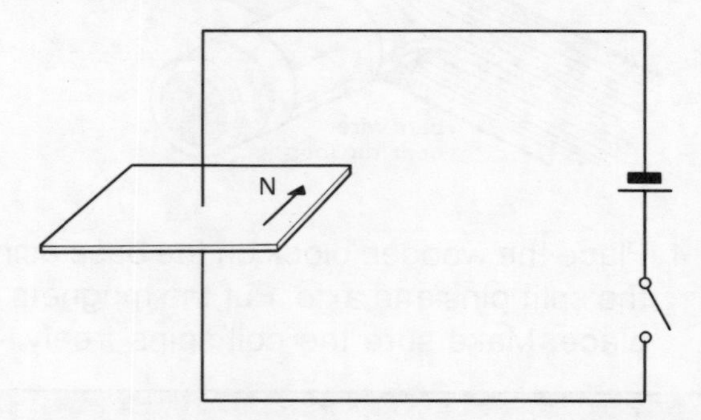

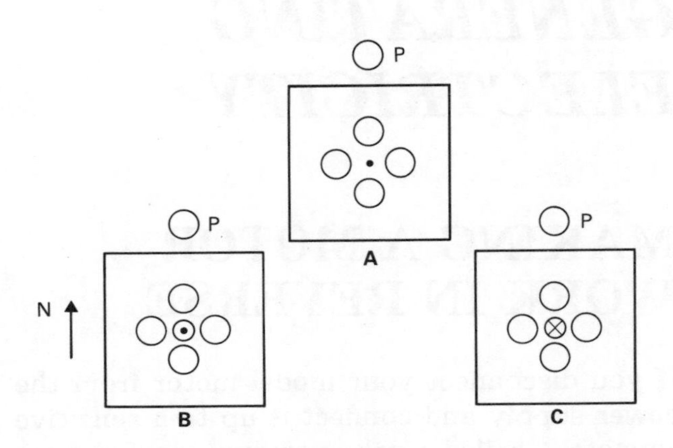

Copy diagrams A, B and C.

(a) Mark on diagram A the compass needles when no current flows.

(b) The switch is closed and a current flows up the wire. Mark on diagram B the compass needles.

(c) The battery is now reversed and a current flows downwards. Mark the new positions of the compass needles on diagram C.

5 (a) You are given a metal rod, a battery, a switch and a length of wire. Describe, with diagrams, how you would make an electromagnet.

(b) Say what would happen if the metal rod was made of (i) soft iron (ii) steel (iii) aluminium.

6 The diagram below shows a simple electric bell. Use the diagram to explain how the bell works.

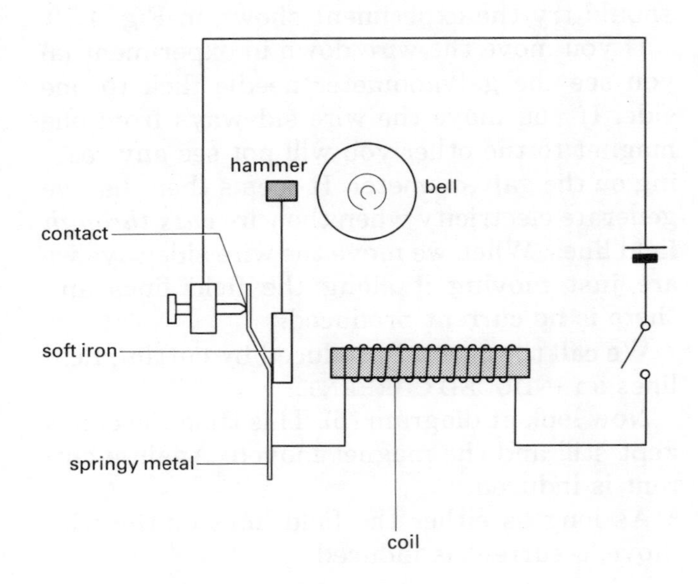

7 The diagram below shows a wire carrying a current between a pair of special block magnets. The magnets have the poles on the sides.

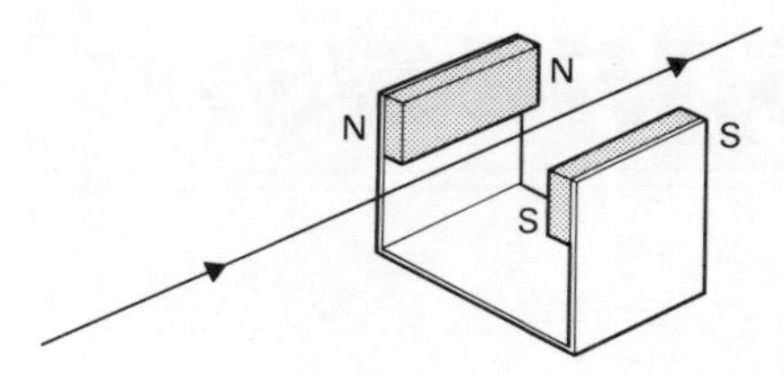

(a) Copy the diagram. Draw a few field lines to show the field due to the permanent magnets.

(b) Draw an arrow and label it F to show the direction of the force on the wire.

(c) What would happen to this force if the direction of the current were reversed?

(d) What would happen if the magnets were reversed?

(e) Give *two* ways of increasing the force on the wire.

8 The diagram below shows magnetic field lines.

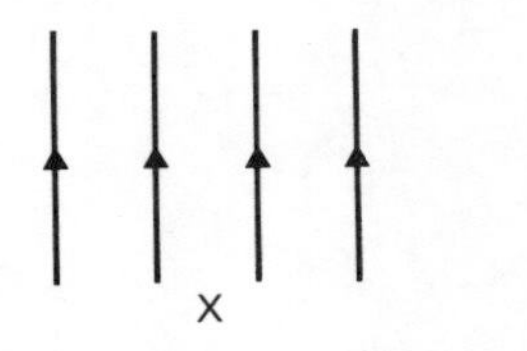

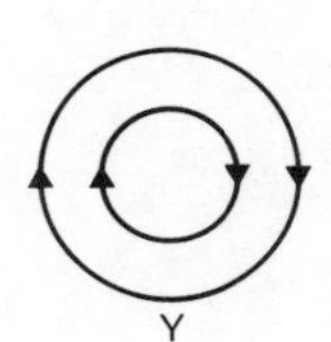

(a) Field X is produced by permanent magnets. Copy the field lines and show the arrangement of magnets that produces them. Be sure to mark the poles.

(b) Field Y is produced by an electric current. Copy the diagram and show how the current is flowing.

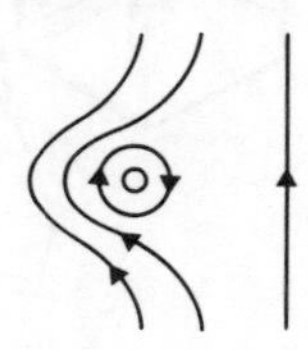

(c) If the wire is put into the magnetic field the above field pattern is produced. On your diagram use an arrow labelled F to show the force on the wire.

(d) If the magnetic field lines in X were reversed so that they went downwards, what would the combined magnetic field now look like? Draw a diagram and mark the force on the wire.

GENERATING ELECTRICITY

MAKING A MOTOR WORK IN REVERSE

If you disconnect your model motor from the power supply and connect it up to a sensitive ammeter - called a galvanometer - you can see what happens when you spin the coil: an electric current is produced and the ammeter needle moves. If you spin the coil in the opposite direction then the current is reversed. The faster the coil spins the greater the current that is produced (Fig. 4.69).

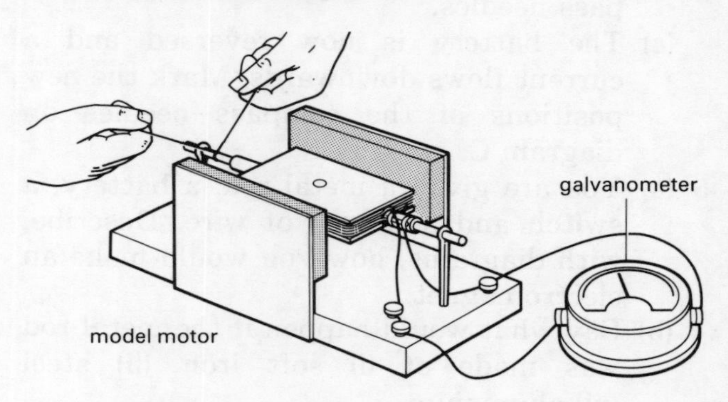

Fig. 4.69 An electric motor can act as a dynamo

The motor is now acting like a DYNAMO. To understand this dynamo effect better you should try the experiment shown in Fig. 4.70.

If you move the wire down in experiment (a) you see the galvanometer needle flick to one side. If you move the wire sideways from one magnet to the other you will not see any reading on the galvanometer. It seems then that we generate electricity when the wire *cuts through* field lines. When we move the wire sideways we are just moving it along the field lines and there is no current produced.

We call the current produced by cutting field lines an INDUCED CURRENT.

Now look at diagram (b). This time the coil is kept still and the magnet moved. Again a current is induced.

As long as either the field lines or the wire move, a current is induced.

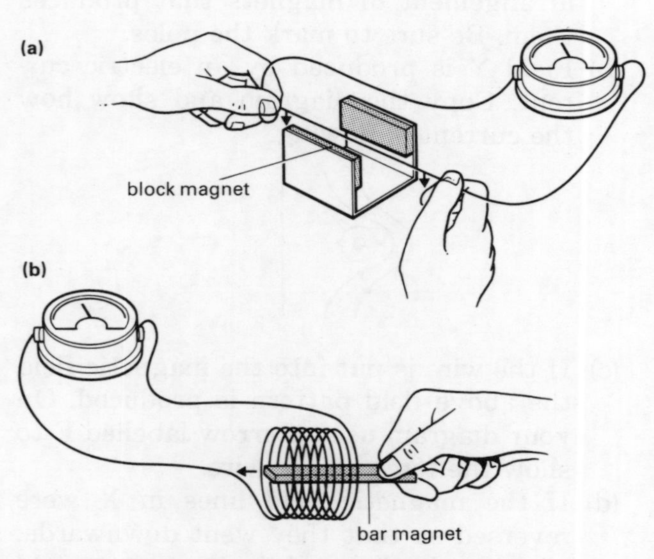

Fig. 4.70 Electric currents are induced when (a) wires cut through field lines or (b) field lines cut through wires

The size of the induced current

Before dynamos can be built, more has to be known about factors affecting the induced current. If you try out the experiments shown in Fig. 4.70 you find the following results.

(i) Reversing the direction of motion *or* the direction of the field reverses the current.

(ii) The longer the wire the greater the induced current.

(iii) The stronger the magnetic field the greater the induced current.

(iv) The faster the movement the greater the induced current.

These rules were first produced by Michael Faraday. He summarised them in an important law.

Induced current is proportional to the rate at which field lines are cut.

Assignment – The microphone and loudspeaker

A microphone is a special kind of generator. It turns sound energy into electrical energy. The diagram below shows how it works.

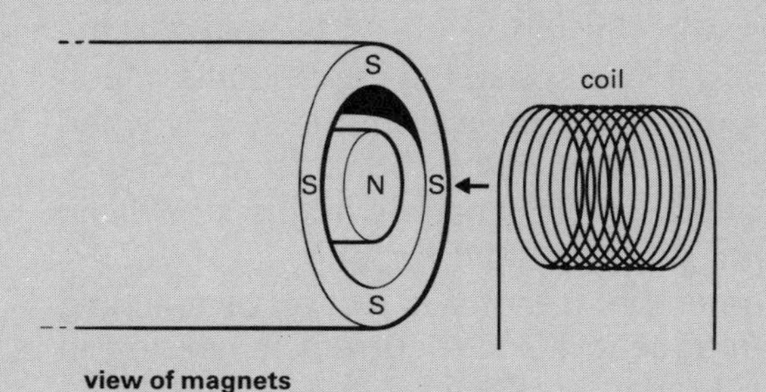

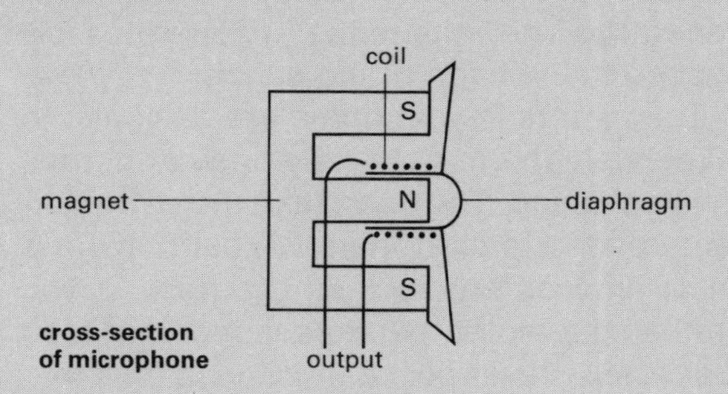

1 Copy the diagram and include the field lines produced by the permanent magnets.

2 The diaphragm is a thin metal sheet. When the sound waves reach the diaphragm they make it vibrate. The coil is joined to the diaphragm and so the coil moves backwards and forwards between the magnets. Use your knowledge of induced currents to explain what sort of current is produced in the coil.

3 If sound of a higher frequency hit the diaphragm how would the current induced be affected?

4 The microphone could be part of a public address system shown below.

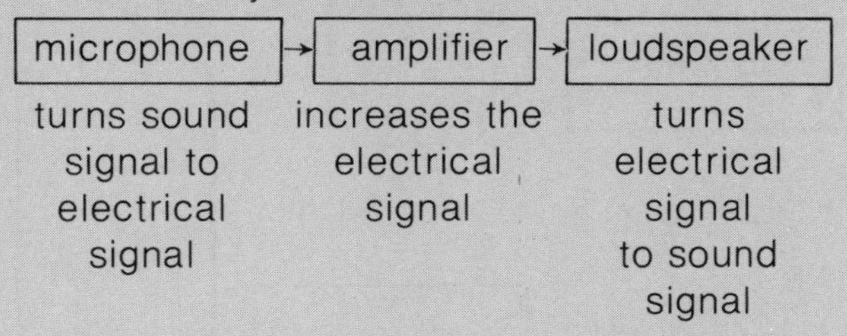

The loudspeaker does the opposite job to the microphone. Look back at the diagram of the microphone. Explain what would happen if an alternating current was sent *into* the coil.

5 The microphone could work in reverse and act as a loudspeaker-but the sound would not be very clear or loud. The diagram below shows a proper loudspeaker. Explain why it works better than the microphone in reverse.

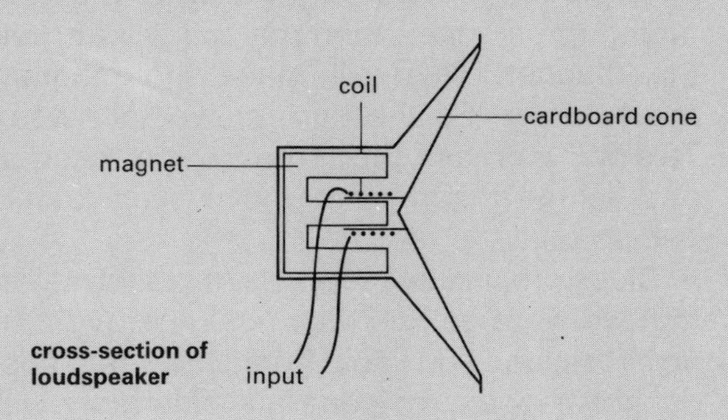

Try to get hold of an old record player or tape recorder that does not work - perhaps from a jumble sale. Take out the loudspeaker and see if you can identify the parts.

Assignment – The direction of the induced current

When a magnet is pushed into a coil, a current is induced. As the current flows it creates a magnetic field which affects the movement of the magnet. Which way will the current flow? Which pole will be produced on the side of the coil nearest to the magnet? The diagram below shows the two possibilities.

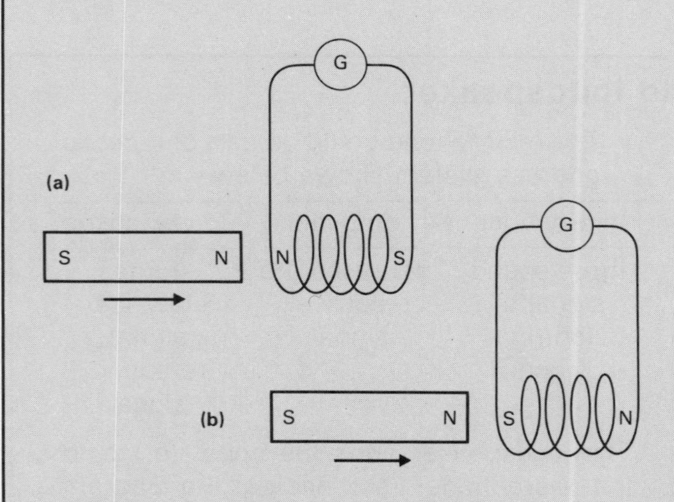

(a) If the current in the coil flows so that an N pole is formed on the side of the coil facing the magnet, then the coil will repel the magnet. A force will be needed to push the magnet into the coil. Work will need to be done on the magnet to produce the electrical energy.

(b) If the current in the coil flows so that an S pole is formed on the side of the coil facing the magnet, then the coil will attract the magnet. This will make the magnet move faster. So the current will increase. This will increase the force on the magnet and so the magnet will move even faster and so on.

Clearly (b) is unacceptable: it would mean that we were getting electrical energy from nowhere, and you know from Unit 3 that this cannot happen. To generate electricity we must do some *work*.

If we think again about the direction of the induced current, it must flow in such a direction that it produces a force that opposes the movement creating it. We can use this fact to understand an important point about electric motors. Look at the diagram below.

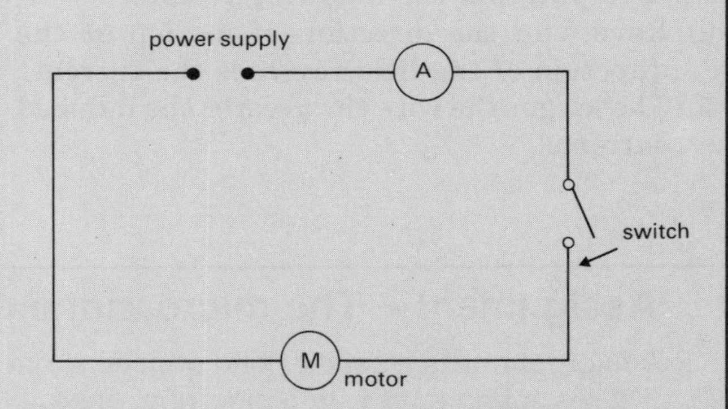

When the switch is closed a large current flows and the motor starts to spin. As the motor gains in speed the current falls. Much less current is drawn from the supply when the motor is spinning rapidly than when it just starts off. This seems a little surprising at first. Why should it happen?

Remember a motor is just a coil spinning in a magnetic field. A current is needed to make it spin, but when it spins it acts as a generator. The voltage that it generates will oppose the voltage of the supply.

The motors in your home are designed to take safely the current they draw at normal running speed. They can take the extra current needed to start them up, but only for a short time otherwise they overheat. If you are using a motor, perhaps in a drill or food processor, never put so much load on it that it jams the motor - otherwise it will burn out the coil.

With large electric motors there is a more serious problem. Special precautions have to be taken to reduce the large starting-up currents.

On snowy mornings some people burn out the windscreen wiper motor in their car by trying to clear large amounts of snow off with the wipers. Explain why you should clear the snow off *before* turning the wipers on. Write a few sentences suitable for someone who is not studying science - perhaps your parents!

ELECTRIC GENERATORS

The model dynamo shown in Fig. 4.69 produces electricity when it is spun round. The galvanometer needle moves just one way. This means that a direct current flows. If you change the motor slightly as shown in Fig. 4.71, you can make a model a.c. dynamo.

The commutator has been replaced by two slip rings, one at each end of the coil. If the dynamo is connected to a galvanometer and spun round, the galvanometer needle rocks backwards and forwards, showing that an alternating current is being generated. If the galvanometer is replaced by an oscilloscope, the spot plots a voltage against time graph. This is shown in Fig. 4.71 (b) - you can see the familiar alternating voltage of an a.c. supply.

The movement of the galvanometer needle in the experiment above is very small. When Faraday first discovered the effect early in the 19th century he was asked 'What use is a tiny little wiggle of a needle in your dusty laboratory?' Perhaps you feel the same! Faraday answered 'What is the use of a newborn child?' Faraday's child has grown up and now supplies us with the electric power for our houses and industry - electric power that makes your life so different from that of your grandparents when they were at school.

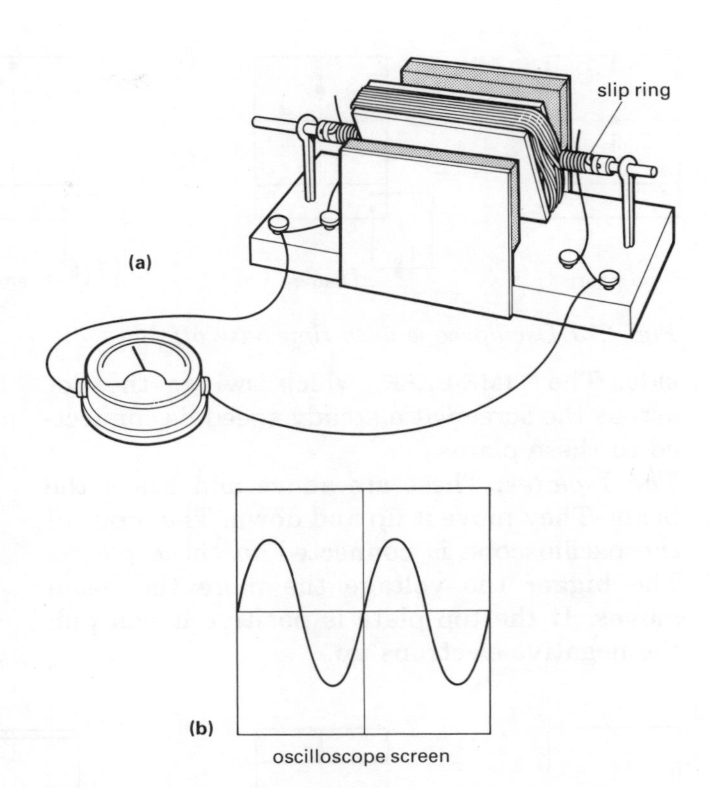

Fig. 4.71 (a) An a.c. generator.
(b) Oscilloscope trace from an a.c. generator

THE OSCILLOSCOPE — A SPECIAL KIND OF VOLTMETER

An oscilloscope is very useful for looking at voltages that change rapidly. The oscilloscope in your laboratory probably looks like Fig. 4.72 (a).

An electron gun shoots a beam of electrons at a screen (Fig. 4.73 (b)). When the electrons hit the screen they make it glow and you see a green dot. There are two sets of plates that move the dot.

The X-plates: These are either side of the electron beam. They move the beam from side to

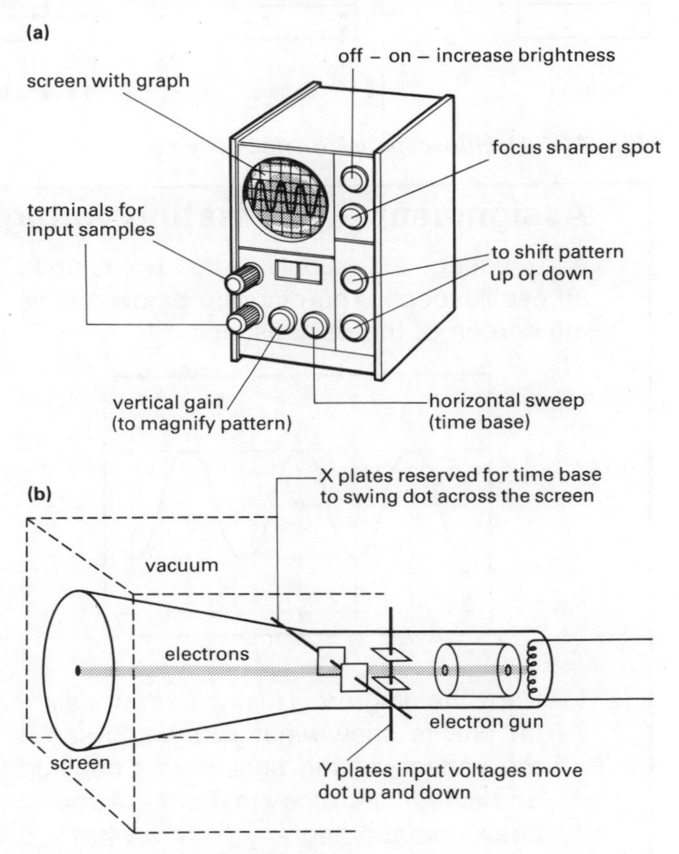

Fig. 4.72 (a) The oscilloscope controls.
(b) Inside the oscilloscope

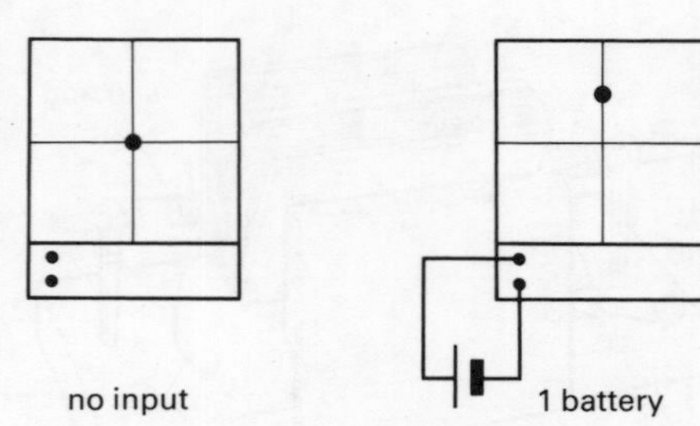

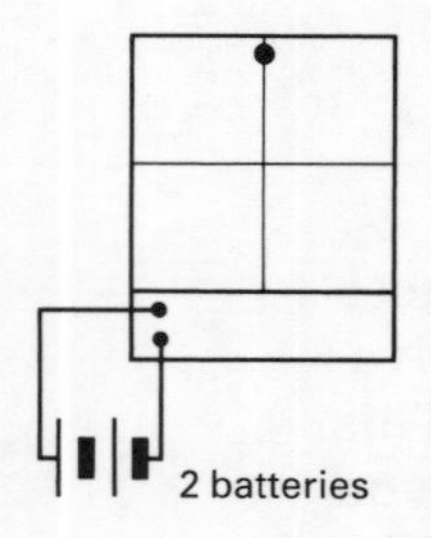

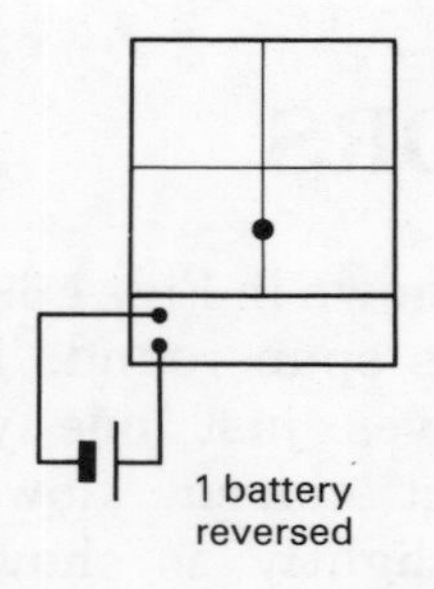

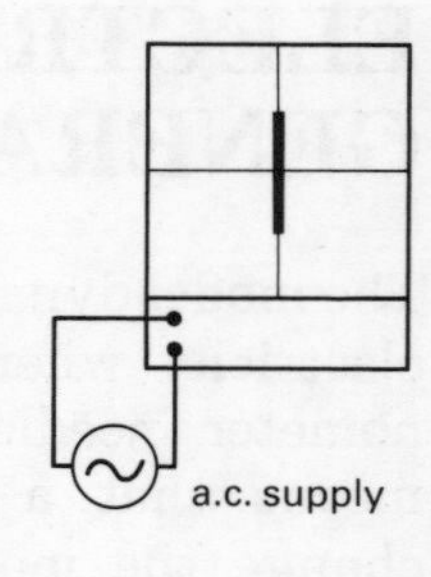

Fig. 4.73 Oscilloscope with time-base off

side. The TIME-BASE, which swings the dot across the screen at a steady speed, is connected to these plates.

The Y-plates: These are above and below the beam. They move it up and down. The input of the oscilloscope is connected to these plates. The bigger the voltage the more the beam moves. If the top plate is positive it will pull the negative electrons up.

Figure 4.73 shows what happens when the time-base is turned *off* and different voltages are applied to the oscilloscope.

You can see that the oscilloscope is acting like a voltmeter. The bigger the voltage the more the dot is moved. The voltage from the a.c. supply moves the dot up and down so quickly it looks like a line. If the time-base is turned on the following traces are obtained (Fig. 4.74).

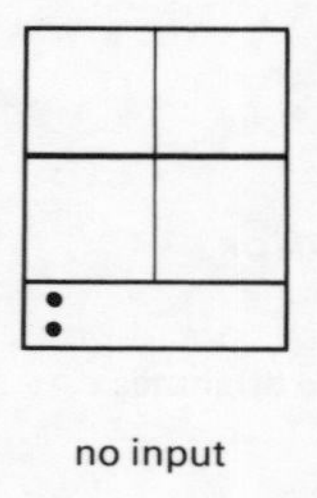

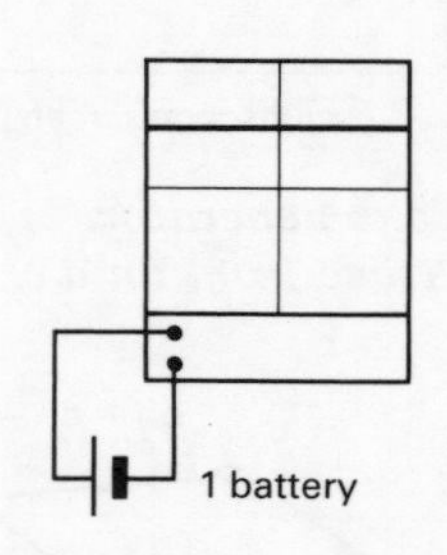

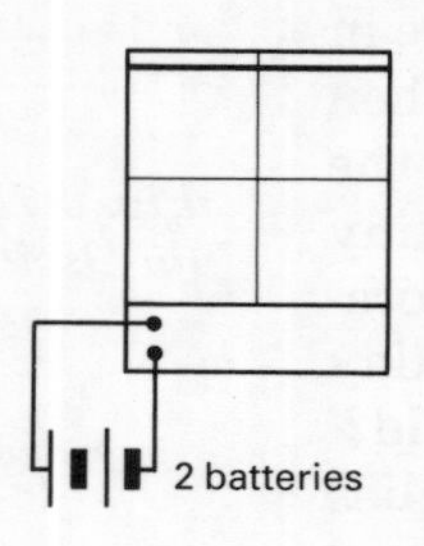

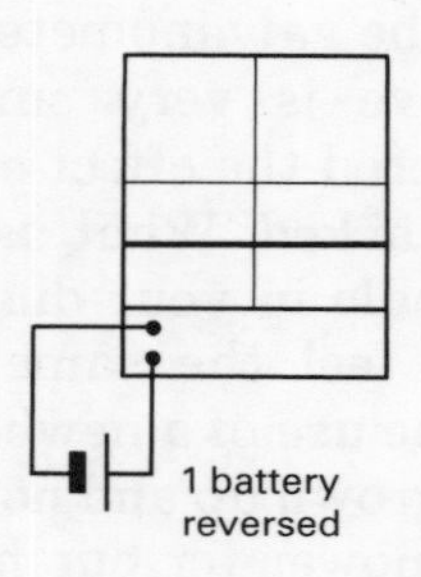

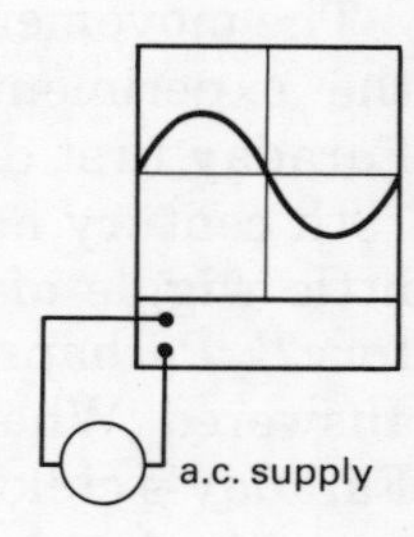

Fig. 4.74 Oscilloscope with time-base on

Assignment – Alternating voltages on the oscilloscope

A low voltage a.c. supply is connected up to an oscilloscope. The diagram below shows the screen of the oscilloscope.

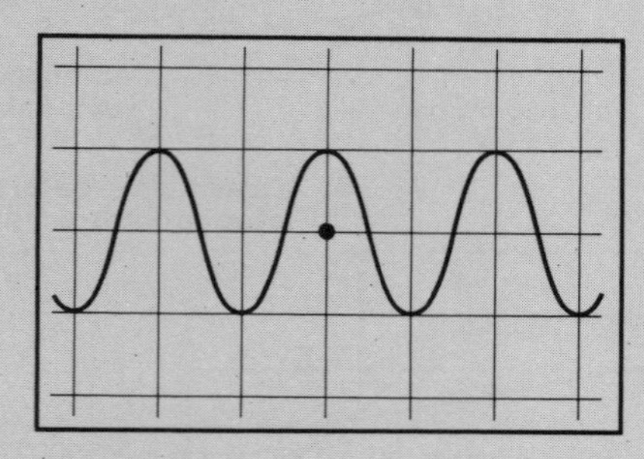

1 Copy the diagram. Next to it draw a similar one to show what you would see if the voltage of the supply was doubled and everything else stayed the same.

2 Draw another diagram to show what you would see if the original voltage was put in and the time-base of the oscilloscope adjusted to twice the speed, i.e. the dot is moved twice as fast across the screen.

3 The original diagram above was obtained when the time-base was set at $\frac{1}{100}$ s per square.

(a) Look carefully at the diagram. How far has the dot moved sideways while the a.c. supply has produced 1 full wave?

(b) How long does 1 wave then take?

(c) How many waves does the a.c. supply produce in 1 second?

(d) What special name is given to the number of waves per second? Does the answer to (c) surprise you?

Look carefully at the a.c. input. The a.c. voltage moves the dot up and down and the time-base moves the dot sideways. The trace on the screen is a sort of voltage-time graph.

POWER STATIONS

A generator in a modern power station supplies about 2000 million watts of power at 25 000 V. The generator works on the same principle as your model but there are important differences. In a real generator there is not just one coil but a large number side by side. These coils are kept stationary and the magnets rotate inside the coils. Figure 4.75 shows the magnets being pushed inside the coils.

Fig. 4.75 The magnets being pushed inside the coils of a generator

Energy transfers in the power station

The energy transfer that takes place in the generator is:

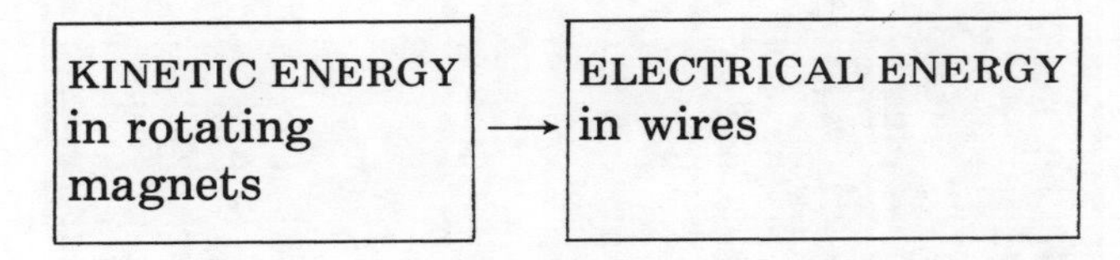

What produces the kinetic energy of the magnets? There are three main ways that this can be done.

(a) *Using chemical energy from fossil fuels* (Fig. 4.76) Coal is burned and boils water to produce steam. The steam is blasted against the fins of a turbine. This spins the magnets in the generator.

(b) *Using nuclear energy* (Fig. 4.77) Nuclear fuel produces heat which is used to boil water. The steam is then used in the same way as in the coal-fired power station.

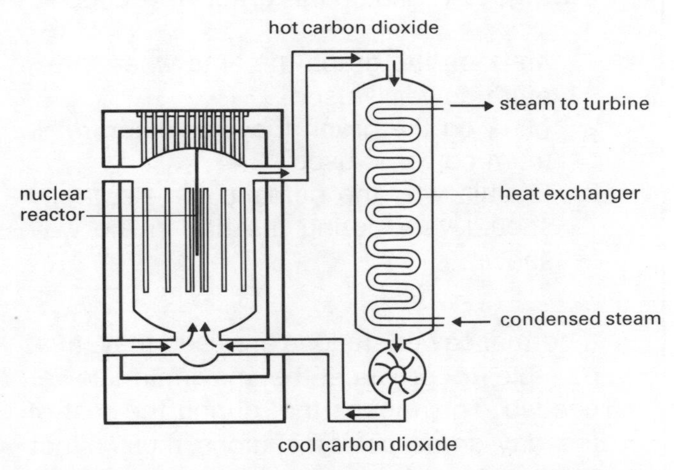

Fig. 4.77 Using nuclear energy - carbon dioxide is used to get the thermal energy from the reactor to the water

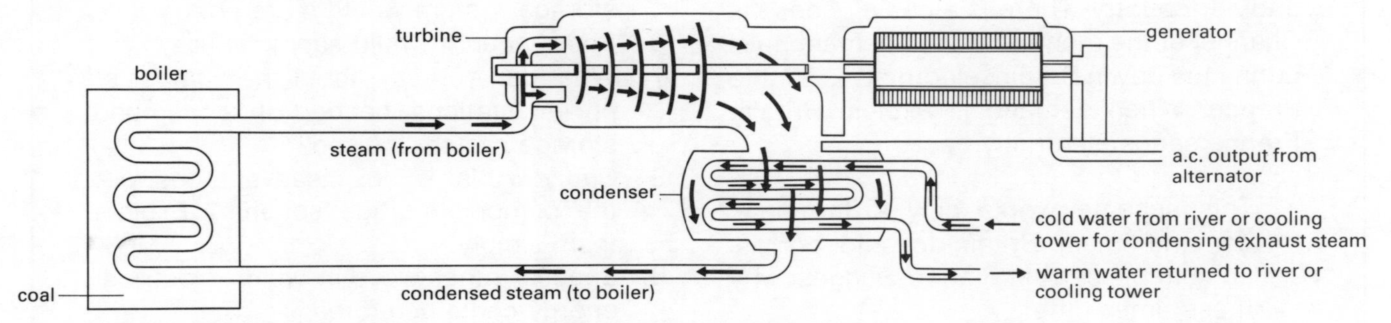

Fig. 4.76 Using chemical energy in coal to produce electrical energy

(c) *Using the potential energy of water* (Fig. 4.78) Water from a high reservoir falls down pipes and then hits the fins of turbines. This spins the magnets in the generator. Electricity generated like this is called hydroelectric power.

You have already met the problem of the energy crisis in Unit 3. Fossil fuels like oil are running out; how will we produce electricity in the future? With very few sites suitable for

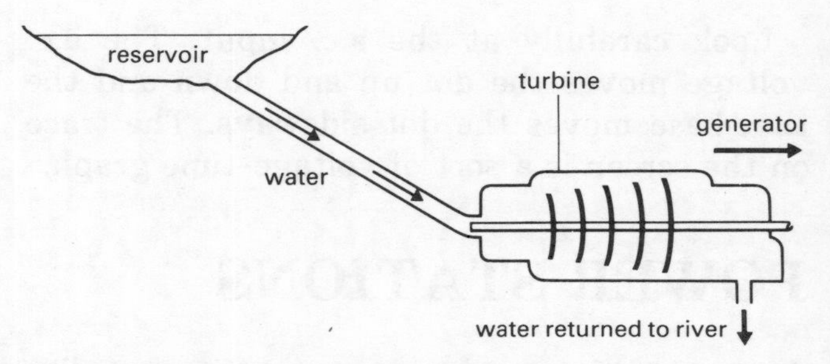

Fig. 4.78 Using the potential energy of water to turn the generator

Assignment – Storing electrical energy

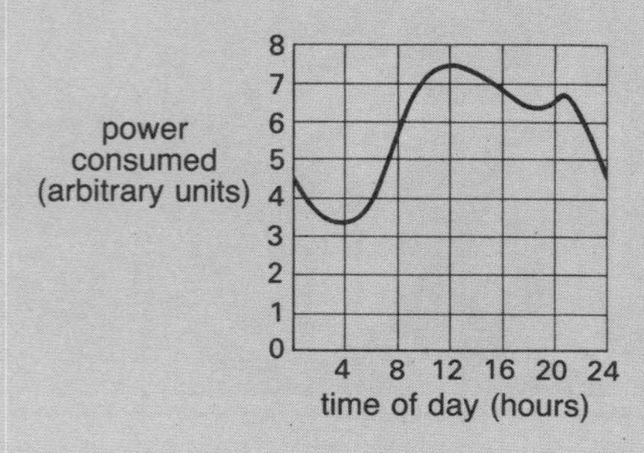

The graph above shows how the amount of power consumed in Britain varies during an average day. Sketch this graph in your book.

1 Mark on the graph the time when maximum power is used.
2 Mark on the graph the time when minimum power is used.
3 Explain why the demand for electricity should vary during the day in the way shown.

If normal power stations are used they must be able to produce the maximum power needed. This means that during the rest of the day some power stations must either produce less electricity or shut down. This is inefficient and expensive. One way round this problem is to swap the electricity with another country. There is a cable under the Channel at the moment. When demand falls during the day in Britain, electricity is sent to France. When demand is high in Britain, France sends electricity back.

4 This scheme works only if the peak demand time in Britain does not correspond to the peak in France. Suggest why these times differ.

Look back at the paragraph on pumped storage schemes in Unit 3, page 130.

The pump/turbine area of the Dinorwig power station, a pumped storage scheme in Wales (courtesy of the CEGB)

5 On your graph, show when a pumped storage system would store energy.
6 Show when it would supply energy.
7 What can you say about the number of power stations needed if a pumped storage system is used?
8 Can you think of any disadvantages with the pumped storage scheme? Explain them simply.
9 Suggest other ways in which electrical energy could be stored.

hydroelectric power stations in countries like Britain, the choice is between continuing with fossil fuels or using nuclear energy. Some people are very worried about using nuclear energy, others think that it is the obvious way forward. You will meet this issue in Unit 5 when you have studied nuclear reactors in more detail.

Alternative energy sources

There are other suggestions for producing electricity that you have met in Unit 3. These methods are being investigated at the moment. They look unlikely to be able to produce enough electricity for a small, highly populated, industrial country like Britain. They may be much more useful in Third World countries.

GENERATING ELECTRICITY USING ELECTROMAGNETS

Look at Fig. 4.79. The C-cores have coils of wire around them. The coil called the *primary* is connected to a battery and switch. The *secondary* is connected to a galvanometer. If you set up this apparatus you should find the following things:

current turned on in primary - momentary flick of galvanometer;

steady current in primary - zero reading in the galvanometer;

current turned off in primary - momentary flick of galvanometer needle in opposite direction.

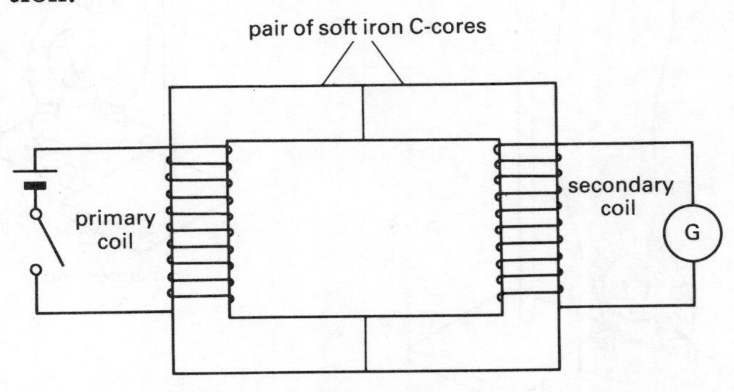

Fig. 4.79 Generating electricity using electromagnets. The galvanometer only gives a reading when the switch in the primary circuit is turned on or off

No current can flow from the primary coil to the secondary coil - the wires are insulated. How is the electricity produced in the secondary coil? Why is it only produced when the primary current is turned on and off? Look at Fig. 4.80.

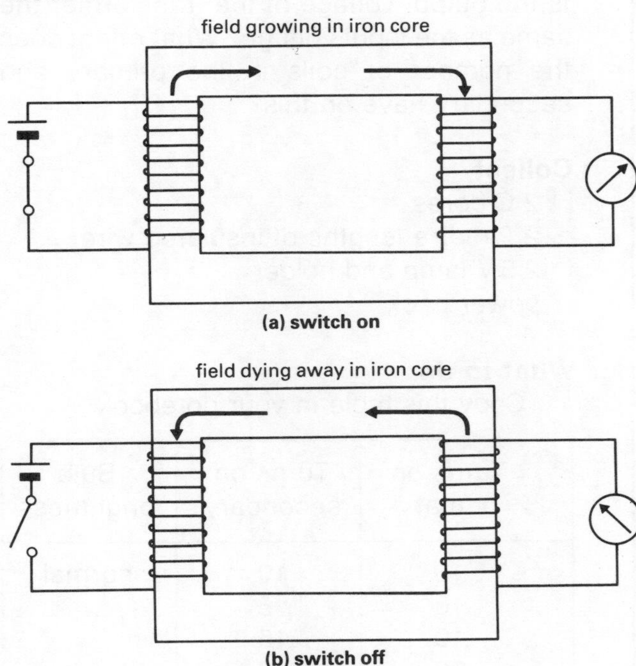

Fig. 4.80 The changing fields in the C-cores when the current is turned on and off

When the current in the primary is turned on, a magnetic field grows in the C-cores. The field lines are trapped in the iron and surge through the secondary coil. Whilst the current is growing, the field is growing and cutting through the secondary coil, inducing a current.

When the current in the primary coil is steady the magnetic field is steady. Nothing is moving so no current is induced. When the current is turned off as in (b) the field lines die away. The magnetic field is moving: it is cutting the coil in the opposite direction. A current is induced in the secondary, but in the opposite direction to the one induced before.

The transformer

In the experiment shown in Fig. 4.80 a current was produced in the secondary only if the current in the primary was changing. If we could have a constantly changing current in the primary coil we would always have a current in the secondary.

Investigation 4.7 How does the number of turns on a transformer coil affect the voltage?

Is the output voltage of the transformer the same as the input voltage? What effect does the number of coils in the primary and secondary have on this?

Collect

2 C-cores
2 2-metre lengths of insulated wire
2.5 V lamp and holder
power pack

What to do

1 Copy this table in your notebook.

Turns on primary	Turns on secondary	Bulb brightness
10	10	normal
10	5	
10	15	
20	20	
20	30	

2 Set up this equipment. Set the power pack at 3 V a.c.

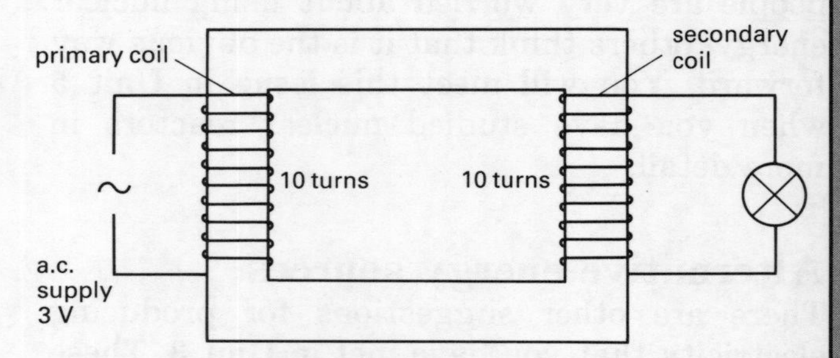

Put 10 turns on the primary, 10 turns on the secondary. Notice the bulb brightness. We will call this 'normal'.

3 Change the number of turns on the primary and secondary coils as shown in the table. Note the bulb brightness. Complete the table using the words 'normal', 'bright' or 'dim'.

4 Copy and complete this paragraph.

If there are more turns on the secondary than the primary, the bulb is __________. This means that if there are more turns on the secondary than on the primary, then the transformer __________ voltages.

If an a.c. supply is connected to the primary coil the current is constantly changing backwards and forwards. This means that the magnetic field lines trapped in the C-cores are also constantly changing direction. This will produce an a.c. in the secondary coil. An alternating current in the primary is transformed into an alternating current in the secondary. The device is called a TRANSFORMER.

Transformers in the home

You probably have a number of transformers in your home. Many devices run at 9 V yet can be plugged into the mains. It is better to have mains electricity than batteries because it is cheaper. The device then must have a transformer to reduce the voltage from 240 V to 9 V. You may be able to spot the transformer in your cassette recorder, record player or computer.

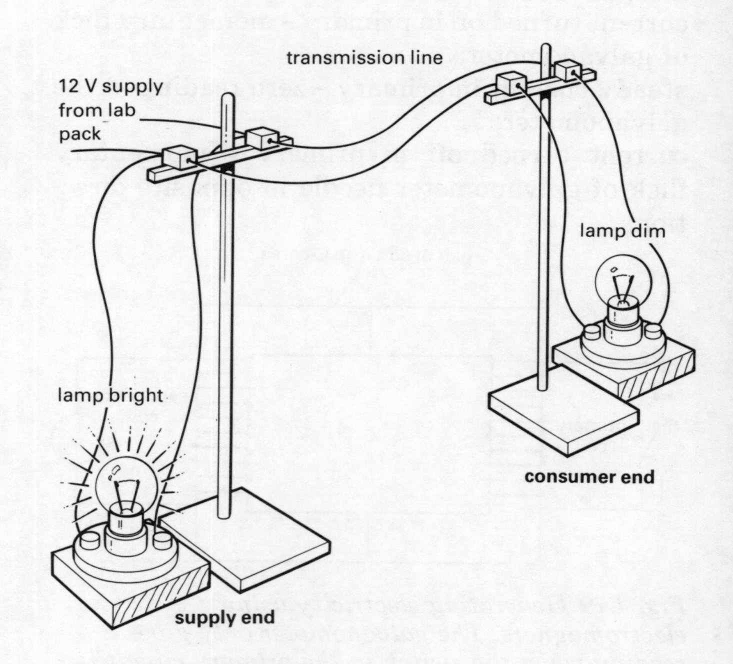

Fig. 4.81 A model power line

Assignment – Transformers

A pupil does Investigation 4.7 but decides she wants to find out more. She uses a voltmeter to measure the input and output voltages to the transformer. She also uses a transformer with more coils and uses a higher voltage than in the original experiment. Her table of results looks like this. (One of the results is incomplete.)

Turns on primary	Turns on secondary	V in	V out
100	100	20 V	20 V
100	50	20 V	10 V
100	150	20 V	30 V
200	200	20 V	20 V
200	300	20 V	30 V
100	200	10 V	20 V
400	600	20 V	

1 What can you say about the input and output voltages if the number of turns on the primary is the same as the number of turns on the secondary?

2 What happens if there are twice as many turns on the secondary as on the primary?

3 The pupil thinks she sees a clear pattern in the results and writes down

$$\frac{\text{volts out}}{\text{volts in}} = \frac{\text{turns on secondary}}{\text{turns on primary}}$$

Check her results to see if they fit this pattern.

4 Write down the pattern and use it to complete the table of results.

Transferring electrical energy efficiently

Cables supported by pylons carry electricity around the country. These cables carry the electricity at a very high voltage - around 250 000 V. Why is electricity transmitted at these very high voltages? You can see what would happen if the electricity was transmitted at low voltage by looking at the model in Fig. 4.81.

The model shows there are two main problems.

(a) The lamp at the far end of the line glows dimly - there is not enough voltage to make it work properly. Energy has been lost in the transmission wires.

(b) The wires get warm. Energy is wasted as heat in the wires.

These problems can be overcome by sending the power down the lines at a higher voltage. By sending electrical energy at high voltage, less energy is lost. Figure 4.82 shows the voltage of the electricity supply from the power station to the consumer.

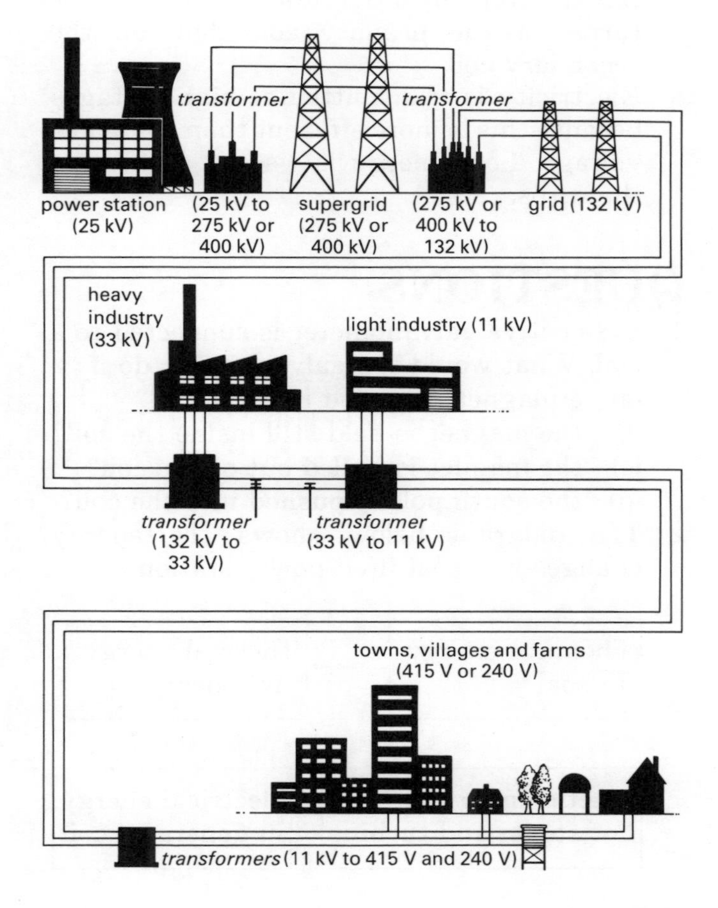

Fig. 4.82 Getting electrical energy to your home - the transformers produce a high voltage so that the energy can be transferred efficiently

WHAT YOU SHOULD KNOW

1 An electric current is induced if a wire cuts through magnetic field lines.
2 The size of the induced current is proportional to the rate at which the field lines are cut.
3 An oscilloscope is a voltmeter. It can measure the voltage of a rapidly changing supply.
4 A transformer is made of two coils called the primary coil and secondary coil, wound on a soft iron core.
5 The alternating current in the primary coil produces a changing magnetic field in the core which induces an alternating current in the secondary coil.
6 A transformer that increases voltage is called a step-up transfer. It has more turns on the secondary coil than on the primary coil.
7 A transformer that decreases voltage is called a step-down transformer. It has more turns on the primary coil than on the secondary coil.
8 Electricity is transmitted at high voltages because this is more efficient than using low voltage. Less energy is wasted as heat in the wires.

QUESTIONS

1 A sensitive galvanometer is connected to a coil. What would the galvanometer do if
  (a) a magnet is moved into the coil?
  (b) the magnet is held still inside the coil?
  (c) the magnet is pulled out of the coil?
  (d) the south pole is pushed into the coil?
2 The diagram below shows the energy changes in a coal fired power station.

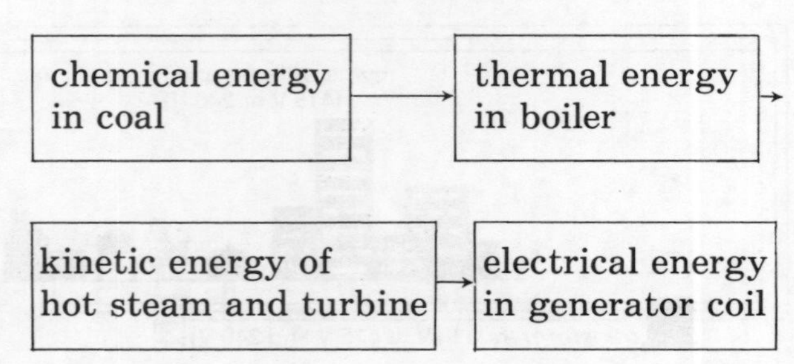

Draw similar diagrams for
  (a) a nuclear power station,
  (b) a hydroelectric power station.
3 Look back at Fig. 4.79 on page 217.
  (a) Describe carefully what is seen on the galvanometer when the switch is closed.
  (b) Explain how a current flows in the secondary coil when no current actually flows around the soft iron core.
  (c) Describe what you would see if the switch was opened again.
  (d) The galvanometer is now replaced by a lamp. What could be done to the primary circuit to make the lamp light?
4 (a) Draw a diagram of a transformer and label the parts.
  (b) What is the difference between a step-up and a step-down transformer?
  (c) A transformer is needed to change 240 V mains supply into a 12 V supply for recharging a battery. The primary coil has 1000 turns. How many turns are there on the secondary?
5 Describe three uses of transformers.
6 Electricity is transmitted around the country at high voltages – up to 400 000 volts. Explain why these high voltages are used.
7 The diagram below shows an oscilloscope with the time-base on. Copy the diagram and next to it draw similar diagrams showing what you would see if (a) a battery and (b) a low voltage a.c. supply were connected to the oscilloscope.

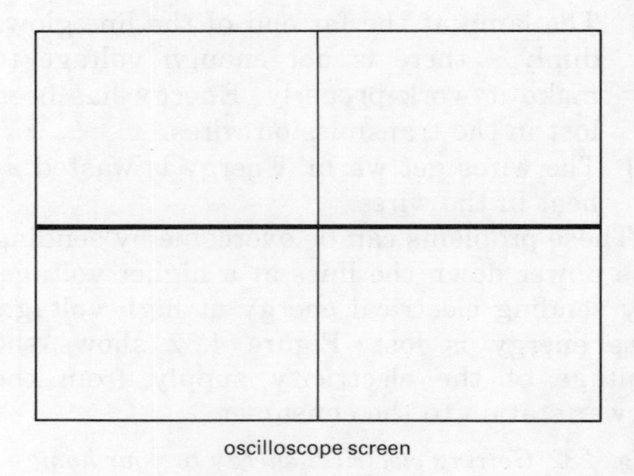

oscilloscope screen

5. What Substances are Made of

Fig. 5.1 Life without man-made chemicals would be difficult - and cold!

ATOMS AND MOLECULES

The two pictures in Fig. 5.1 show what life might be like without man-made chemicals. Think how you would be affected. Do you wear plastic shoes? Do you wear clothes made from synthetic fibres? Do you use records or tapes or electric cables? We take these things for granted, but we only have them because we have learnt how to make and use them.

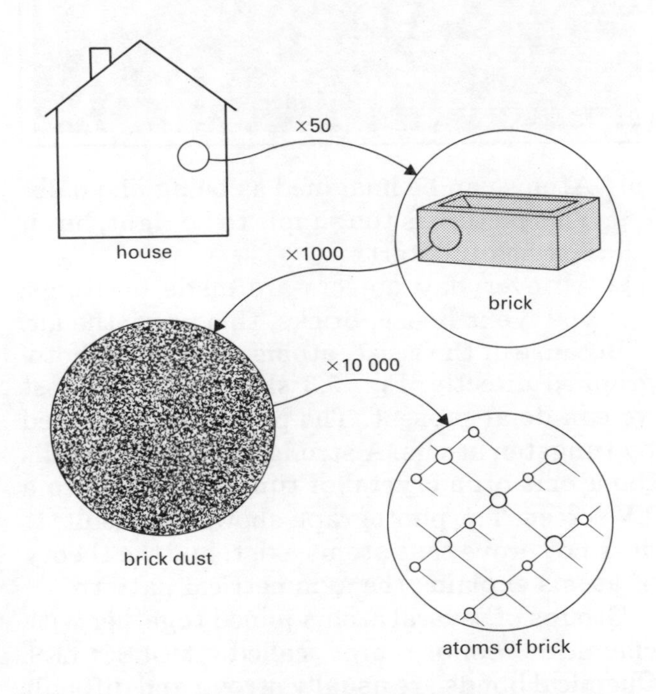

Fig. 5.2 How you would need to magnify a house to see the atoms it is made of

Scientists and engineers need to know what substances are made from. In this unit you will find out the story behind substances like plastics - what they are made from, how they differ, what we know about them.

Atoms and molecules are the building blocks of all substances. ATOMS are very tiny. Look at Fig. 5.2. The total magnification needed to see single atoms is enormous.

Here is some information about atoms:

(a) They are very small. Two million sulphur atoms in a row would measure about 1 mm. The brick in the diagram would contain about 5 000 000 000 000 000 000 000 000 atoms.

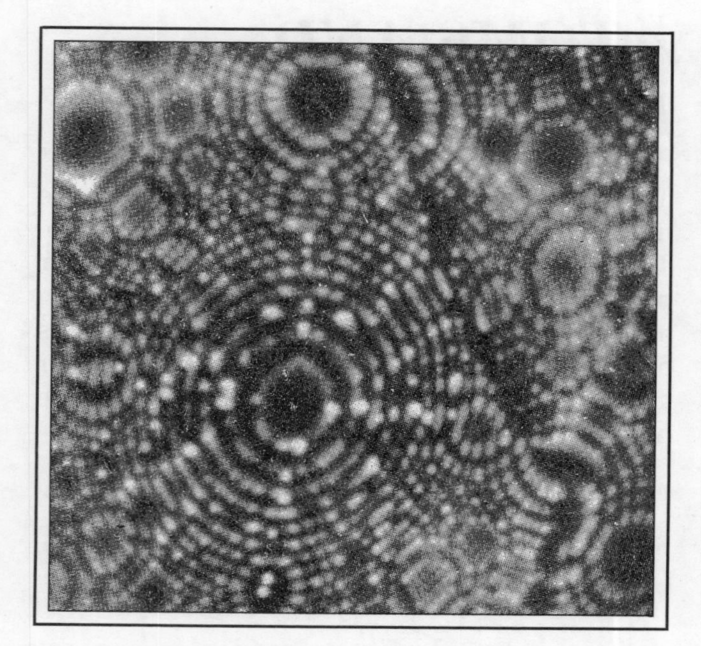

Fig. 5.3 Field ion microscope picture of a crystal of tungsten

(b) Atoms can be imagined as being like balls. This picture is too simple to be right, but it is a helpful start.

(c) All everyday objects are made of atoms: you, your house, bricks, this page, the air.

Because of their size, atoms cannot be photographed directly. Fig. 5.3 shows you the best we can do at present. The pattern is produced by tungsten atoms. A strong electric field pulls the atoms off a crystal of tungsten and onto a TV screen. The photograph shows the result. It does not prove that atoms exist, but the theory of atoms explains the symmetrical pattern.

Groups of several atoms joined together with chemical bonds are called MOLECULES. Chemical bonds are usually strong and difficult to break. Most atoms do not exist on their own; they are joined together as molecules. For example, sugar, oxygen and cornflakes are all made of molecules which contain two or more atoms.

EVIDENCE FOR SMALL PARTICLES

Atoms and molecules are so small that it is hard to find evidence that they really exist. This page looks solid. There is no sign that it is made of lots of small particles. However there are some observations that are best explained by using the ideas of atoms and molecules.

Investigation 5.1 The movement of smoke particles

Collect

power pack
smoke cell
microscope
wax straw
cover slip

What to do

1. Connect the smoke cell to the power pack.
2. Adjust the microscope to give a low magnification (× 40 is best).
3. Put the smoke cell on the stage of the microscope. Adjust the focus control so that the lens is about 1 cm above the cell.
4. Put smoke in the cell by lighting one end of the drinking straw and letting the smoke fall from the other end into the cell. Put the cover slip over the cell.
5. Quickly move the cell under the microscope lens and adjust the focus. You are looking for tiny moving flashes of light.
6. When you have the smoke in focus, watch the movement of one speck carefully. If possible, try this with a more powerful objective lens. What you can see is light reflected off the smoke particles.

Questions

1. Describe the movement of one smoke particle.
2. You should have seen two movements: one is the particles drifting around. This is caused by air currents. What is the other movement? What could cause it?
3. A brick is hung on a string in a room with no air currents. Will the brick move? Explain your answer.

Investigation 5.2 Diffusion

When a person wearing perfume enters a room, the smell gradually spreads round the room. This investigation will help you to explain why.

Collect

a beaker (250 cm^3)
some coloured liquid (ink or potassium manganate(VII).
a syringe (10 cm^3)

What to do

Half-fill the beaker with water. Very carefully squeeze the coloured liquid from the syringe into the bottom of the beaker. Try not to disturb the water at all. Observe what it looks like. Leave the beaker on one side for a few days and then note what changes have occurred.

Your teacher may demonstrate some other examples of diffusion to you.

Questions

1 When two gases mix, what happens to the particles which the gases are made from?
2 Why do the particles of liquids mix?
3 Suggest three ways that you could speed up the mixing of the liquids.
4 Gases diffuse or mix faster than liquids. Suggest a reason for this.

Investigation 5.3 Crystal growth

One reason why crystals are so attractive is that they have regular shapes. You can grow them quite easily, but you will have to be patient if you want to grow a large crystal.

One substance that forms good crystals is copper(II) sulphate.

Collect

250 cm^3 beaker
copper(II) sulphate crystals
thread

What to do

1 Put about 60 cm^3 of hot water in the beaker.
2 Stir in copper(II) sulphate until no more will dissolve.

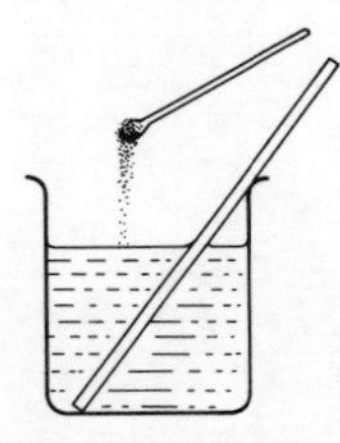

3 Filter the hot solution.

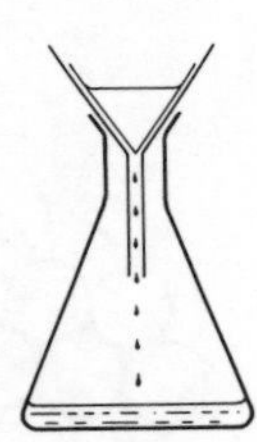

4 Leave the solution in the beaker to cool.
5 Pick out a regular shaped crystal to use as a 'seed' crystal, and tie it on a thread. Hang it in the solution. Cover the top of the beaker to reduce evaporation.

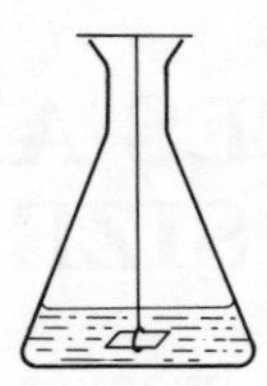

6 After two or three days, remove the crystal and repeat steps 2 to 5 until the crystal is as large as you want.

Different crystals have different shapes. Figure 5.4 shows the shapes of three types. The shape of the crystals can be explained by the arrangement of the particles. The models (Fig. 5.5) show two possible shapes produced by round particles.

The cubic arrangement on the left gives a cubic crystal like sodium chloride (common salt). The sloping shape on the right gives crystals like calcite or copper(II) sulphate. The actual shape is determined by the size and the type of atoms from which the crystal is made.

These three experiments are part of the evidence for particles. The results can be explained using two ideas or theories:

1 All substances are made of small particles.
2 The particles in liquids and gases are always moving.

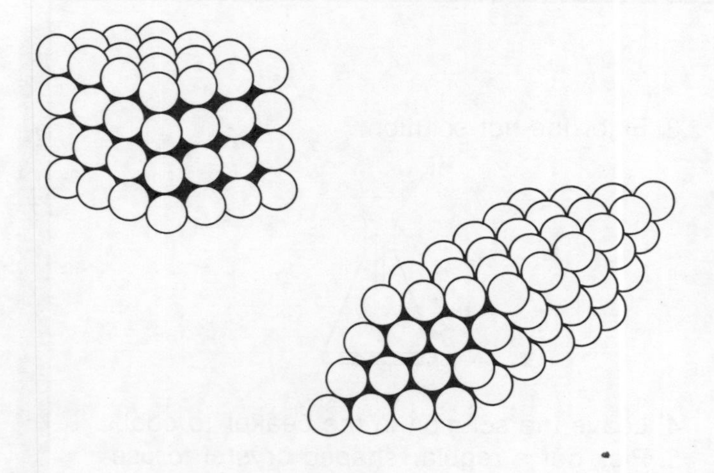

Fig. 5.5 These models show how the particles are arranged in two types of crystal

SUBSTANCES AND PARTICLE SIZE

The way a substance behaves depends on how its particles are arranged. The size of the particles also makes a difference. Substances with smaller particles are more likely to be gases and to have a low density. Those with large particles are more likely to be solid and to have a high density.

Fig. 5.4 Crystals of (a) copper sulphate, (b) sodium chloride, (c) calcite

Assignment – Carbon atoms

The pictures show two types of carbon. Beside each picture is a model of the arrangement of the atoms. All the atoms are the same; only the way that they are arranged is different.

1. What is the graphite form of carbon used for?
2. Write a short description of graphite.
3. The atoms in graphite slide over each other easily. Look at the model, then explain how this happens.
4. Suggest why graphite is used as a lubricant in oil.
5. What is the diamond form of carbon used for?
6. Write a short description of diamond.
7. Look at the model of the atoms in diamond. Why is a diamond very hard?
8. Study these figures:

substance	mass of 1 cm^3
diamond	3.5 g
graphite	2.3 g

   Which form of carbon has more atoms in 1 cm^3? Explain your answer.

9. How does your answer to question 8 help to explain the differences in the way diamond and graphite behave?
10. When a diamond is polished, carbon atoms are removed from the surface of the crystal. It is easier to do this in the 'soft' direction than the 'hard' direction. Explain why.

A diamond tool being used to produce a high-quality polished surface

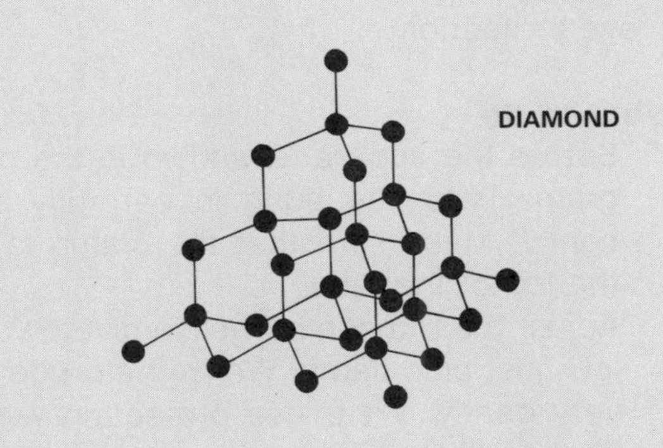

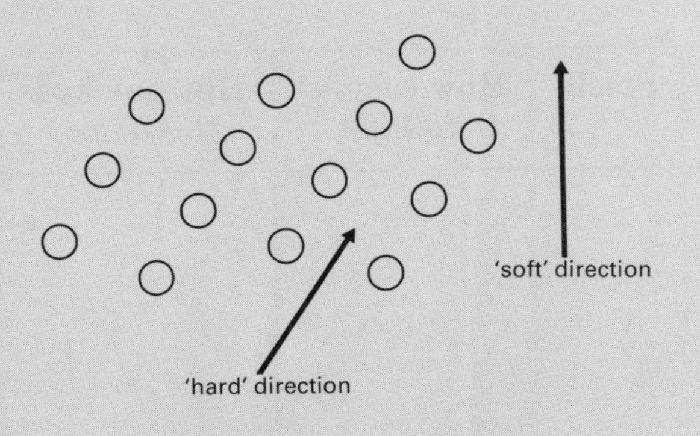

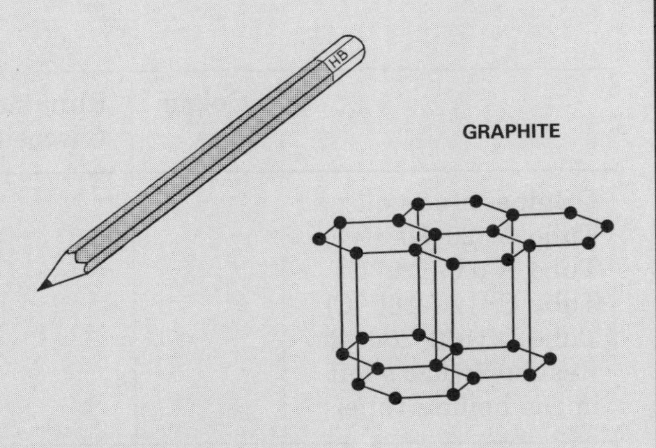

Investigation 5.4 Separating particles of different sizes

Crude oil is a natural substance which contains a mixture of particles of different sizes. You can read about the formation of oil in Unit 1, page 69. Without treatment, oil is not of much use. But it can be made into some important substances simply by separating the particles into similar size groups. This is done by a process called fractional distillation that you can try for yourself. Remember, crude oil produces highly flammable gases when heated. Wear eye protection until everybody has finished the investigation.

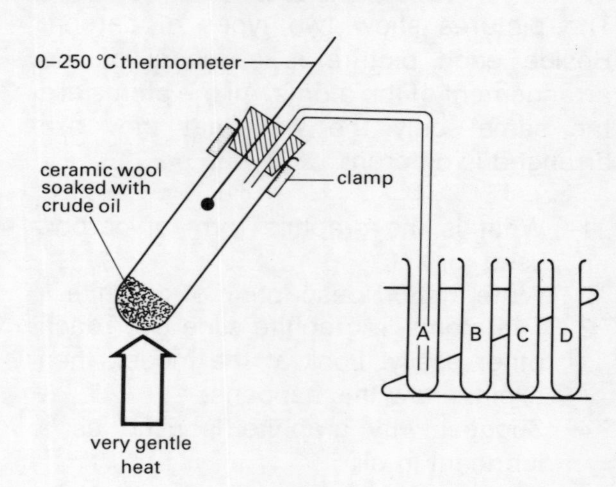

Collect

- boiling tube containing ceramic wool soaked with crude oil
- bung with 2 holes
- 5 test tubes
- test tube rack
- bent glass tubing
- thermometer (0-250°C)
- Bunsen burner
- clamp stand
- eye protection

What to do

1. Set up the apparatus shown in the diagram. Twist the bung in carefully but tightly. Make sure that the clamp is at the top of the tube.
2. Adjust the Bunsen burner to give a very low, just blue flame. Warm the crude oil very gently. Wear eye protection while heating the oil.
3. When the temperature reaches 80°C stop heating. Move the glass tube from test tube A to test tube B.
4. Heat again until the temperature reaches 120°C. Stop heating, and move the glass tube on to test tube C.
5. Repeat the process up to 160°C.
6. Finally move on to test tube D. Stop heating when the temperature reaches 200°C.
7. Copy the table below into your book and fill in the first three columns. Take care with the residue, it will be hot.
8. Test each tube in turn to see how the sample burns. The safest way to do this is to pour a little of the liquid onto some ceramic paper in a tin tray or crucible. Stand the container on a heatproof mat. Light the sample with a spill.

	Colour	Runniness (viscosity)	Smell	How easy is it to light?	How smoky is the flame?
Original crude oil Tube A (20-80°C) Tube B (80-120°C) Tube C (120-160°C) Tube D (160-200°C) Residue (what's left in the boiling tube)					

▶

Questions

1 What differences can you see between the samples you collected?
2 How is the residue different from the other samples?
3 Give three reasons why the crude oil that you started with would not be a good fuel.
4 One problem with the method you have used is that a lot of vapour can be wasted. Explain why this happens.
5 Sketch an improved apparatus which would stop most of the vapour from being lost.
6 Which of the tubes you collected contained the biggest particles? Explain your answer.
7 Which of the tubes contained the smallest particles? Explain your answer.

Assignment – Particles from crude oil

Crude oil contains many different molecules. Here is a table of the melting and boiling points of some of the simplest molecules. Each molecule contains carbon atoms joined together. The first substance, methane, contains only one carbon atom in each molecule. The rest increase in size up to decane which has chains of ten carbon atoms.

Name of substance	Common usage	Number of carbon atoms in chain	Melting point/ °C	Boiling point/ °C
Methane	Natural gas	1	−182	−161
Propane	Camping gas	3	−188	−42
Butane	Calor gas	4	−138	0
Hexane		6	−95	69
Octane		8	−57	126
Decane		10	−30	174

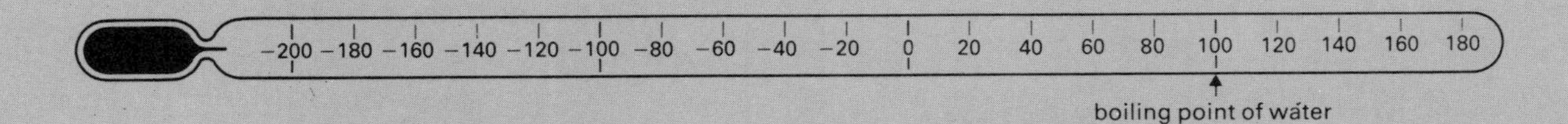

1 Copy the temperature scale. Mark on it the following:
 the boiling point of butane
 the boiling point of decane
 the melting point of decane
 the melting point of methane
 room temperature (approximately)
 the boiling point of water
The last one has been done for you.
2 Why are the melting points always below the boiling points?
3 Which of the substances are gases at room temperature? (Hint: they boil below room temperature.)
4 Which of the substances are liquids at room temperature?
5 Write down a pattern that connects the melting point and the size of the molecules.
6 If you had a mixture of butane, octane and decane in a sealed can, how could you separate them? Explain your method, with diagrams.

SULPHUR PARTICLES

Sulphur is a yellow powder found underground in the United States and Poland. Sulphur is made of sulphur atoms. The atoms are all identical, but they can be joined together in different ways. These different arrangements give different types of sulphur, which you can make in the laboratory.

Investigation 5.5
Making rhombic sulphur

Collect

pestle and mortar
small piece of roll sulphur
a little dimethylbenzene (xylene) in a test tube **(WARNING - flammable)**
microscope and microscope slide
dropper
beaker
eye protection

What to do

1 Put on your eye protection.
2 Crush a small piece of roll sulphur.
3 Warm the dimethylbenzene by putting the test tube in a beaker of hot water.
4 Dissolve some of the sulphur in the dimethylbenzene

BE VERY CAREFUL: dimethylbenzene is flammable

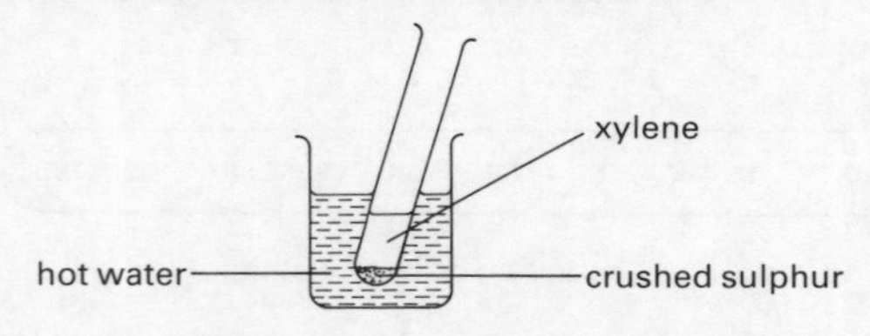

5 Using the dropper, put a few drops of the solution on a microscope slide.
6 After 15 minutes, observe the slide through the microscope.

Questions

1 What is the job of the dimethylbenzene in this experiment?
2 Why is the dimethylbenzene heated by warm water instead of a Bunsen burner?
3 Draw a sketch of one or two of the rhombic sulphur crystals that you have made.

Investigation 5.6
Making monoclinic sulphur

Collect

an old test tube
some pieces of roll sulphur
test tube holder
filter paper
paper towel
Bunsen burner
heatproof mat
tongs
hand lens or low power microscope
eye protection

What to do

1 Put on your eye protection.
2 Half fill the test tube with small pieces of roll sulphur.
3 Fold a filter paper to pour the hot sulphur into.
4 Warm the sulphur gently. Keep shaking the tube.

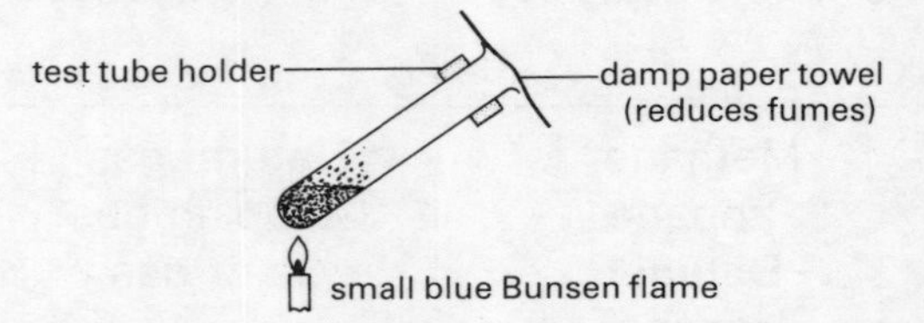

5 When all the sulphur has just melted and is a pale yellow colour, pour it quickly into the paper. (Hold the paper with the tongs, on the heatproof mat.)
BE CAREFUL - the sulphur gets less runny as it gets hotter.

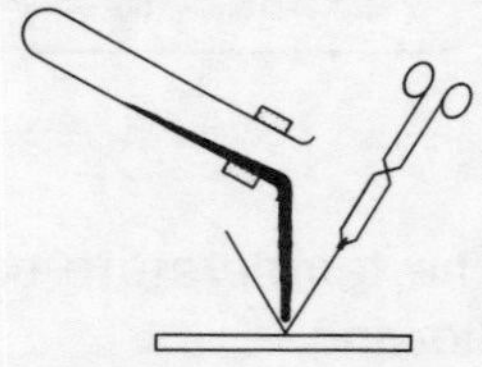

6 When a crust has formed, open up the paper carefully (the sulphur will still be molten inside). Carefully split the lump in half.

Questions

1 Draw one or two of the monoclinic sulphur crystals that you made.
2 Describe what the crystals are like after a day or so.

Investigation 5.7
Making plastic sulphur

WARNING - This experiment is very smelly. It produces a poisonous gas. You should work in a fume cupboard and wear safety spectacles. Have a damp cloth available to put over the mouth of the test tube if the sulphur starts to burn.

Collect

an old test tube
test tube holder
piece of Universal indicator paper
Bunsen burner
beaker
heatproof mat
eye protection

What to do

1. Put on your eye protection.
2. Half fill the test tube with small pieces of roll sulphur.
3. Half fill the beaker with cold water and place it on the heatproof mat.
4. Heat the test tube strongly. Shake it as you heat it.
5. When the sulphur starts to boil, hold a piece of *damp* indicator paper in the vapour.
6. Carefully pour the boiling sulphur into the cold water in the beaker. Remove the plastic sulphur and examine it.

Questions

1. Describe the plastic sulphur that you made.
2. Is the gas produced acid, neutral or alkaline? How do you know?
3. After a day or two, describe what the plastic sulphur is like.

Both rhombic and monoclinic sulphur are made of small molecules. Each molecule contains eight sulphur atoms arranged in a ring.

In rhombic sulphur the rings fit into each other. This gives rhombus-shaped crystals. In monoclinic sulphur, the rings fit edge to edge. This produces needle-shaped crystals. Monoclinic sulphur is less stable than rhombic sulphur. It turns back into rhombic sulphur at room temperature.

When plastic sulphur is made, the rings of sulphur atoms are broken up by the heat into chains of atoms. Sudden cooling in cold water 'freezes' the molecules into these long chains. The long chains make the sulphur flexible. Plastic sulphur changes back into rhombic sulphur if left for a few days.

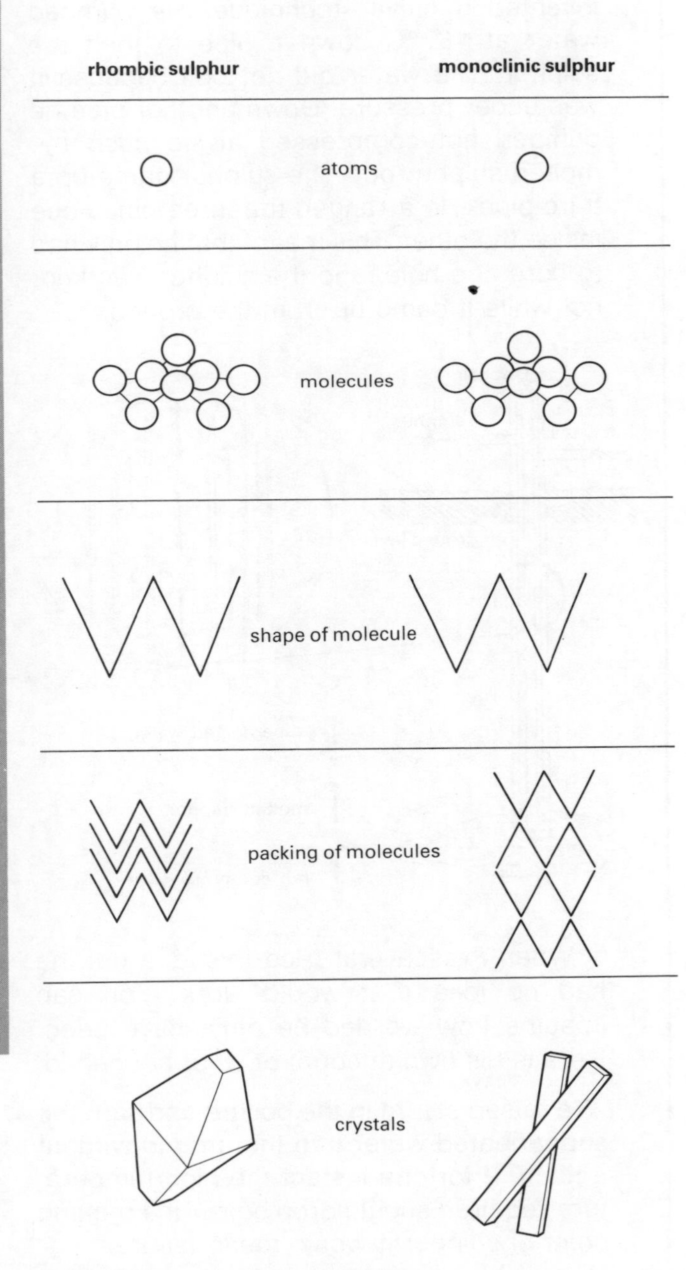

Fig. 5.6 Two forms of sulphur

Assignment – Mining sulphur

Most of our supply of sulphur comes from the United States. It was discovered there over 100 years ago by men who were prospecting for oil. They used their experience of pumping oil to pump the sulphur out of the ground. It would have been impossible to mine it like coal, because it was buried under thick layers of unstable sand – 'quicksand'.

One of the prospectors, Herman Frasch, invented a clever technique. He pumped water at 167°C down a pipe to melt the sulphur. The water did not boil because it was under pressure. Down another pipe he pumped hot compressed air to push the molten sulphur out. The sulphur came up a third pipe. He arranged the three pipes one inside the other. This meant that he only had to bore one hole, and the sulphur was kept hot while it came up from the ground.

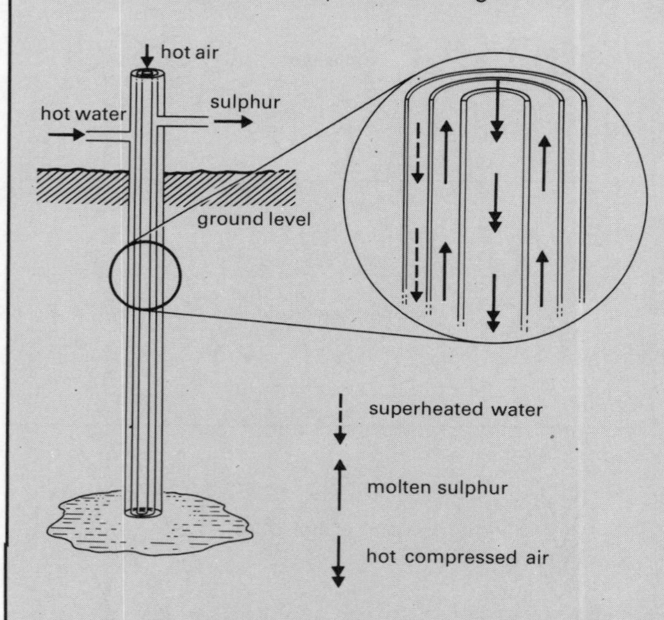

When Frasch first tried this idea out, he had no idea if it would work. You can imagine how worried he must have been. Here is his own account of what happened:

'We raised steam in the boilers and sent the super-heated water into the ground without a hitch. If for one instant the high temperature required should drop below the melting point of sulphur it would mean failure.

'After permitting the melting fluid to go into the ground for 24 hours, I decided that sufficient material must have been melted to produce sulphur. The pumping engine was started on the sulphur line. More and more slowly went the engine, more steam was supplied, until the man at the throttle sang out at the top of his voice, "She's pumping". A liquid appeared, and when I wiped it with my finger I found my finger covered with sulphur.

'After pumping for about 15 minutes, the 40 barrels we had supplied were found to be inadequate. Quickly we threw up embankments and lined them with boards to receive the sulphur that was gushing forth. When everything had been finished, the sulphur all piled up in one heap, and the men had departed, I mounted the pile and seated myself on the very top. It pleased me to hear the slight noise caused by the contraction of the warm sulphur which was like a greeting from below.'

(from *J. Soc. Chem. Ind.*, 1912, Vol. 168)

1 Why was it important that the people who discovered the sulphur had experience of extracting oil?
2 Why does Frasch call the water 'super-heated' water?
3 Give two reasons why Frasch arranged the three pipes inside each other.
4 Why was it important to keep the sulphur hot on its way up to the surface?
5 Why do you think Frasch was so pleased about sitting on the pile of warm sulphur?
6 What sort of sulphur did he sit on? Explain your answer.
7 Why was it important that Frasch did not superheat the water much above 167°C? Look back at Investigation 5.6 if you are not sure.
8 Draw a sketch to show how the different forms of sulphur – rhombic, plastic and monoclinic – can be converted into each other. Label it carefully.

ELEMENTS AND COMPOUNDS

We have found that all chemical substances are made up of atoms. A substance in which all the atoms are the same is called an ELEMENT. The element carbon is made up entirely of carbon atoms and the element sulphur contains only sulphur atoms. Carbon atoms are different from sulphur atoms. There are about 100 elements, so there are about 100 different types of atom. However only about 20 of the elements are common. Ask your teacher to show you some samples of common elements.

Atoms are very small, so there are many millions of them in even a small amount of an element. A diamond on an engagement ring contains about 100 000 000 000 000 000 000 000 atoms. This large number is not easy to imagine. Here is another way of looking at it. Imagine covering Britain with sand so that every part is covered with a layer 100 metres deep. There would be about as many grains of sand as there are atoms in a small diamond.

Elements can be made of atoms on their own, or joined in small groups as molecules, or in very large groups as giant structures. Table 5.1 gives some examples.

COMPOUNDS are also pure substances, but they contain more than one element. Water is an example. It is made of two elements, hydrogen and oxygen. It is a compound because there are always two parts of hydrogen to one part of oxygen, and nothing else. If you could take one single molecule from a bucket of pure water, it would contain two atoms of hydrogen and one atom of oxygen.

Compounds may be made from small molecules, like water, or they may be giant structures with the particles in a regular pattern, like common salt. Table 5.2 gives some examples.

Table 5.1 Some elements.

What the element is made of	Where it may be found	Some examples
single atoms	some gases	neon, argon, helium
molecules	many gases, some solids	hydrogen, oxygen, sulphur, phosphorus
giant structures	nearly all metals, some other solids	copper, iron, sodium, carbon

Table 5.2 Some compounds

What the compound is made of	Where it may be found	Some examples
small molecules	gases, many liquids	steam, water, petrol
medium size molecules	some liquids, many solids	sugar, fat
giant structures	crystals, some other solids	common salt, copper(II) sulphate, starch

Assignment – Chemical symbols

Chemical symbols are a kind of scientists' shorthand. Every one of the 100 or so elements is given its own symbol. This makes it much easier to write down the elements which are found in a compound, or in a chemical reaction.

Normally, each symbol represents an element and also tells you how much of that element you are dealing with. For the moment, we are just going to use the symbols to represent one atom of the element. Here are some symbols:

hydrogen	H
carbon	C
nitrogen	N
oxygen	O
chlorine	Cl
sulphur	S
magnesium	Mg
copper	Cu

Symbols for other elements are shown in the table on p.273.

Molecules are made by combining the symbols for the atoms. A molecule of hydrogen gas contains two atoms of hydrogen, so the symbol or formula for hydrogen gas is H_2. A molecule of water contains two atoms of hydrogen and one of oxygen, so the formula of water is H_2O.

You may be able to make models of some of the particles above. You can certainly try to draw a model of each one. Here is a sketch of a water molecule:

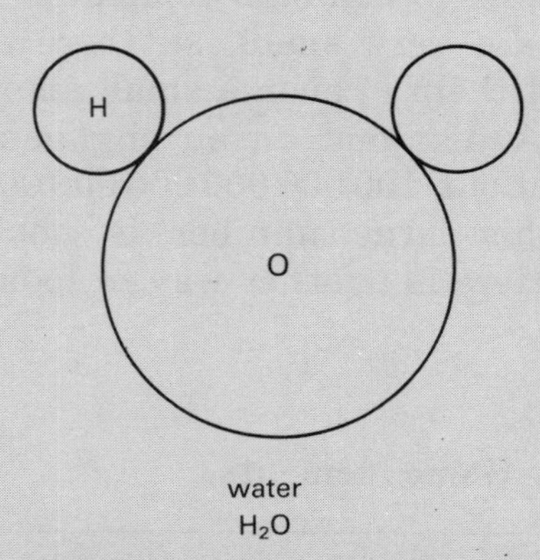

water
H_2O

Draw and label a sketch of each of the molecules in the table. If possible, make a key and represent each atom that you use by a different colour.

Copy the table below and fill in the gaps. The first one has been done for you.

Substance	Chemical formula	How many types of atom	How many atoms in total	Element or compound?
hydrogen gas	H_2	1	2	element
water	H_2O	2		
sulphur	S_8			
methane (natural gas)	CH_4			
carbon dioxide				
vinegar (ethanoic acid)	H_3CCO_2H			
sulphuric acid	H_2SO_4			
propanone	H_3CCOCH_3			
sulphur (IV) oxide	SO_2			
nitrogen gas				
alcohol (ethanol)	H_3CCH_2OH			

Compounds usually behave very differently from the elements that they are made from. Water is a good example. It is made from hydrogen, a flammable gas, and oxygen, a gas that is needed for burning and breathing. Yet water is a liquid that puts fires out.

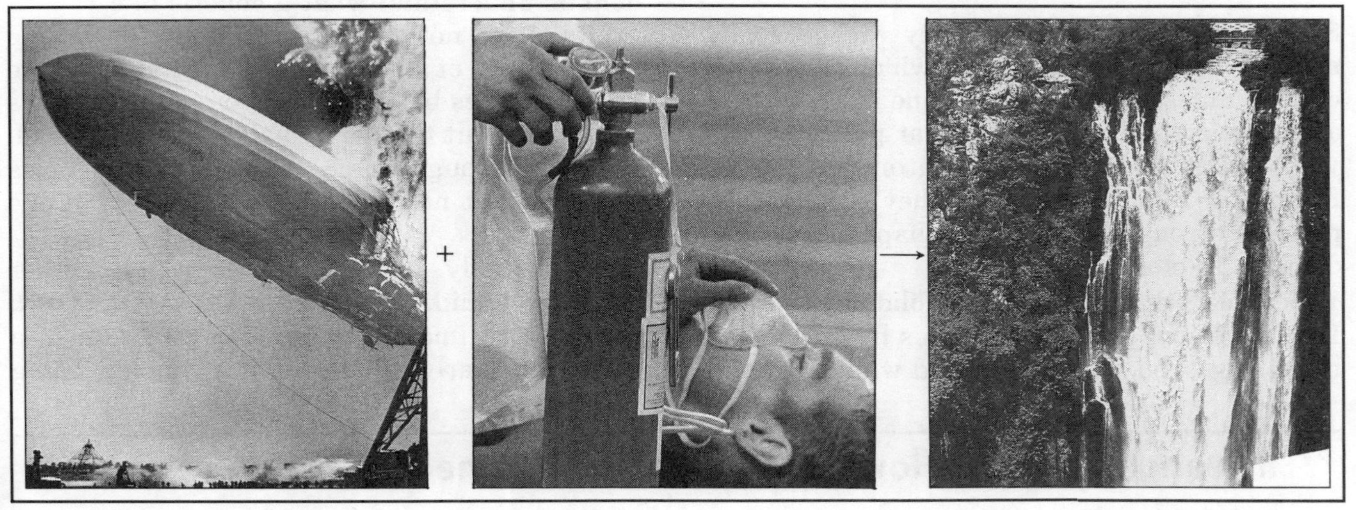

Fig. 5.7 Hydrogen + oxygen ⟶ water

Investigation 5.8 To make a compound from its elements

Collect

test tube and test tube clamp
10 cm of copper wire, coiled
2 spatulas of sulphur powder
Bunsen burner
heatproof mat
eye protection

WARNING - It is best to do this experiment in a fume cupboard and wear eye protection. Burning sulphur produces a poisonous gas. Have a damp cloth available to put over the mouth of the tube if the sulphur starts to burn in air.

What to do

1 Set up the equipment as shown in the diagram. Before you start heating the tube, read through the questions. Put on eye protection.

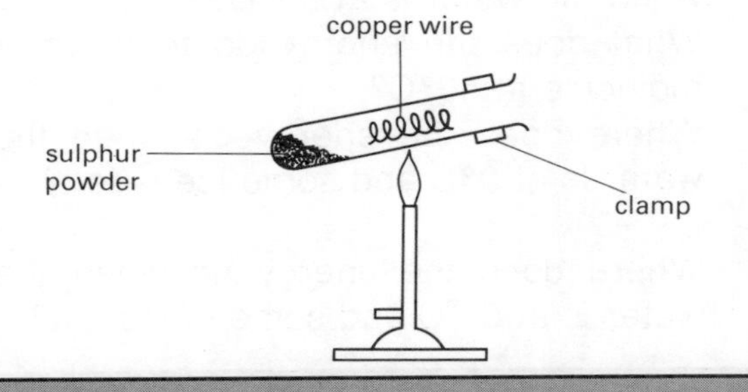

2 Heat the copper wire (not the sulphur). Use a roaring flame.
3 When the wire glows, move the flame quickly on to the sulphur. Try not to breathe any fumes!
4 When nothing more happens, turn the Bunsen off. Leave the tube to cool. You can answer some of the questions while you are waiting.
5 Tip the residue out of the test tube. It is called copper sulphide. Try to bend it.
6 Tidy your equipment away.

Questions

1 What is the sulphur like at the start?
2 What is the copper like at the start?
3 What do you see when you heat the sulphur?
4 What is the copper sulphide like after the experiment?
5 How is the compound different from its elements?
6 Two other common compounds are carbon dioxide (CO_2) and sulphuric acid (H_2SO_4). How do these two compounds differ from the elements that they are made from?

SOLIDS, LIQUIDS AND GASES

Solids are all around us. They are generally strong, and have a shape which does not change unless something is done to them. The atoms in a solid are in a regular pattern. It is very difficult for one of the atoms to escape from this regular pattern. They are held in place by strong forces. So the shape of a solid is difficult to change.

However, the particles in a solid can vibrate. If you can imagine the particles held together by springs you will understand what the solid is like. They cannot easily break out of the structure, but they can move around. But the springs do not actually exist; they are just to help us understand what a solid is like.

Liquids do not have a fixed shape. They take up the shape of anything they are put in. There are still forces between the atoms or molecules of a liquid, but they are smaller than in a solid. There is enough force to hold the particles together, but not enough to keep them in one place.

Occasionally particles escape altogether from the liquid. This is called EVAPORATION. One way to imagine a liquid is to think of a small box nearly full of marbles which is being

Investigation 5.9 How ice changes as it is heated

In this investigation you are going to change a solid into a liquid and then into a gas. Try to imagine what is happening to the particles as you do the experiment.

Collect

100 cm^3 beaker
Bunsen burner
gauze
heatproof mat
thermometer (0–110°C)
ice
clock or watch

What to do

1 Fill the beaker with ice, and add a little cold water.
2 Copy the table into your book.

Time/min	Temperature/°C	Observations
0 (start)		
1		
2		
3		
etc.		

3 Adjust your Bunsen burner to get the smallest possible just-blue flame. This is very important. Do not put the flame under the beaker yet.
4 Stir the ice carefully with the thermometer and read the temperature.
5 Start heating the ice. Do not adjust the Bunsen at all. Every minute, gently stir the mixture, then read the temperature. Look after the thermometer carefully when you are not taking the temperature! The more you stir the mixture, the better your results should be.
6 Continue taking readings until the water has kept boiling for five minutes (but stop if it boils dry).

Questions

1 Plot a graph of your results. Make time the horizontal (*x*) axis, and temperature the vertical (*y*) axis.
2 What do you notice about the temperature while the water was boiling?
3 What do you notice about the temperature while the ice was melting? You may have to look carefully.
4 The Bunsen burner gives out energy at a steady rate. Where does this energy go when the water is at 50°C?
5 What does the energy do to a water molecule at 50°C?
6 Where does the energy go when the water is at 0°C and some ice is left?

7 Where does the energy go when the water is at 0 °C and some ice is left?

shaken. The marbles are not arranged in any particular way, and every now and then, a marble may fly out of the box. The particles of a liquid are like the marbles.

Gases do not have any shape either. The particles of a gas are further apart and they travel all over the place. They do not have much effect on each other. Imagine a snooker table on which a few snooker balls are being thrown around at high speeds. The particles in a gas are like the snooker balls, only they are millions of times smaller and travel much faster and in all directions.

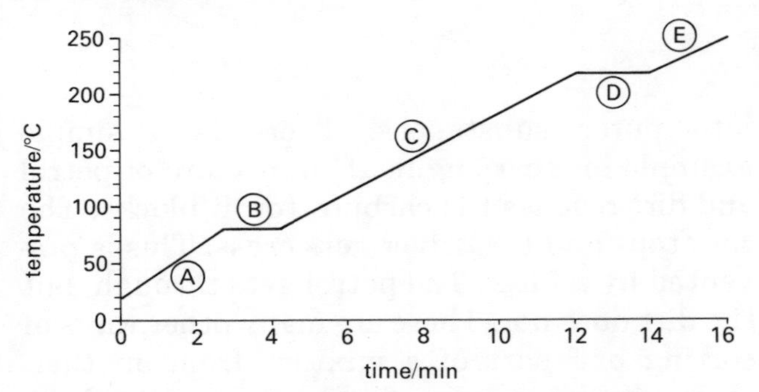

Fig. 5.8 How the temperature changes when naphthalene is heated

Figure 5.8 shows the result of doing Investigation 5.9 with naphthalene. Naphthalene is used to make mothballs. Try to work out from the graph what the melting and boiling points of naphthalene are.

A: solid naphthalene. The heat energy makes the molecules vibrate faster. The shape of the naphthalene does not alter.

B: melting. Heat energy breaks the bonds holding the molecules together. The solid becomes a liquid.

C: liquid naphthalene. The heat makes the molecules move faster, but there is still a weak attraction between them. One or two molecules have enough energy to escape or evaporate.

D: boiling. Heat energy breaks the weak bonds between the molecules. They escape from the liquid.

E: naphthalene gas. The molecules are all separated. They travel very fast. The heat energy makes them travel faster still.

At stages A, C and E, the heat energy makes the molecules vibrate or move faster in all directions. This means that the temperature rises. You can find a more detailed explanation in Unit 7. At stages B and D, the temperature does not rise. The naphthalene is melting or boiling. These changes are called CHANGES OF STATE.

MELTING : solid ⟶ liquid
BOILING : liquid ⟶ gas

At these points the energy from the Bunsen burner is breaking the bonds between the molecules. It is not making them move faster.

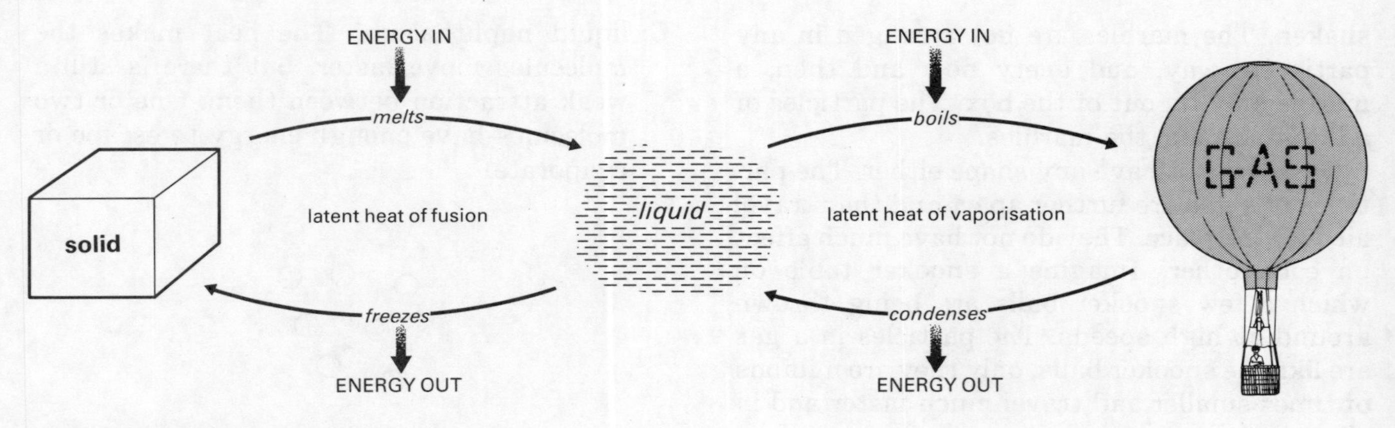

Fig. 5.9 Changes of state

LATENT HEAT

LATENT HEAT is the amount of heat needed to turn a solid into a liquid, or a liquid into a gas. Latent means something which is there but cannot be seen. Latent heat is energy you cannot see because the temperature does not rise.

The latent heat of *fusion* is the energy needed to turn a solid into a liquid. It is also the energy given out when a liquid becomes a solid.

The latent heat of *vaporisation* is the energy needed to change a liquid into a gas. It is also the energy given out when a gas becomes a liquid. For water, the latent heats are:

$$\text{fusion} : 334\,\text{kJ kg}^{-1}$$
$$\text{vaporisation} : 2260\,\text{kJ kg}^{-1}$$

So, it takes 334 kJ to change 1 kg of ice at 0 °C into water at 0 °C, and 2260 kJ to turn 1 kg of water at 100 °C into steam at 100 °C.

PURE SUBSTANCES

Pure is a confusing word. In science a pure substance is one that is the same throughout. All of its particles are the same. So all elements are pure, and so are all compounds. Some examples of pure substances are gold, sulphur, soldium chloride (common salt) and distilled water. Things which are not pure are called mixtures. Sea water, soil, a cup of tea, ink and air are all examples of mixtures.

It is often necessary to separate mixtures into purer substances. There is a simple example in a car engine. If a mixture of petrol and dirt reaches the carburettor, it blocks. The car stops and the driver gets cross. This is prevented by a filter. The petrol gets through, but the dirt does not. There are many other ways of sorting one part of a mixture from another. Here are some common ones.

1 Filtration

This separates a solid which is not dissolved from a liquid.

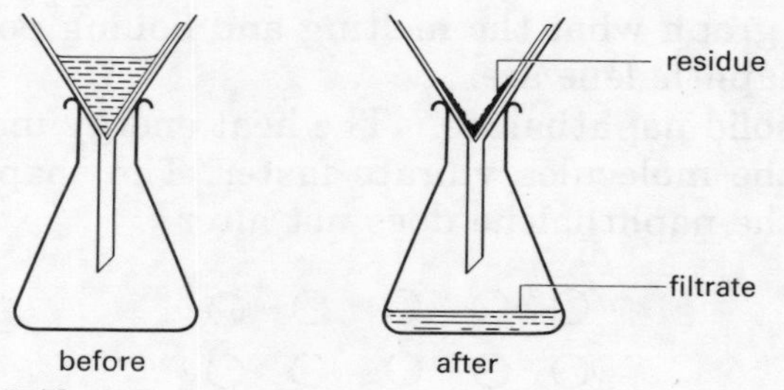

Fig. 5.10

2 Evaporation

This removes the liquid from a solution, leaving behind the dissolved solid.

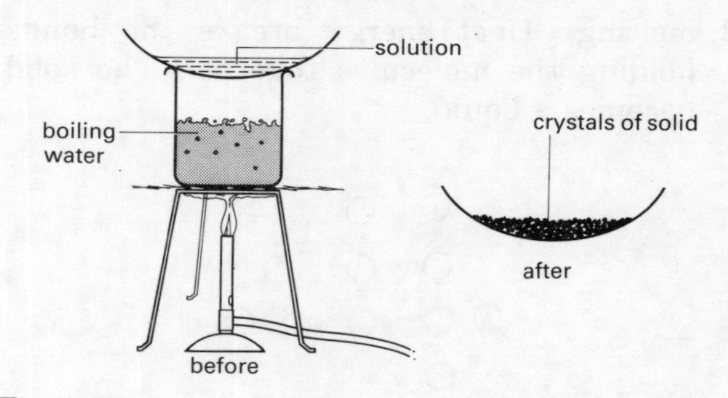

Fig. 5.11

3 Centrifuging

This separates a solid which is not dissolved from a liquid. It works by whirling the mixture around, very fast, so that the solid particles settle firmly to the bottom of the tube. It is like filtering, but quicker though less efficient.

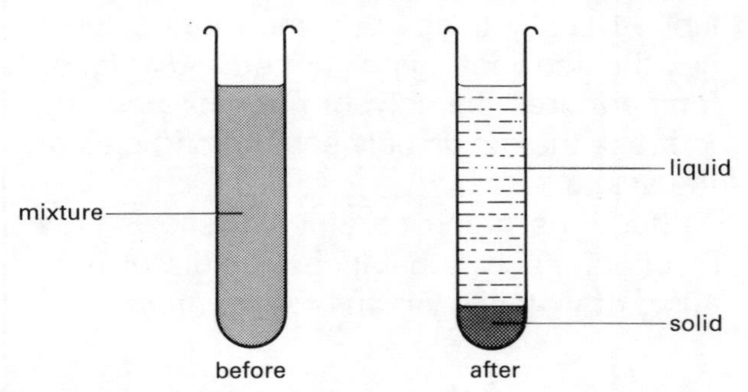

Fig. 5.12

4 Distillation

This separates a liquid from a solution or mixture of liquids. The technique was used in Investigation 5.4 to separate the fractions in crude oil.

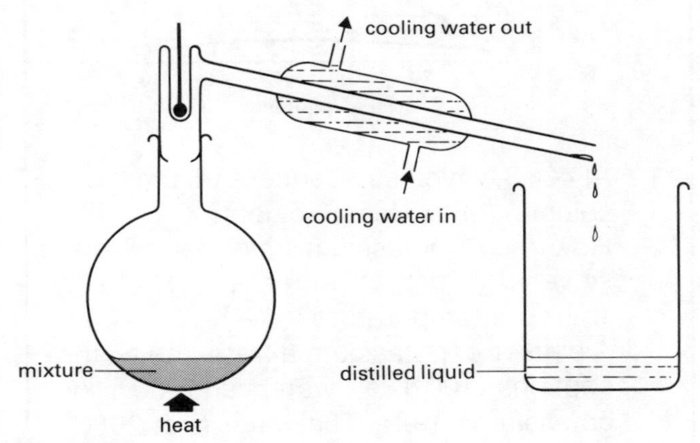

Fig. 5.13

5 Separating funnel

This can be used to separate a liquid from a mixture of liquids which do not dissolve in each other.

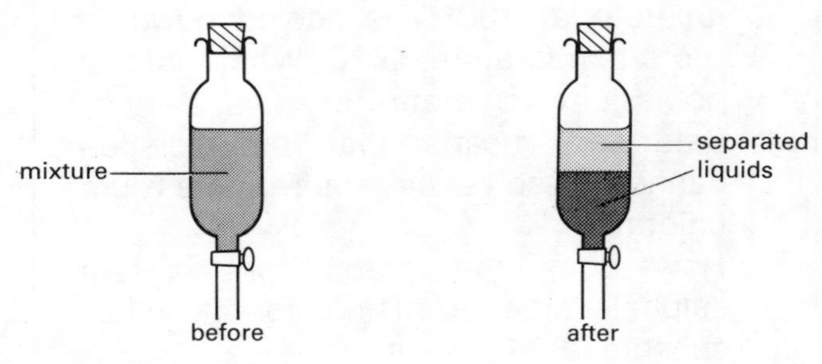

Fig. 5.14

6 Chromatography

This is used to separate and identify different substances in a mixture.

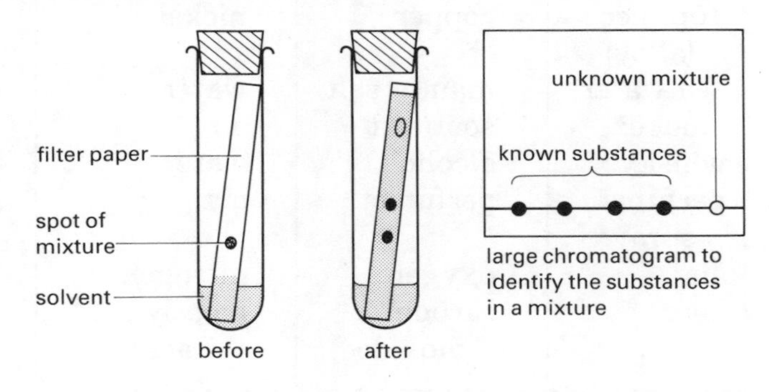

Assignment – Separating mixtures

Use sketches with labels to show how you would separate the following mixtures:

1 Sand and water, to get the sand back.
2 Oil and water, to get the oil back.
3 Ink and water, to get the ink back.
4 Salt and sand, to get both the salt and the sand back.
5 Iron filings and pepper, to get the pepper back.
6 Salt dissolved in water, to get both the salt and the water back.
7 Black ink, which is a mixture of different inks, to find out which colours are in it.

SOLUTIONS

A SOLUTION is one substance dissolved in another. In a solution, the bonds between the particles of the substance being dissolved are broken, and the particles are spread out evenly into the solution.

When a crystal of common salt dissolves in water, the forces holding the particles of the crystal together are broken by the water molecules. The particles of salt split up and spread out into the water. The substance that does the dissolving (water in this case) is called the SOLVENT. The substance that is dissolved is called the SOLUTE.

The most common solutions are solids dissolved in liquids, but there are many others.

Solution	Solute	Solvent
10p piece (alloy)	copper	nickel
sea water	mainly salt	water
smoke*	soot/dirt	air
whisky	alcohol	water
perfume spray*	perfume	air
air	oxygen	nitrogen
beer	carbon dioxide	mostly water

The solutions marked * are poor examples, as the solute may not be evenly spread in the solvent.

Solutions do not always behave as you would expect them to. Normally, $10 + 10 = 20$. If you add $10\,cm^3$ of one substance to $10\,cm^3$ of another, you would usually get $20\,cm^3$, but not in the next investigation!

Assignment – Home experiments

WARNING – These may be a little messy. Ask your parents first please.

1 Fill a small glass with warm water until it almost overflows. Very carefully, add salt, a little at a time. It should dissolve. The glass will not overflow until you have added quite a lot of salt. Try to explain where the salt goes. Give a reason why the glass does not overflow.

2 Find a cold water tap that will give a slow thin stream of water. Rub a plastic comb on a cloth. Hold the comb as near as possible to the stream of water without touching it. What happens?

Find out if any other objects when rubbed will affect the stream. Can you find anything that will have the opposite effect? Look back at the first few pages of Unit 4 and the last question of Investigation 5.10. Explain why the stream of water can be moved. Have a guess why it is always attracted and never repelled.

Assignment – Solubilities

How well one thing will dissolve in another is called its SOLUBILITY. Some things, like sugar in water, are very soluble. Others, like sand in sea water, will not dissolve and are insoluble. Usually things become more soluble as the solvent gets hotter. At higher temperatures the solvent has more energy to break the bonds between the particles of the solute.

Study the graph carefully. It shows how much of each substance will dissolve in 100 g of water at various temperatures.

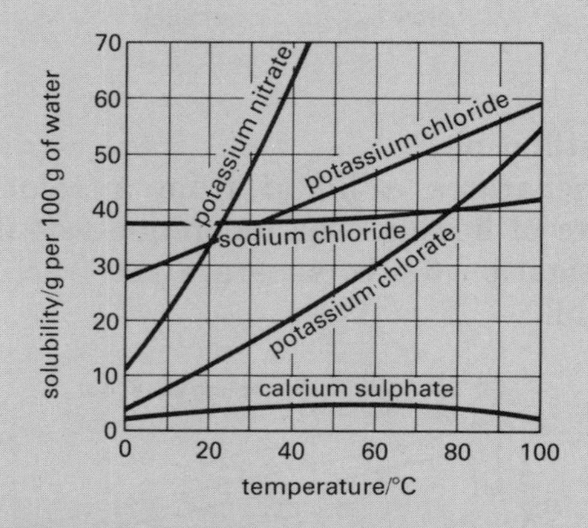

1 At 40°C, which substance is (a) the most soluble, (b) the least soluble?

2 How much potassium chloride will dissolve in 100 g of water at (a) 100°C, (b) room temperature?

3 A beaker of potassium nitrate and water contains 100 g of water and 60 g of potassium nitrate. The water is at 20°C. How much solid would there be in the beaker?

4 If you heated the beaker in question 5, at what temperature would all the potassium nitrate dissolve?

5 A saturated solution of potassium chlorate at 100°C is cooled down to room temperature, 20°C. What mass of potassium chlorate crystals forms? Saturated means that the solution cannot dissolve any more potassium chlorate.

6 How does the graph for calcium sulphate differ from that for the other substances?

Investigation 5.10 Mixing ethanol and water

Collect

5 graduated test tubes or measuring cylinders
test tube rack
You will need access to burettes containing water and dry ethanol.

What to do

1 Copy this table

Experiment	1	2	3	4	5	6	7
Volume of water/cm^3	20	18	15	10	5	2	0
Volume of ethanol/cm^3	0	2	5	10	15	18	20
Expected volume/cm^3							
Actual volume/cm^3							

2 Before starting the investigation, fill in the answers you expect to get when you add the ethanol to the water. You should also be able to fill in the actual volume for experiments 1 and 7 without doing the experiment.

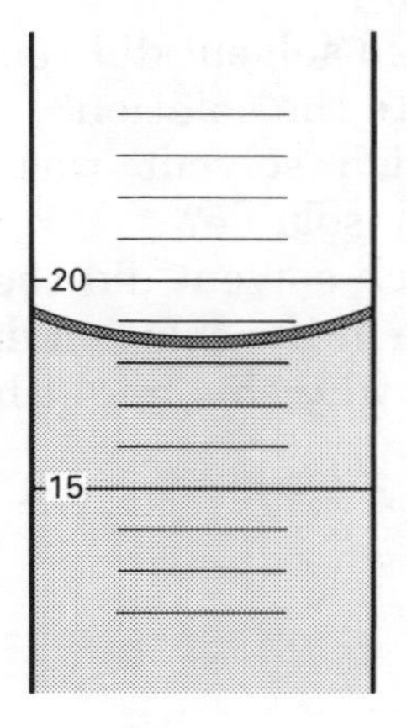

3 To get reasonable results you need to be able to use and read the burette. Here are some hints. When you turn the tap, keep it pushed in. If you do not, it may leak. You should be able to use the tap without watching it.

When reading the burette, or the test tubes, you must measure at the lowest part of the curve. This is called a MENISCUS. The picture shows you a test tube which contains 18.5 cm^3 of liquid. It is worth practising this with your test tubes before starting the experiment.

4 Make sure your test tubes are dry. Stand them in a test tube rack so that they cannot get out of order.

5 Measure exactly the right amount of water from the burette into each tube. Put 18 cm^3 into the first, 15 cm^3 into the second and so on.

6 Add the right amount of ethanol from the burette into each tube: 2 cm^3 into the first tube, and so on.

7 Leave the tubes to stand for a minute, then very carefully read the volumes in each tube. Get other people in your group to check your readings.

Questions

1 Plot a graph of your results. Make the *x*-axis the volume of ethanol used, and the *y*-axis the actual volume of the solution.

2 Sketch on the graph, in a different colour if possible, the line for your expected results.

3 Is the actual volume more or less than the expected volume?

4 Which volumes of ethanol and water gave the largest difference from the expected result?

5 The molecules of ethanol and water contain small electric charges. When they are mixed the + charges attract the – charges. This pulls the molecules together. Try to explain why the volume does not stay at 20 cm^3 in your experiment.

WHAT YOU SHOULD KNOW

1 Atoms are tiny individual particles that make up everything we know.
2 Molecules are two or more atoms joined together.
3 Atoms and molecules are always moving.
4 Particles of different sizes can be separated by fractional distillation.
5 The arrangement of the atoms or molecules alters the way a substance behaves (e.g. sulphur and carbon).
6 Elements are pure substances made from atoms of one type.
7 Compounds are pure substances which contain two or more different elements in a fixed proportion.
8 A symbol can be used to represent each element.
9 A compound is often totally different from the elements of which it is made.
10 Substances are either solid, liquid or gas. Solids have a shape and a rigid structure. Liquids have no shape, but the particles are held together. Gases have no shape and there are no forces between the particles.
11 Heat is needed to turn a solid into a liquid, or a liquid into a gas. This is called latent heat.
12 Mixtures can be separated by a range of techniques such as filtering, evaporation, chromatography, distillation, centrifuging.
13 A solution is a mixture of two substances. One substance, the solute, is spread out evenly throughout the other, the solvent.
14 There are sometimes forces between the particles of solute and solvent that increase or reduce the final volume of the solution.

QUESTIONS

1 5 g of sodium chloride (common salt) was stirred into 100 g of water until it all disappeared.
(a) Write a sentence to describe the experiment using the words solute, solvent and dissolve.
(b) What was the *mass* of the solution after the experiment? Explain your answer.

2 The graph shows what happens when sodium chloride (common salt) is dissolved in five different solvents. 100 cm³ of solvent is used each time and different amounts of salt are added. The final volume of the solution is measured, ignoring any solid. Each line on the graph represents one solvent.

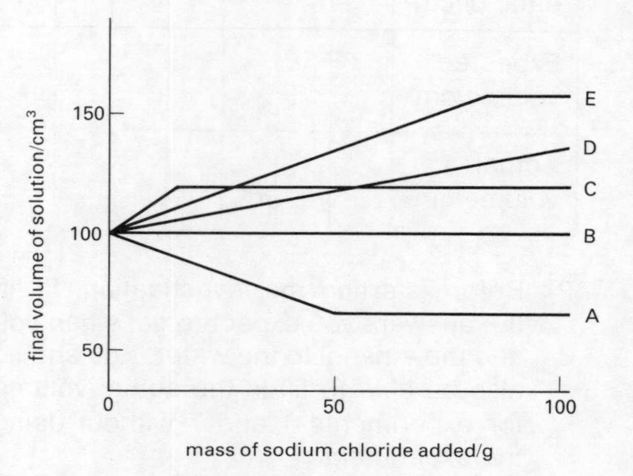

(a) In which solvent did the salt not dissolve?
(b) In which solvent did 100 g of salt fail to saturate the solution?
(c) In which solvent was the salt only slightly soluble?
(d) In which solvent did the salt make the volume of the solution drop? Suggest a reason why this might have happened.

3 The graph shows how the solubility of calcium hydroxide changes with temperature. A saturated solution of calcium hydroxide is made at 80 °C. An extra spatula measure of calcium hydroxide is added. The solution is then left to cool down to 30 °C.

What would you see? Explain your answer.

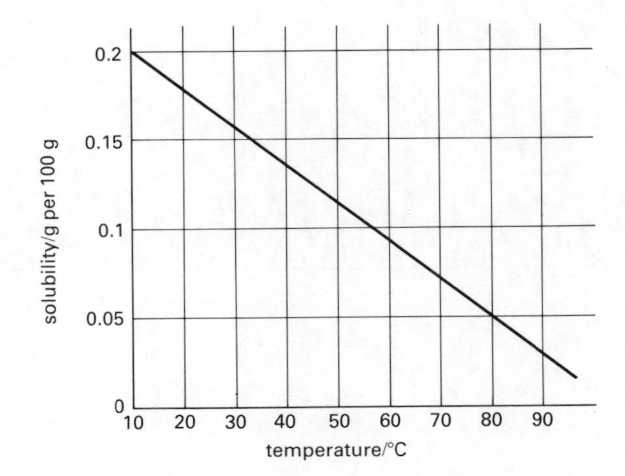

4 Here is some information about the particles in five different substances.

(a) Which substance could be a lump of lead? Explain your choice.

(b) Which substance could be water in a beaker at room temperature? Explain your choice.

(c) Which substance is unlikely to exist? Give two reasons for your answer.

Substance	Distance between the particles	Arrangement of the particles	Movement of the particles
A	close together	haphazard	random
B	close together	regular	in straight lines
C	far apart	haphazard	random
D	far apart	regular	in straight lines
E	close together	regular	vibrating a little

What Substances are Made of

INSIDE THE ATOM

So far we have only thought of atoms as small hard balls. This is useful for explaining many things, like the behaviour of solids, liquids and gases. However, atoms can do things which small hard balls cannot do. In the last section we did not need to look inside the atom; in this section we must.

WHAT ATOMS ARE MADE OF

A hundred years ago, scientists knew nothing about the inside of atoms. Now they know a great deal. The assignment on page 243 describes how some of these early experiments were carried out. The experiments led to the picture of the atom shown in Fig. 5.16. The nucleus in the centre is very small, but it makes up almost the entire mass of the atom. In orbit around the nucleus are even smaller particles called electrons. Imagine magnifying an atom about one hundred thousand million times so that it is the size of a tennis court. The nucleus would be a pin-head in the centre of the court. The electrons would be orbiting somewhere in the rest of the court, but they would still be too small to see. Most of the atom is just empty space.

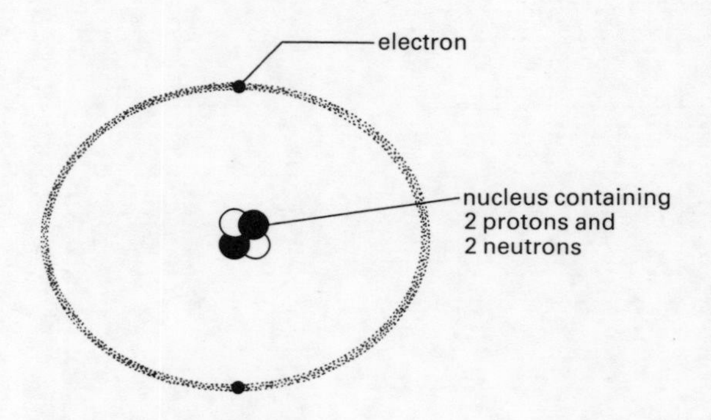

Fig. 5.16 A helium atom

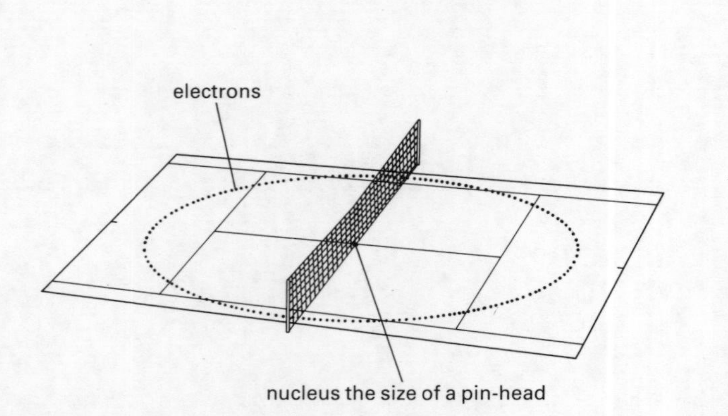

Fig. 5.17 A tennis-court size atom

As you saw in Unit 4, electrons have a negative charge. An atom is electrically neutral, so the negative charge of the electrons has to be balanced by positive charge in the nucleus. The particles which carry the positive charge are

Particle	Where it is	Mass of particle	Electrical charge
proton	nucleus	1 atomic mass unit	+1 atomic charge unit
neutron	nucleus	1 atomic mass unit	0
electron	in orbit	almost nil (1/1840 a.m.u.)	−1 atomic charge unit

Table 5.3 Parts of the atom.

called PROTONS. A proton has an equal but opposite charge to the electron. Also in the nucleus are particles called NEUTRONS. A neutron has no charge, but has about the same mass as the proton. An electron has a mass about 1800 times smaller than a proton or a neutron.

Table 5.3 gives a summary of this picture of the atom.

1×10^{23} atomic mass units make 1 gram
6×10^{18} atomic charge units flowing in 1 second is 1 amp of current.

With this picture, most of the things that atoms do can be explained. However, even this is too simple. Modern physicists have a much more sophisticated view of the inside of an atom. The assignment on page 245 gives you an idea of their work, and of how much still remains unknown.

Assignment – Finding out about atoms

The secrets of the atom have only really been discovered in the last hundred years. Many people were involved in the work. This assignment describes a little of what they did. If you are interested you can find out more from a library.

In the 18th and early 19th centuries several men, including Michael Faraday, passed electricity through gases. They found that gases conducted better at low pressures. They noticed that the gases glowed, and that the glow was caused by rays from the cathode.

1 What uses do we make now of electricity making gases glow?
2 Why do you think that gases at low pressure conduct better than gases at atmospheric pressure? (Hint: when a gas conducts, its particles are charged. If a current is to flow, the particles have to be able to move.)

Crookes, Lenard and Perrin, working independently in the 1890s, found that these 'cathode rays' travelled in straight lines. They could be bent away from a straight line by an electric or magnetic field. Perrin showed that the rays had a negative charge.

3 Why were the rays called 'cathode rays'?

The 'Maltese Cross' tube used by Crookes for an experiment with cathode rays - the rays cast a shadow of the cross on the end of the tube. What did this show?

4 How could you demonstrate that cathode rays are bent by both electric and magnetic fields? Use a sketch to explain your answer.
5 What would this experiment tell you about the rays?

A few years later J.J. Thomson, a Cambridge physicist, deflected a narrow beam of cathode rays with an electric field. He saw that they all followed the same path. He said that the ratio of charge to mass for the particles must be the same. At the same time, he showed that 'cathode rays' were really a stream of identical negatively charged particles called electrons. Thomson guessed that an atom was mostly solid (a small, hardish ball) but with electrons dotted in it. He said that the electrons were like the currants in a plum pudding.

6 Why did Thomson have to use a *narrow* beam of rays?

7 Draw a sketch to show what Thomson's experiment would have looked like if the rays did not all have the same charge. Label the rays which have the most charge, and the rays that have the least charge.

8 Imagine you are J.J. Thomson. Write a scientific paper (remember, in 1897!) to explain your model of the atom. Describe your evidence.

Between 1902 and 1920, Ernest Rutherford put all this information together with other ideas about atoms. With help, he carried out several experiments which led to the model of the atom with a nucleus and electrons.

Perhaps the most important step came when the nuclei of helium atoms were fired at a thin sheet of gold. A thin sheet of gold is about 1000 atoms thick. A helium nucleus can be detected with a standard radioactivity detector. On Thomson's model, the helium particles should have stuck in the gold atoms (making a very stodgy pudding?!). Rutherford's assistants Geiger and Marsden found that this was wrong. Most of the nuclei went straight through the gold atoms. Even more surprising, a few came straight back towards the source of the nuclei. Rutherford described this as 'almost as incredible as if you had fired a 15 inch shell at a piece of tissue paper and it had come back and hit you.'

In 1932, Chadwick discovered a new particle in the nucleus. He found it had no charge, but the same mass as a proton. He called it the neutron.

9 Describe with a sketch how you might carry out the experiment with the helium nuclei and the gold foil. What would you measure?

10 Some helium nuclei went straight through the foil. What did this tell Rutherford about the nucleus of a gold atom? Explain your answer.

11 How did Rutherford explain the helium nuclei that came back towards the source? (Hint: work out what charge a helium nucleus has, and look back at the start of Unit 4).

ATOMIC STRUCTURE AND THE ELEMENTS

The table shows the 20 lightest elements. It is part of the Periodic Table of the elements. (The full version is given at the back at the book.)

$^{1}_{1}$H *1*							$^{4}_{2}$He *4*
$^{7}_{3}$Li *7*	$^{9}_{4}$Be *9*	$^{11}_{5}$B *11*	$^{12}_{6}$C *12*	$^{14}_{7}$N *14*	$^{16}_{8}$O *16*	$^{19}_{9}$F *19*	$^{20}_{10}$Ne *21*
$^{23}_{11}$Na *23*	$^{24}_{12}$Mg *20*	$^{27}_{13}$Al *27*	$^{28}_{14}$Si *28*	$^{31}_{15}$P *31*	$^{32}_{16}$S *32*	$^{35}_{17}$Cl *35·5*	$^{40}_{18}$Ar *40*

There are two numbers beside each element.

This is the MASS NUMBER. It is the number of protons in the nucleus added to the number of neutrons in the nucleus. Helium has 2 protons and 2 neutrons so its mass number is 4.

$^{4}_{2}$**He**

This is the ATOMIC NUMBER. It is the number of protons in the nucleus of the atom.

Atoms are neutral; they have the same number of electrons and protons. So helium, which has an atomic number of 2, also has 2 electrons.

The elements in the table on p.244 are in order of increasing atomic number. Each element has one more proton than the element before it.

In some elements the number of neutrons in the nucleus varies. This does not make the elements different but it does affect their mass. Carbon is an example. Most carbon atoms contain 6 protons, 6 neutrons and 6 electrons. These atoms are $^{12}_{6}C$. Some carbon atoms have 8 neutrons, but still have 6 protons and 6 electrons. These atoms are $^{14}_{6}C$. The two types of atoms are called ISOTOPES. Isotopes always contain the same number of protons, but different numbers of neutrons.

The RELATIVE ATOMIC MASS (often abbreviated as A_r) of an element is the mass of a normal set of atoms of the element. This sounds like the mass number, but it is not quite the same. With most elements, a normal set of atoms contains some isotopes. These make the relative atomic mass different from the mass number. Chlorine is an example. There are two common isotopes of chlorine, $^{35}_{17}Cl$ and $^{37}_{17}Cl$. A normal set of chlorine atoms contains about three mass 35 atoms for every one mass 37 atom. This makes the relative atomic mass of chlorine between 35 and 37. It is about 35.5.

A standard element has to be used to compare the masses of the other elements against. The modern standard is atoms of $^{12}_{6}C$, with no isotopes present. This is called carbon-12.

Assignment – The frontiers of physics

Modern nuclear physics is concerned with forces. This includes all sorts of force, from gravity to the forces inside an atomic nucleus. At the moment there seem to be four basic types of force.

The *strong force* is the force which holds the nucleus of an atom together. It stops the protons and the neutrons from flying apart. The particle which transmits this force is called a *gluon*. This name was chosen because gluons seem to stick the protons and the neutrons together like glue.

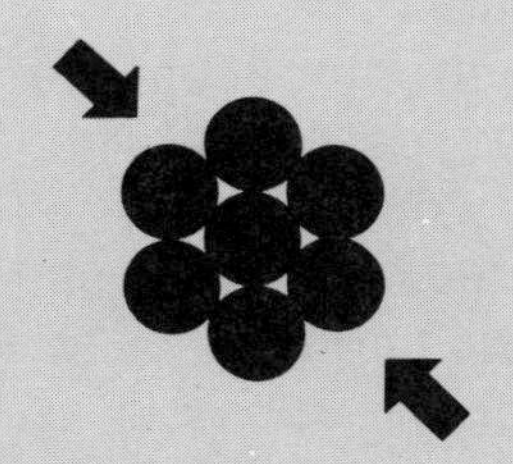

strong force binds the nucleus

The *weak force* is the force that causes radioactivity. It pushes particles like electrons out of the nucleus when a neutron breaks up. The weak force is transmitted by particles called *bosons*. It is about one million times weaker than the strong force.

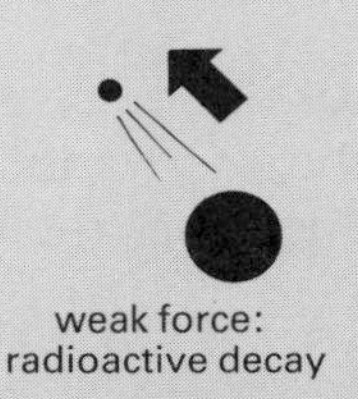

weak force: radioactive decay

The *electromagnetic force* acts between particles that have an electric charge. It is the force that keeps the cloud of negative electrons around the positive nucleus of an atom. It is transmitted by photons, which are the particles that transmit light. Physicists think that the weak force and the electromagnetic force are linked. Modern theories combine the two and call them the '*electroweak force*'. The electromagnetic force is about one thousand times weaker than the strong force.

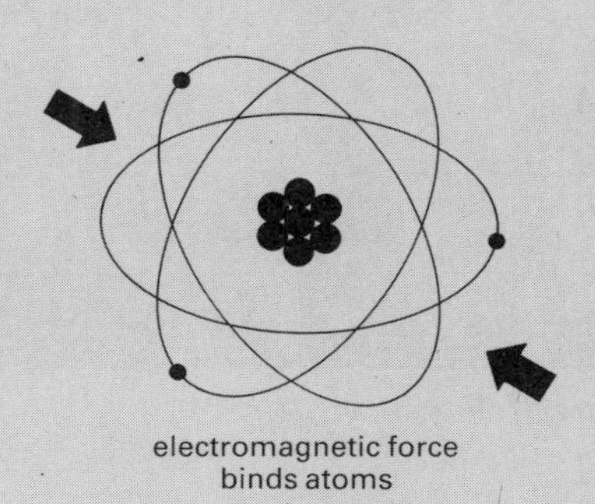

electromagnetic force binds atoms

Gravity is the force that acts between all particles. It is so weak that it only really affects large objects. It pulls things to the ground, and keeps the planets orbiting the sun. Physicists do not know how the gravitational force is transmitted. They think that there must be some particles called *gravitons* that do it, but they are not sure. The gravitational force is 1 and 38 zeroes (1×10^{38}) weaker than the strong force.

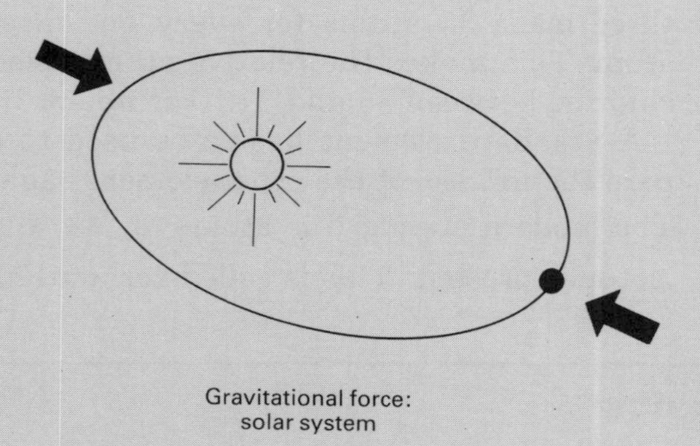

Gravitational force: solar system

To find out more about these forces and the nucleus of the atom, nuclei have to be split open. Huge particle 'guns' or accelerators are used. These give the protons or neutrons (which are the 'bullets') very high energies. They are fired at targets, or at other beams of particles. The collisions are observed very closely. This can provide a lot of information about the nuclear forces. Unfortunately, the particle guns are very expensive, and it takes a long time to get reliable results.

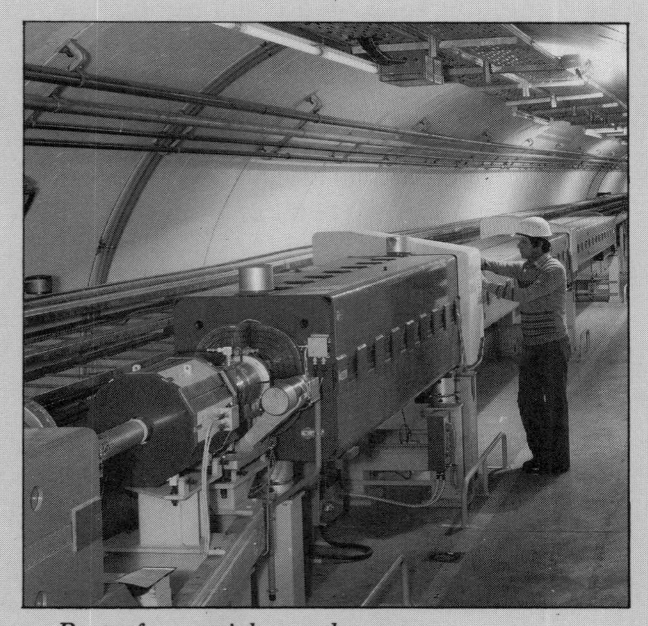

Part of a particle accelerator

Physicists using these particle accelerators have discovered some remarkable things. They now believe that protons and neutrons are made of even smaller particles called *quarks*. Quarks have never been seen on their own, but three quarks together make particles like the neutron. A second type of particle is the *lepton*. Leptons are particles with no size, which are not affected by the strong force. An electron is one example of a lepton.

At the moment, there are more questions than answers. How can particles have no size? How is the force of gravity transmitted? Why can't quarks be found on their own? Maybe the answers will be found in the next few years.

1 Make a table which shows the four types of force. In the table put the strength of each force, how each is transmitted, and one example of each.

2 Which force is responsible for each of the observations below?
   (a) A pencil falls to the ground when it is dropped.
   (b) Uranium produces particles that can be detected with a Geiger-Müller counter.
   (c) A rubbed balloon will stick to a wall.
   (d) Oxygen particles do not fly apart and break into lots of hydrogen particles.

3 'Taxpayers' money should go to improve everyday life, not on scientific research.' Write an essay to explain your views on this comment, paying attention in particular to why you think money should be spent (or not spent!) on research into nuclear physics.

IONS

An ION is an atom which has lost or gained electrons. Atoms that gain electrons form negative ions called ANIONS. For example:

$$\underset{\text{iodine atom}}{I} + \underset{\text{electron}}{e^-} \longrightarrow \underset{\text{iodine ion}}{I^-}$$

When an atom loses electrons, it forms a positive ion called a CATION. For example:

$$\underset{\text{magnesium atom}}{Mg} \longrightarrow \underset{\text{magnesium ion}}{Mg^{2+}} + \underset{\text{2 electrons}}{2e^-}$$

Assignment – Ions and elements

When an atom gains electrons, it forms an anion or negative ion. Anions are larger than the atom that they are made from because they have more electrons. Cations have fewer electrons than the atom that they are made from and so are smaller. In this assignment you can work out which elements form anions, and which form cations.

1. Copy the Periodic Table of the first 20 elements on page 273 into your book. You only need to fill in the symbol for each element, not all the other information.
2. Study the information in the table below. Work out which of the elements form positive ions, or cations. Shade all these elements in with the same colour on your Periodic Table.
3. Work out which elements form anions. Shade them all in with a different colour to show the anions.
4. Write down any patterns you can see in your table from the shading.
5. Under each element that forms an ion in your table, write the charge of the ion that it forms.
6. Write a few sentences to describe any patterns that you can see from the charges that you have just added to your table.

Element	Symbol	Size of atom ($m \times 10^{-10}$)	Size of ion ($m \times 10^{-10}$)	Charge on ion (a.c.u.)
Hydrogen	H	0.4	1.5	1
Helium	He	0.9	none	
Lithium	Li	1.3	0.7	1
Beryllium	Be	0.9	0.4	2
Boron	B	0.8	0.2	3
Carbon	C	0.8	2.6 & 0.2	4
Nitrogen	N	0.8	1.7	3
Oxygen	O	0.7	1.3	2
Fluorine	F	0.7	1.3	1
Neon	Ne	1.3	none	
Sodium	Na	1.5	1.0	1
Magnesium	Mg	1.2	0.7	2
Aluminium	Al	1.2	0.5	3
Silicon	Si	1.2	0.4	4
Phosphorus	P	1.1	2.1	3
Sulphur	S	1.0	1.8	2
Chlorine	Cl	1.0	1.8	1
Argon	Ar	1.8	none	
Potassium	K	2.0	1.3	1
Calcium	Ca	1.4	1.0	2

WHERE THE ELECTRONS ARE

The electrons in an atom are not all the same. They have different amounts of energy. The electrons nearest the nucleus have least energy, while those furthest away have most.

Around the nucleus are several different ENERGY LEVELS or shells which can contain electrons. The shell nearest the nucleus has the lowest energy electrons in it. But it can only hold two electrons. The second shell is further from the nucleus. It can hold up to eight higher energy electrons. When it is complete, the electrons start filling the next shell. Elements with many electrons, like uranium, may fill six or seven different shells.

The first 20 elements fill the shells in order. Table 5.4 gives some examples.

The number of electrons in the outermost shell has a big effect on the way that the element behaves. In the table above, helium, neon and argon all have full outer shells. This makes them very unreactive. Hydrogen, lithium and sodium all have one electron in their outer shell. This tends to make them very reactive.

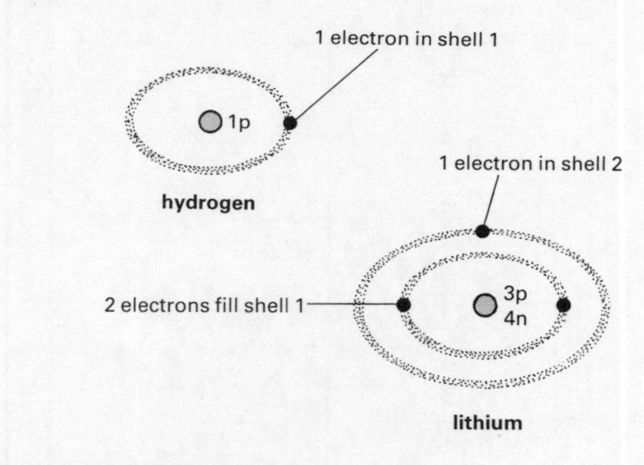

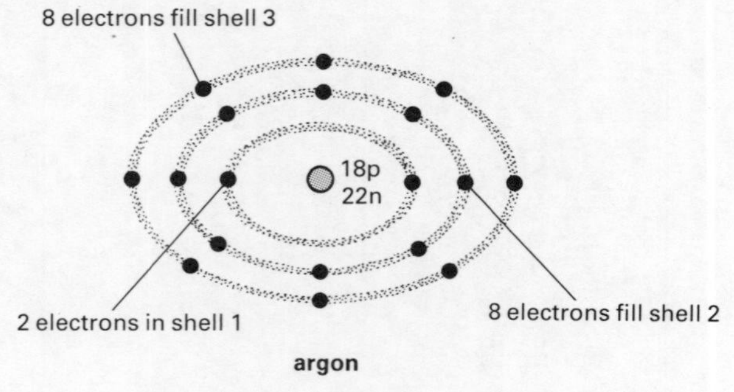

Fig. 5.18

Table 5.4

Element	Atomic number (or number of electrons)	Numbers of electrons in shell 1	2	3
Hydrogen	1	1		
Helium	2	2		
Lithium	3	2	1	
Fluorine	9	2	7	
Neon	10	2	8	
Sodium	11	2	8	1
Chlorine	17	2	8	7
Argon	18	2	8	8

The Periodic Table lists the elements so that those with the same number of outside shell electrons are found in the same column. You can see how this works out for the first 20 elements in the table on page 273. The columns are called GROUPS. Group I, the first column, contains elements like sodium which have only one electron in their outer shell. Group 0, the last column, has elements with full outer shells like helium and neon.

The rows in the Periodic Table also have a name. They are called PERIODS. All the elements in a period have the same number of shells containing electrons.

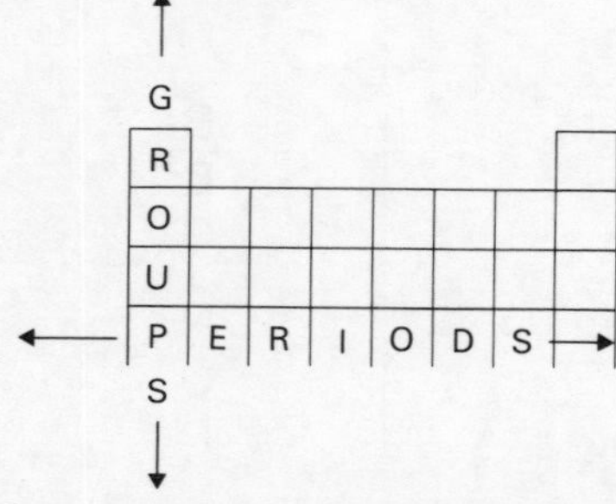

Fig. 5.19 Groups and periods in the Periodic Table

HOW COMPOUNDS ARE FORMED

A compound is made when two or more elements react together. There are two main types of compound: electrovalent and covalent.

ELECTROVALENT COMPOUNDS are formed by elements losing or gaining electrons. Figure 5.20 shows an example.

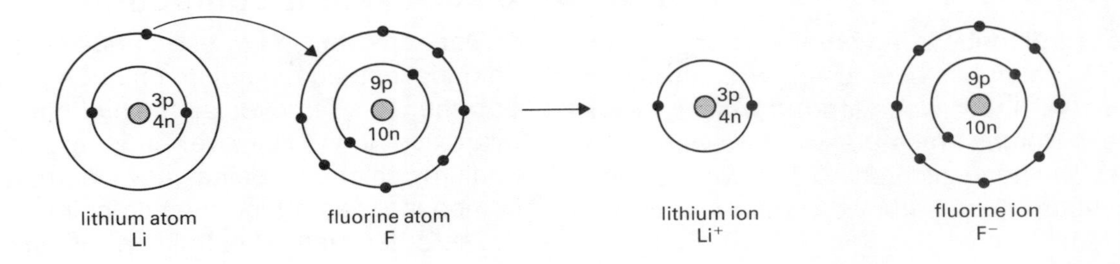

Fig. 5.20

Lithium fluoride can be made by putting lithium metal in fluorine gas. White crystals of lithium fluoride are produced. Why does this happen? Lithium has one outer shell electron. If it can lose this it will be left with a full inner shell. This is a stable and unreactive state. So lithium *loses* an electron easily.

Fluorine has seven outer shell electrons. If it can gain one electron it will have a full outer shell. This is also a stable state. Fluorine therefore readily *gains* an electron.

When lithium and fluorine react, an electron passes from each lithium atom to each fluorine atom. The compound formed contains positive lithium ions and negative fluoride ions.

Sometimes an atom may need to exchange two or more electrons. Oxygen is an example. It has six outer shell electrons. If it gains two more electrons it will have a complete outer shell. Figure 5.21 shows the reaction between lithium and oxygen.

This needs more energy to start it than the reaction between lithium and fluorine. The lithium has to be heated first. Then it will burn in oxygen, or air, to form white lithium oxide.

Elements from the first three Groups in the Periodic Table tend to lose electrons easily. They form positive ions. Elements in Groups 6 and 7 gain electrons easily and form negative ions. It is these elements which are usually found in electrovalent compounds.

Electrovalent compounds are sometimes known as ionic compounds. They have the following properties:

(a) they usually form crystals;
(b) the solid crystals do not conduct electricity;
(c) they usually melt and boil at high temperatures;
(d) when melted they conduct electricity;
(e) when dissolved in water they conduct electricity;
(f) the ions are packed in a regular pattern called a LATTICE.

Fig. 5.22 A model of the crystal lattice of sodium chloride, an ionic compound

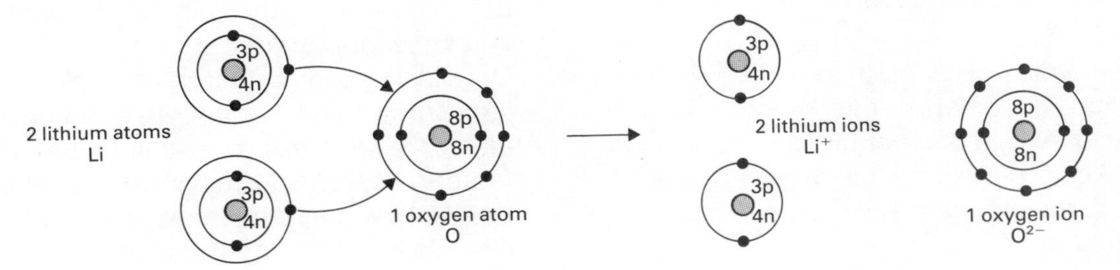

Fig. 5.21

Assignment – Sodium chloride: an electrovalent compound

Sodium chloride is a very important chemical. As salt it has been used for centuries to preserve and flavour food. It was so valuable in Roman times that workers were paid in salt instead of money. The word salary comes from the Latin word *salarium*, meaning 'salt'.

Nowadays salt is used as a raw material for making other chemicals. The most important of these are sodium hydroxide and chlorine. Sodium hydroxide is used to make chemicals for glass, in artifical fibres, and in paper making. Chlorine is used in making plastics and dry-cleaning solvents and to kill bacteria in water.

There are two major methods for getting chlorine and sodium hydroxide from salt. Both use electricity to split the sodium and chloride ions apart. The basis of the method is shown in the diagram. Chloride ions have a negative charge, and tend to move to the anode. Sodium ions, and hydrogen ions from the water, move to the cathode as they both have a positive charge. The hydrogen ions become hydrogen gas. Sodium ions remain in the solution with hydroxide (OH^-) ions.

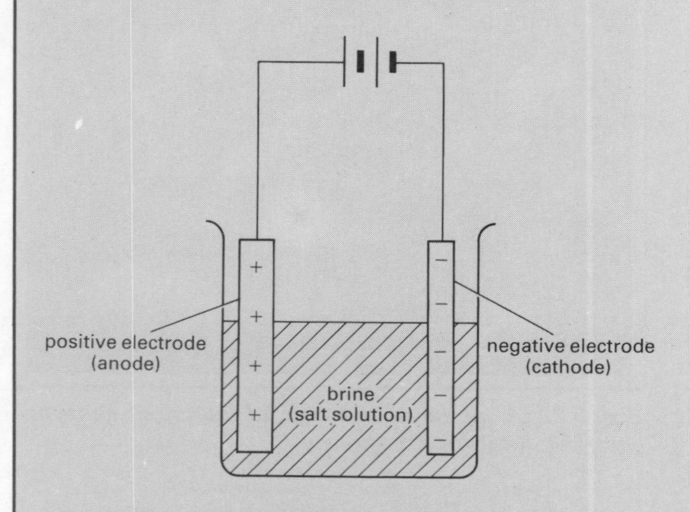

The hydroxide ions make the solution alkaline. The chlorine gas reacts with the alkali, so this simple method cannot be used to make large quantities of chemicals. A method has to be used which keeps the products apart.

One answer is to use molten sodium chloride instead of sodium chloride solution or brine. This removes all the hydrogen and hydroxide ions. However it is expensive. Sodium chloride, being an electrovalent compound, has a high melting point.

The other method is to keep the chlorine gas away from the hydroxide ions. In a DIAPHRAGM CELL this is done with an asbestos sheet. The hydroxide ions and chlorine gas are separated by the diaphragm, so they cannot react. In a MERCURY CELL mercury is used as the negative electrode or cathode. When the sodium ions reach the cathode they react with it to make an amalgam. This can be run out of the cell to a 'decomposer'. Here, water is added to the amalgam, producing sodium hydroxide, hydrogen, and mercury which is recycled.

1 Why do you think salt was so important in Roman times?
2 Make a list of the uses of salt.
3 A non-scientific businessman says that he could make sodium and chlorine much more easily by heating salt. Explain to him why this would not work.
4 Look back at the simple diagram.
   (a) How would you show that the solution becomes alkaline?
   (b) If you had to dispose of large amounts of the solution, how could you neutralise it?
5 Do a simple design for a diaphragm cell. Try to do it from your own ideas without looking it up. It does not matter if you make mistakes! Explain carefully
   (a) what the raw materials are,
   (b) how the cell works,
   (c) what the products are,
   (d) any safety precautions needed.
6 What do you think are the main problems of using a mercury cell? (Hint: what do you know about mercury?)
7 Do a design for a factory to make table salt from brine. Again, use your own ideas. Take care to explain how your factory works. You may find it useful to draw a flow diagram showing the order in which the factory does things.

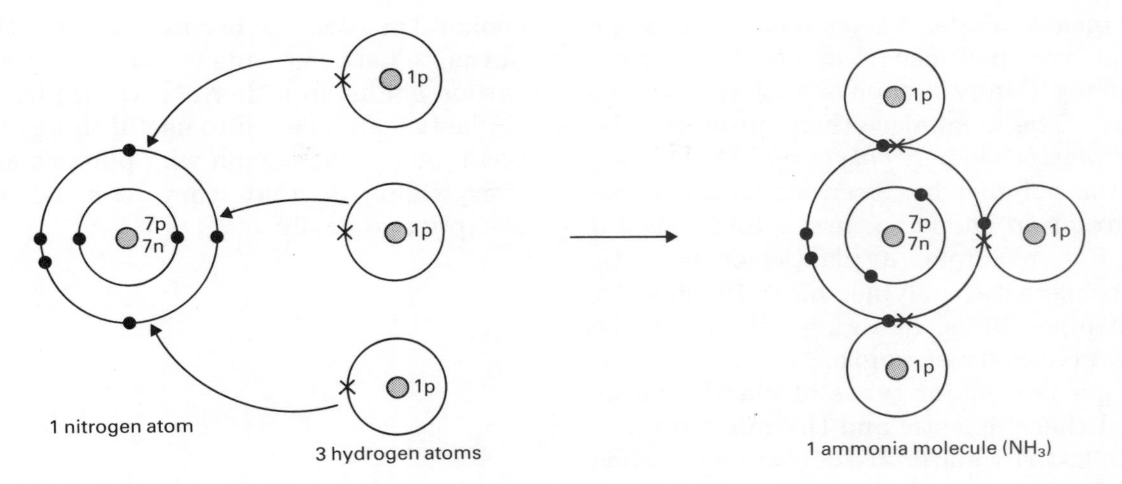

Fig. 5.23 The formation of an ammonia molecule. The hydrogen electrons have been drawn with a cross so that you can see where they have come from. They are not different from the nitrogen electrons

In a COVALENT COMPOUND, electrons are shared between the elements in the compound. Ammonia, made from nitrogen and hydrogen, is an example of a covalent compound.

By sharing the electrons, all the atoms in ammonia have full outer shells and the four atoms are firmly held together into a molecule. This means that they are stable. Covalent compounds do not have a rigid structure like electrovalent compounds. They usually consist of many molecules, attracted to each other by weak forces. This gives covalent compounds different properties from electrovalent compounds:

(a) they are often liquids or gases;
(b) they usually melt or boil at low temperatures;
(c) they often have a smell (like ammonia!);
(d) as they do not contain ions, they do not conduct electricity when melted;
(e) they often will not dissolve in water.

PLASTICS

One important group of covalent compounds is plastics. Plastics are different from simple covalent molecules like ammonia. They are made of very long chains of carbon atoms joined by covalent bonds. A typical plastic molecule contains several thousand carbon atoms. The covalent bonds give each carbon atom a full electron shell. This makes the long chains very stable. Plastics are not easily broken down or attacked by chemicals.

All plastics are made from simple carbon compounds called monomers. One example of a monomer is ethene. An ethene molecule contains two carbon atoms and four hydrogen atoms, all joined by covalent bonds. When ethene molecules are heated under special conditions, the carbon atoms join to form a long chain. This is called polymerisation, and the long chains of molecules form a POLYMER.

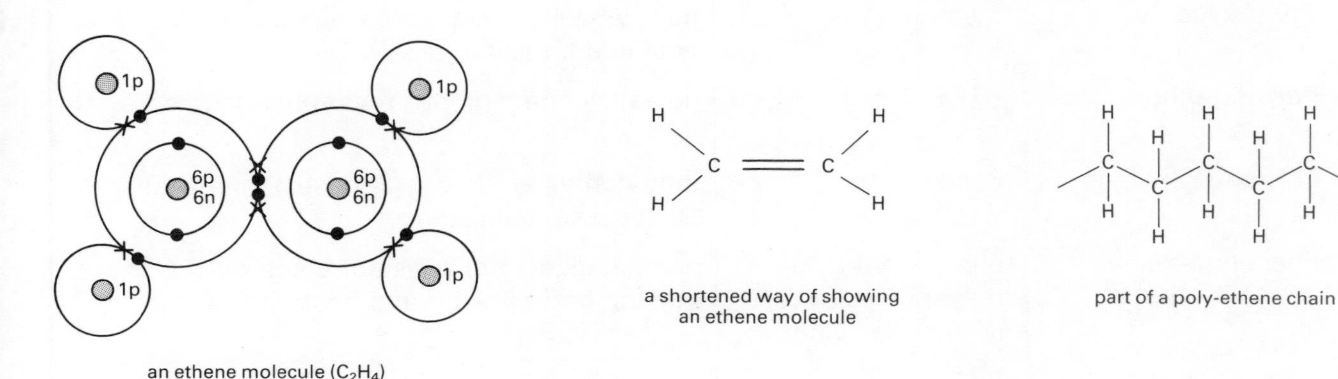

Fig. 5.24

Mono means single. A monomer is the single unit that the polymer is made from. *Poly* means many. Many monomer molecules make a polymer. The monomer ethene produces the polymer poly-ethene, or *polythene* (Fig. 5.24).

Once the polymer has been made, other materials are often added to turn it into a useful plastic. For example, a plasticiser may be added to make the polymer more flexible. Or strengthening fibres, like glass fibre, may be added to give extra strength.

There are two major types of plastic. These are called thermoplastic and thermosetting.

THERMOPLASTIC materials become soft when they are heated and set again when they cool. There are few links between the different carbon chains.

THERMOSETTING materials cannot be altered by heating. Like a boiled egg, the more you cook it, the harder it becomes until in the end it burns. There are many links between the carbon chains in a thermosetting plastic.

Plastics are made into useful things by many techniques. One reason why plastics are found everywhere is that they can be moulded cheaply and easily.

Assignment - Plastics

The table gives the names of some plastics and their uses.

Chemical name	Common name	Some uses
Poly-ethene	polythene	bottles, bags, packaging, household goods
Poly-propene	polypropylene	crates, chairs, pipes, crisp packets
Poly-ester	polyester film	recording tapes, films
Poly-chloroethene	PVC	pipes, gutters, bottles, records, cable insulation
Poly-methyl 2-methyl propanoate	Perspex	light fittings, sheets for signs and displays, furniture, used instead of glass
Poly-amide	nylon	gear wheels, curtain rails, electrical equipment casing
Poly-tetrafluoro ethene	PTFE, Teflon	non-stick coating, low-friction bearings
Poly-phenylethene	polystyrene	food containers (e.g. egg boxes), model kits, flower pots, packaging
Phenol methanal	phenol-formaldehyde polyester resin (often used with glass fibre)	pan handles, heat resistant electrical fittings, car, caravan and boat bodies

Plastics are usually:

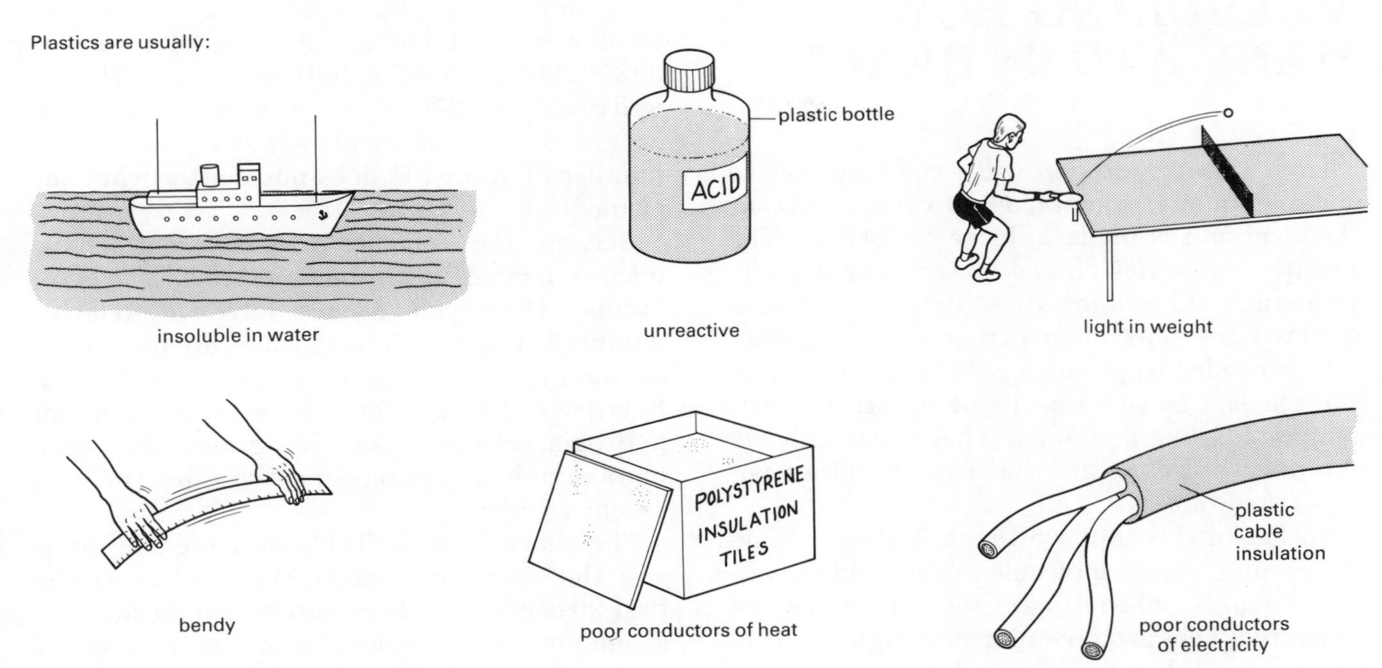

Fig. 5.25 Some properties of plastics

▶ 1 Copy the table below into your book. Fill in the gaps. The first one has been done for you.

Need	Properties needed	Best plastic to use from the list
to cover food	light, bendy, transparent, cheap, clingy	polythene
a bottle for lemonade		
a long-playing record		
coating a non-stick frying pan		
packaging a computer		
a gutter		
an electric light fitting.		

2 For each 'need' in the table, say what material could be used if we had no plastics. In each case, say why a plastic is used instead.

3 Make a list of problems that we have, or may have with plastics. (Some hints: what are they made from? What happens to plastic we have finished with?)

4 The first plastics were displayed over 100 years ago in 1862. Polythene was discovered in 1933. Use a library or reference book to find out how polythene was discovered, by whom, and for what it was first used.

MEASURING HOW MANY ATOMS REACT

Nitrogen and hydrogen will react together to make ammonia. Ammonia is a very important chemical and is made in large quantities. The scientists who design the factories that make ammonia need to know its structure. They also need to know how much nitrogen and hydrogen will be needed to produce each tonne of ammonia. The last section was about the structure of compounds like ammonia. This section shows you how the quantities that react together can be worked out.

Earlier in this unit we thought about single atoms and single molecules. These ideas are not enough when quantities have to be measured. One atom cannot be weighed - it is far too small. Instead we use large numbers of atoms, the sort of number that can be weighed.

You do the same thing when you buy something very small. You do not ask for 1 326 413 grains of sand when you want to mix concrete. Nor do you ask for 2 343 199 277 grains of flour for a cake. You simply ask for 50 kg or 1 kg or whatever is appropriate. It is easier and much more convenient.

Atoms are not measured in kilograms. They are measured in a unit called the MOLE. A chemical mole is not a small furry animal - it is a very large number. It is rather like a thousand or a million, only much bigger. If you went into a chemical factory and asked for a mole of sulphur atoms you would get 32 grams of sulphur exactly (and probably some very odd looks!). A mole of atoms is 602 000 000 000 000 000 000 000 atoms. This is written as 6.02×10^{23} for short. A mole of atoms of any element always means this number of atoms. It does not matter what the element is.

Because the elements have different atomic masses, a mole does *not* mean the same mass of atoms. Hydrogen atoms have a Relative Atomic Mass of 1. Helium atoms are four times as heavy as hydrogen atoms: they have a Relative Atomic Mass of 4. If a mole of hydrogen atoms weighs 1 gram, the same number of helium atoms must weigh 4 times as much: 4 grams.

The short Periodic Table on page 248 gives you the Relative Atomic Masses (A_r) of the first 20 elements. The Relative Atomic Mass in grams of each element contains a mole of atoms. So a mole of magnesium (A_r 24) atoms weighs 24 g. The further you go across and down the table, the heavier becomes a mole of atoms.

As a mole is just a big number, you can have a mole of anything. A mole of elephants is a little hard to imagine. A mole of moles would be more moles than have ever existed. But it is useful to know what a mole of the particles in a compound weighs.

Look again at the ammonia molecule in Fig. 5.23. It is easy to work out what a mole of ammonia weighs. There are three hydrogen atoms in an ammonia molecule, so there are three moles of hydrogen atoms in a mole of ammonia molecules. The three moles of

Fig. 5.26 (left) A biological mole and (right) a chemical mole (of sulphur)

hydrogen weigh 3g. In the same way, there is 1 mole of nitrogen atoms in a mole of ammonia. The Relative Atomic Mass of nitrogen is 14, so 1 mole of nitrogen weighs 14g. A mole of ammonia weighs 3g + 14g which is 17g. This is called the RELATIVE MOLECULAR MASS of ammonia.

Here is another example. The formula of anhydrous copper(II) sulphate is $CuSO_4$. What will be the mass of 1 mole of it?

First, look up the Relative Atomic Masses:
copper 63.5
sulphur 32
oxygen 16

Next, work out the number of moles of each element in 1 mole of copper(II) sulphate:
copper 1 mole Cu_1
sulphur 1 mole S_1
oxygen 4 moles O_4
Now add up the masses

1 mole of copper	: 1 × 63.5	=	63.5g
1 mole of sulphur	: 1 × 32	=	32 g
4 moles of oxygen	: 4 × 16	=	64 g
			159.5g

So a mole of copper(II) sulphate has a mass of 159.5 g. This is the Relative Molecular Mass of copper(II) sulphate.

Assignment – Do you understand the mole?

To answer these questions you will need to have read pages 245 to 254. Use these Relative Atomic Masses to answer them:

hydrogen: 1	sulphur: 32
carbon: 12	chlorine: 35.5
nitrogen: 14	copper: 63.5
oxygen: 16	bromine: 80
magnesium: 24	mercury: 200

Type 1: Moles to mass

Example: What is the mass of 0.25 mol of nitrogen atoms? (mol is the abbreviation for mole).

Working: 1 mol of nitrogen atoms has a mass of 14 g. So 0.25 mol of nitrogen atoms has a mass of 14 × 0.25 = 3.5 g

Questions: What is the mass of the following? Show all your working.

1 1 mol of mercury atoms
2 1 mol of magnesium atoms
3 4 mol of hydrogen atoms
4 10 mol of oxygen atoms
5 0.5 mol of bromine atoms
6 0.25 mol of oxygen atoms

Type 2: Mass to moles

Example: How many moles of atoms are there in 24 g carbon?

Working: 1 mol of carbon atoms has a mass of 12 g so 24 g of carbon contains 24 ÷ 12 = 2 mol.

Questions: How many moles of atoms are there in each of the following? Show all your working.

7 64 g of oxygen
8 48 g of magnesium
9 127 g of copper
10 10 g of hydrogen
11 3 g of carbon (careful!)
12 20 g of bromine

Type 3: Comparing atoms

Example: Which contains more atoms, 12 g of hydrogen or 96 g of oxygen?

Working: 1 mol of hydrogen atoms has a mass of 1 g. So 12 g of hydrogen contains 12 ÷ 1 = 12 mol.
1 mol of oxygen atoms has a mass of 16 g. So 96 g of oxygen atoms contains 96 ÷ 16 = 6 mol.
12 mol is more atoms than 6 mol, so 12 g of hydrogen contains more atoms than 96 g of oxygen.

Questions: Which element in each pair below contains more atoms? Show all your working out,

13 12 g of carbon or 28 g of nitrogen
14 32 g oxygen or 3 g of hydrogen ▶

15 100 g of mercury or 80 g of bromine
16 16 g of sulphur or 18 g of carbon
17 7 g of nitrogen or 20 g of bromine
18 40 g of oxygen or 30 g of carbon
19 2 mol of copper or 21 g of nitrogen (care!)
20 10 mol of bromine or 10 mol of chlorine?

Type 4: Compounds, moles to mass
Example: What is the mass of 2 mol of nitrogen dioxide (NO_2)?

Working: NO_2 means N_1O_2
So NO_2 contains 1 mol of nitrogen atoms and 2 mol of oxygen atoms. It is easiest to use a table for the working out:

Element	Moles of atoms	Relative Atomic Mass	Total mass
N O	1 2	14 16	1 × 14 = 14 g 2 × 16 = 32 g
Mass of 1 mol of NO_2			= 46 g

So 2 mol of NO_2 have a mass of 2 × 46 g = 92 g.

Questions: What is the mass of the following? Show all your working out.
21 1 mol of methane (CH_4)
22 4 mol of water (H_2O)
23 10 mol of chlorine molecules (Cl_2)
24 0.5 mol of copper(II) chloride ($CuCl_2$)
25 0.25 mol of magnesium sulphate ($MgSO_4$)
26 3 mol of ammonium carbonate ($(NH_4)_2 CO_3$)
(Hint: work out the bracket first, then multiply by two. Add in the carbonate CO_3 last.)

Type 5: Compounds, mass to moles
Example: how many moles is 66 g of carbon dioxide (CO_2)?

Working: Use a table as in type 4.

Element	Moles of atoms	Relative Atomic Mass	Total mass
C O	1 2	12 16	1 × 12 = 12 g 2 × 16 = 32 g
	Mass of 1 mol of CO_2		= 44 g

1 mol of CO_2 has a mass of 44 g, so 66 g is 66 ÷ 44 = 1.5 mol.

Questions: How many moles is each of the following? Show all your working out.
27 128 g of sulphur dioxide (SO_2)
28 9 g of steam (H_2O)
29 142.5 g of magnesium chloride ($MgCl_2$)
30 49 g of sulphuric acid (H_2SO_4)
31 182.5 g of hydrogen bromide (HBr)
32 57 g of octane (C_8H_{18})

FORMULAE FROM EXPERIMENTS

One important reason for using the mole is to help work out the formulae of compounds. Although you have already met and used the formulae of many chemicals, they all had to be worked out by doing experiments in the first place. The next investigation shows you an example of how this is done.

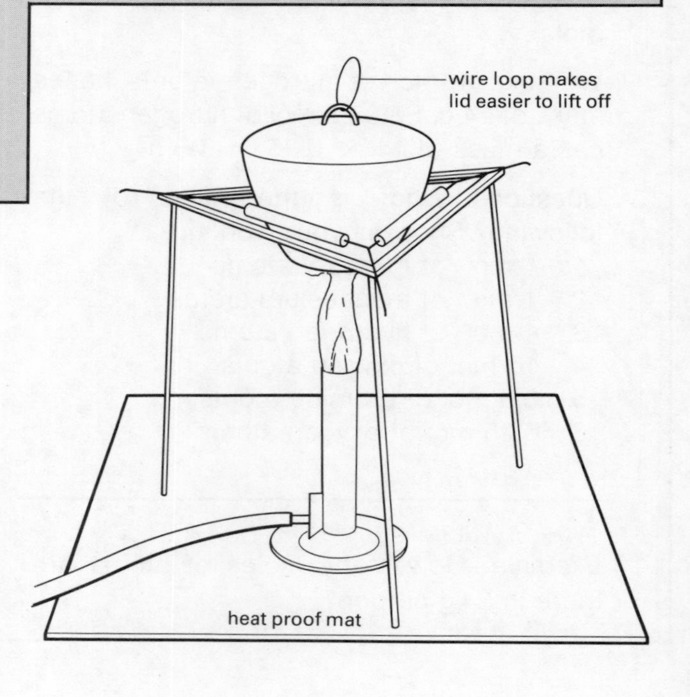

Investigation 5.11 To find the formula of magnesium oxide

When magnesium is heated in air it reacts with the oxygen to form magnesium oxide.

Collect

- crucible and lid
- pipe-clay triangle
- tongs
- 10 cm of magnesium ribbon
- piece of emery paper
- small square of heatproof ceramic paper
- Bunsen burner and heatproof mat
- access to a balance that reads to 0.01 g
- eye protection

What to do

1. Put the ceramic paper in the crucible. Put on eye protection. Heat the two with a blue flame to dry them. Leave to cool.
2. Clean the magnesium thoroughly with emery paper. Coil it loosely so that it will fit into the crucible.
3. Copy this table down
   - (1) mass of crucible without magnesium ________ g
   - (2) mass of crucible + magnesium ________ g
   - (3) mass of crucible + magnesium oxide ________ g
4. Find the mass of the crucible with the ceramic paper in, but without the lid. Do it very carefully. Put the mass in space (1) in your table.
5. Put the magnesium in the crucible. Find its mass again carefully. Put the result in space (2) in your table.
6. Put the lid on the crucible. Heat it with a hot blue Bunsen flame. After 5 minutes, lift the lid for a moment to let some air in. Try not to let any smoke or ash escape. Repeat this several times until all the magnesium has become ash. Then take the lid off and put it on the heatproof mat (Care!). Heat the crucible for 5 minutes more.
7. Let the crucible cool. When it is cool enough to hold in your hand (Care!) find its mass. Put the result in (3) in your table.
8. Tidy up. Take care, the tripod will be hot.

Calculations

Put these calculations down as you do them.

1. Work out how much magnesium you started with. Subtract mass (1) from mass (2).
2. Work out how much magnesium oxide was made. Take mass (1) from mass (3).
3. Work out how much oxygen reacted with the magnesium. Take the mass of magnesium from the mass of magnesium oxide to get this.
4. How many moles of magnesium did you use? 1 mol of magnesium weighs 24 g. Divide your mass of magnesium by 24. If you use a calculator, 3 significant figures will do.
5. How many moles of oxygen atoms reacted? Divide your mass of oxygen by 16, the mass of 1 mol of oxygen.
6. Work out how many moles of oxygen atoms combined with 1 mole of magnesium atoms. Simply divide your answer to 5 by your answer to 4.
7. Write down the formula of magnesium oxide:
   Mg_1O_x ← fill in your answer from 6.

Here is a short summary of what you have done:

work out the mass of each element
↓
work out how many moles this is
↓
write this down as a ratio.

Questions

1. What did you see during the experiment?
2. Why did you dry the crucible first?
3. Why was it important to clean the magnesium?
4. Why did you have to let some air in?
5. Why was it important not to let any smoke out?
6. What could you do to improve your result?
7. Write down a word equation that describes the reaction.

Assignment – Working out a formula

This assignment gives you a chance to work out some formulae from the results of experiments.

Example: Mercury(II) chloride is a white powder. When it is heated with a weak acid it is converted into mercury. In an experiment, 5.42 g of mercury(II) chloride produced 4.00 g of mercury metal. A_r (Hg) = 200; A_r (Cl) = 35.5.

Working: mass of mercury used = 4.00 g
mass of chlorine in mercury(II) chloride = 5.42 g − 4.00 g = 1.42 g
mass of 1 mole of mercury = 200 g
so moles of mercury used = 4.00 ÷ 200 = 0.02 mol
mass of 1 mole of chlorine = 35.5 g
so moles of chlorine = 1.42 g ÷ 35.5 = 0.04 mol
moles of chlorine combined with 1 mol of mercury: 0.04   0.02 = 2
So the formula of mercury(II) chloride is Hg_1Cl_2 or Hg $C1_2$

1. Copper(II) oxide is a black powder. When it is heated with hydrogen passing over it, it is converted into copper metal and water. 3.82 g of copper(II) oxide became 2.54 g of copper in an experiment. Work out the formula of copper(II) oxide. Show all your working.
   A_r(Cu) = 63.5; A_r(O) = 16.
2. An oxide of lead weighs 6.85 g. It is heated with hydrogen passing over it. The lead left at the end of the experiment has a mass of 6.21 g. Work out the formula of the oxide of lead.
   A_r (Pb) = 207; A_r (O) = 16.

WHAT YOU SHOULD KNOW

1. Atoms consist of shells of electrons around a nucleus.
2. The nucleus contains protons and neutrons. It makes up nearly all the mass of the atom.
3. Protons have a positive charge, electrons have a negative charge.
4. The mass number of an atom is the number of protons added to the number of neutrons.
5. The atomic number of an atom is the number of protons in the nucleus.
6. In a neutral atom, there are equal numbers of protons and electrons.
7. The relative atomic mass of an element is the average mass of its atoms compared to carbon atoms.
8. Isotopes are atoms with the same number of protons, but with more or less neutrons.
9. An ion is formed when an atom loses or gains electrons.
10. The number of electrons in the outside shell of an atom determines its behaviour. Atoms with the same number of outside shell electrons are found in the same Group in the Periodic Table.
11. Electrovalent compounds are formed when one atom gives electrons to another.
12. Covalent compounds are formed when atoms share electrons.
13. Electrovalent and covalent compounds have different properties.
14. Plastics are giant chains of carbon atoms. They have many useful properties.
15. The mole is a very large number. It is used to make the measuring out of atoms easier.
16. The relative molecular mass of a compound is the mass of 1 mole of that compound.
17. The formula of compounds is worked out from experiments.

QUESTIONS

1 In each of the following, say which element contains more atoms. Show how you reached your answer. Relative atomic masses can be found on the Periodic Table on page 273.
  (a) 35.5 g of chlorine or 60 g of calcium?
  (b) 32 g of oxygen or 4 g of hydrogen?
  (c) 36 g of magnesium or 25 g of fluorine?
  (d) 20 g of mercury or 6.5 g of zinc?
  (e) 60 g of carbon or 2.5 mol of magnesium?
  (f) 40 g of potassium or 39 g of calcium?

2 In each of the following, say which element would weigh more. Show all your working.
  (a) 2 mol of sodium atoms or 1 mol of calcium atoms?
  (b) 5 mol of hydrogen atoms or 0.5 mol of oxygen atoms?
  (c) 12 mol of helium atoms or 1 mol of iron atoms?
  (d) 4 mol of oxygen atoms or 1 mol of copper atoms?
  (e) 3 mol of carbon atoms or 1.5 mol of magnesium atoms?
  (f) 0.1 mol of uranium (A_r (U) = 238) atoms, or 1 mol of sodium atoms?

3 Draw diagrams to show the arrangement of the electrons in
  (a) $^{23}_{11}Na$ (b) $^{14}_{7}N$ (c) $^{37}_{17}Cl$

4 The compound formed when sodium reacts with chlorine has a high melting point. Answer these questions:
  (a) What is the name of the compound?
  (b) Is the compound electrovalent or covalent? Explain your choice.
  (c) Is the compound likely to have a giant or a molecule structure? Explain your choice.
  (d) Using one particle of chlorine and one particle of sodium, show how the electrons are arranged in the compound.
  (e) What would the chemical formula of the compound be? Include any charges on the particles.
  (f) Make a list of the properties you would expect the compound to have.

5 Copy and complete the table. The Periodic Table on p.273 may help you. Lithium has been done for you.

Element	Lithium	Oxygen	Sodium	Chlorine	Potassium
Symbol	Li				
Atomic number	3	8			
Mass number	7	16	23	35	39
Protons	3				
Neutrons	4				
Electrons	3				

6 Methane is formed when 1 atom of carbon reacts with 4 atoms of hydrogen.
  (a) Draw the electron arrangement for hydrogen ($^{1}_{1}H$).
  (b) Draw the electron arrangement for carbon ($^{12}_{6}C$).

  Methane has a low melting point and does not conduct electricity when liquid.
  (c) Is methane an electrovalent or covalent compound? Explain your answer.
  (d) Draw the electron arrangement in one particle of methane.
  (e) Write down the chemical formula of methane.
  (f) Write down the mass of 1 mole of methane.
  (g) What other properties would you expect methane to have?

7 1.35 g of aluminium powder was heated strongly in a stream of oxygen. A white powder, aluminium oxide, was produced which had a mass of 2.55 g. What is the formula of aluminium oxide?

8 7.73 g of dried rust (iron oxide) was heated with hydrogen passing over. The iron left at the end had a mass of 5.60 g. What is the chemical formula of the rust?

9 The formula of ammonia is NH_3.
  (a) How many atoms are there in 1 molecule of ammonia?
  (b) Draw the arrangement of the electrons in a molecule of ammonia.
  (c) In an experiment, 2.8 g of nitrogen reacted completely with hydrogen to make ammonia. What mass of hydrogen reacted?
  (d) In the experiment in (c), what mass of ammonia was produced?

ENERGY FROM THE NUCLEUS

Most of the energy that we use comes from moving atoms around. For example, a lump of coal contains many carbon and hydrogen atoms. When coal is burnt, the bonds between these atoms are broken. The carbon and hydrogen atoms join up with oxygen atoms instead. This process gives out heat energy.

To get energy from the nucleus of an atom, moving the atom around is not enough. Nuclear energy is obtained by removing bits from the nucleus of an atom, or adding bits to it. It is not easy to do this. However, if you succeed, the amount of energy that you can obtain is enormous. This is because there is a very large amount of energy stored in the nuclei of atoms.

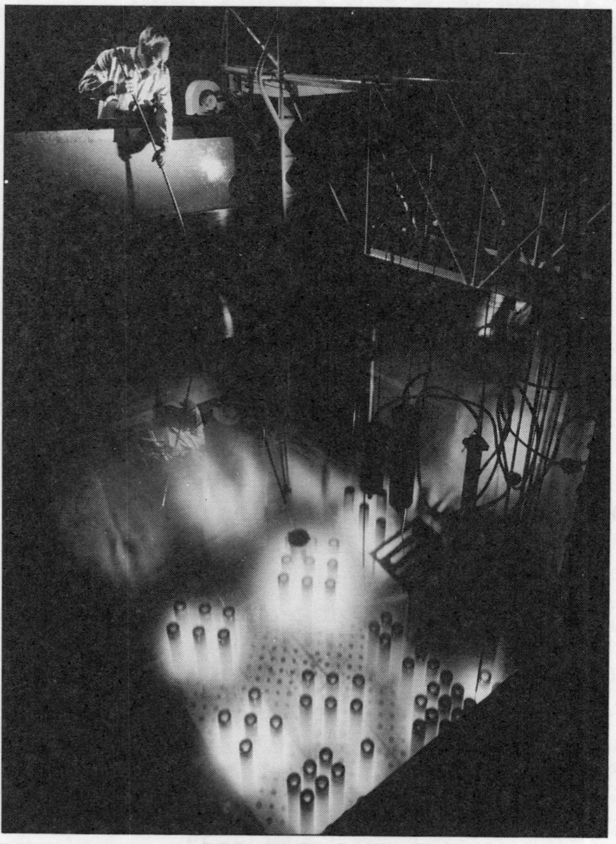

Fig. 5.27 *Inside a nuclear reactor*

EVIDENCE FOR NUCLEAR ENERGY

Nuclear energy is hard to understand because you cannot see it. Heat energy is much easier. You can both see and feel a fire! However, with special instruments, energy from the nucleus can be detected. Unfortunately much of the energy is dangerous and hard to detect, so complicated equipment has to be used.

Some natural substances give off particles whose energy comes from the nucleus. These are called RADIOACTIVE substances. Henri Becquerel discovered this in 1896. He found, quite by chance, that elements like uranium and radium blackened photographic paper. This is one way of detecting some radioactive particles. Nowadays we have other methods of detecting nuclear radiations.

The electroscope

This is a device that detects charges. You can find out more about it in Unit 4, page 163. If an electroscope is charged up, the gold leaf rises. A radioactive element near the leaf will gradually discharge the electroscope. The leaf moves back down.

The cloud chamber

If radioactive particles pass through a saturated vapour, a trail is made where each particle passes. The trails are similar to those made by high flying aircraft on a clear day. In a cloud chamber, a saturated vapour is produced. Radioactive particles are then allowed into the chamber.

The spark counter

Radioactive particles produce sparks between a wire mesh at a high potential and a wire at ground potential (zero volts).

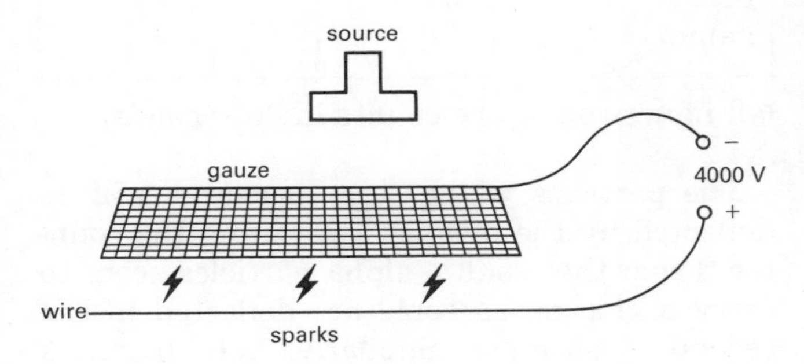

Fig. 5.29 A spark counter

These three detectors work because the radioactive particles have enough energy to knock electrons off the air particles. This produces ions in the air. The ions will conduct electricity better than normal air. They also provide a place for the vapour in the cloud chamber to condense.

The standard detector used in simple radioactive measurements is a Geiger-Müller tube. This is really just a sophisticated spark counter. A radioactive particle that passes through the thin 'window' produces some ions in the tube. These ions allow a pulse of current to flow in the tube circuit. This pulse is amplified and passed to a counter or a loudspeaker.

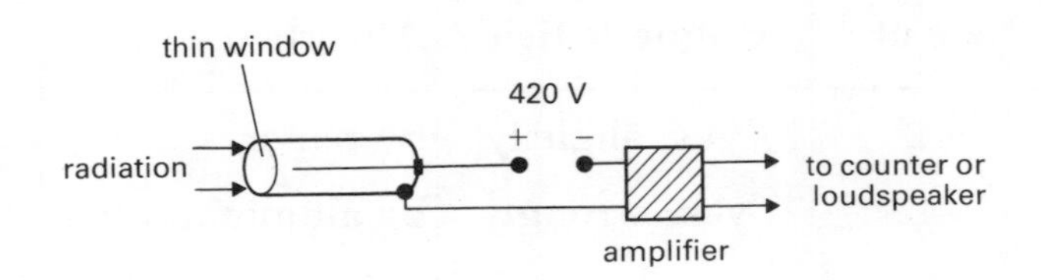

Fig. 5.30 A Geiger-Müller tube

Fig. 5.28 Tracks made by radioactive particles in a cloud chamber

TYPES OF PARTICLE

A Geiger-Müller tube can be used to show that there are three types of natural radioactive particle. These are called alpha (α), beta (β) and gamma (γ) particles. The experiments below show some of their properties.

How penetrating are the radiations?

A source of each type of radiation is put in turn in front of a counter. The number of counts in 30 seconds is measured. A control experiment is done without a source present. The experiment is repeated with different materials between the source and the counter to see which materials each radiation can get through.

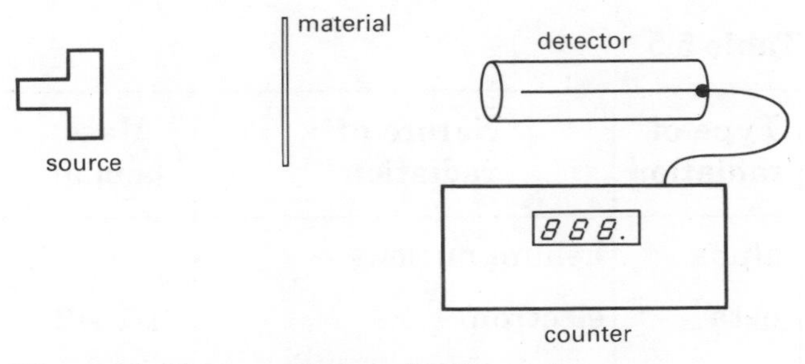

Fig. 5.31 Using a Geiger-Müller tube to count radiation

Here is a set of results.

Source	Material between source and counter			
	nothing	paper	aluminium foil	lead foil
alpha	521	13	7	2
beta	134	121	10	4
gamma	388	395	343	121
no source	8	6	5	1

(all the figures show the number of counts in 30 seconds)

The results show that alpha particles are stopped by paper, beta particles are stopped by aluminium foil and gamma particles are partly stopped by lead. The control experiment readings are caused by natural radiation. This comes from natural radioactive elements and cosmic rays from outer space.

Do the radiations carry charges?

A source of each type of radiation is aimed at a counter as before. The number of counts in 30 seconds is measured. A large magnet is then put between the source and the counter and the counts are measured again. This is repeated for each of the three types of radiation. Here is a set of results.

Source	Counts with no magnet present	Counts with a magnet present
alpha	500	410
beta	175	30
gamma	410	415

(all figures show the counts in 30 seconds)

The particles which carry a charge will be deflected by the magnet away from the counter. From the results, alpha particles seem to carry a charge, as some are deflected in the second experiment. Similarly, beta particles are charged. However gamma particles are not significantly affected, so they have no charge. Table 5.5 gives a summary of the properties of the three types of radiation.

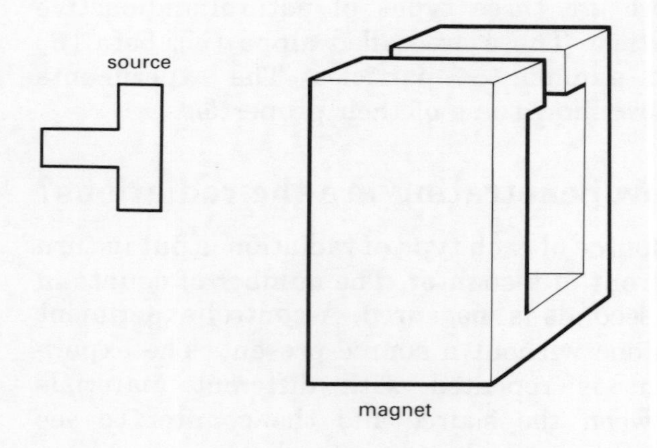

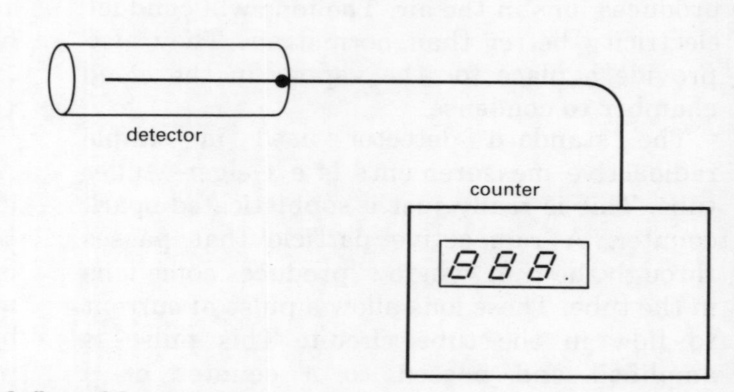

Fig. 5.32 Measuring the amount of radiation not deflected by a magnet

Table 5.5

Type of radiation	Nature of radiation	Mass (a.m.u)	Charge (a.c.u)	Deflected by a magnetic field?	Absorbed
alpha	helium nucleus	4	+2	yes, slightly	by paper
beta	electron	1/1840	−1	yes, strongly	by aluminium foil
gamma	electromagnetic wave	0	0	no	a little by lead

Assignment – Using radiations

Radioactive elements are used to produce nuclear power and in nuclear weapons. They are also very important in many aspects of everyday life. This assignment outlines some of the everyday uses of radiations.

1. Using radiations to kill harmful organisms

Intense radiation kills off living things. Medical instruments can be treated with these radiations (usually gamma rays) to make them sterile. Radiations are also used in the treatment of cancer. Intense beams are aimed at the malignant cells, which are killed off. The difficulty is to avoid damage to healthy cells.

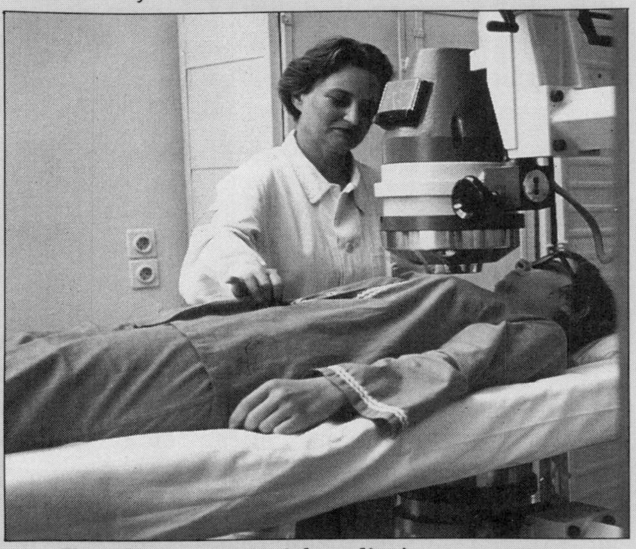

Treating cancer with radiation

2. Using radiation to detect things

Radioactive substances can be used to find or follow the position of something. For example, a weak radioactive liquid can be put into a leaky underground pipe. Radiation will build up where the leak is. This radiation can then be detected with a Geiger-Müller counter.

In hospitals, patients may be given a weak radioactive drink or injection. A scan of the patient with a counter can show if the digestion, or the blood system, is working properly. Tumours can sometimes be detected, as they tend to absorb the radioactive material.

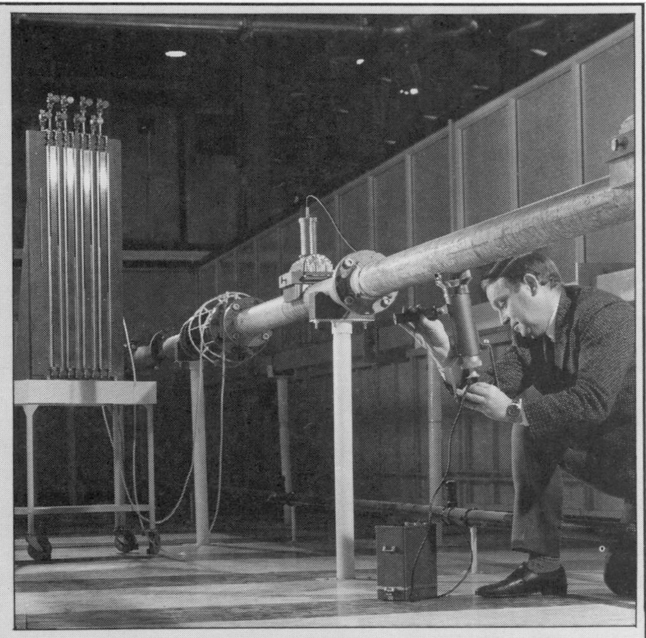

Using radiation to measure the thickness of a pipe wall

3. Using radiation to measure things

An object that is difficult to measure, like the thickness of a ship's hull, can be measured with radiation. Normally a source is placed on one side of the object, and a detector on the other. There is no need for either to touch the object being measured. Sometimes the detector is put on the same side as the source. In this case the detector reads the amount of radiation reflected back by the object. The denser and the thicker it is, the more radiation will be reflected back.

1 Would you use an alpha source, a beta source or a gamma source to *kill* cancer cells? Look back at page 262 if you need help. Explain your choice.

2 Food could be sterilised in the same way that medical syringes are.
(a) Why could this be a good idea?
(b) Would it be dangerous? Explain!

3 (a) What are the problems of using radiations to kill cancer cells? (Write down at least two).
(b) What can be done to reduce the problems?

4 Would you use an alpha source, a beta source, or a gamma source to *find* a tumour in a living person? Explain your choice.

▶

5. The manager in a cornflakes factory wants a method of knowing that her packets are filled to the right level. Do labelled sketches to show *two* ways she could do this. One method should use a radioactive material. Assume that the packets are sealed. (Hint: using an employee to check each packet is no good - it is too expensive, and very boring anyway).

6. (a) If you had to measure the thickness of the walls of an old metal sewer pipe, how could you do it without opening it up? Use diagrams to explain your answer.
   (b) If you used a radioactive source to do the job, would the readings be highest or lowest where the pipe was thick? Explain.

SAFETY

One of the major problems of using nuclear energy is that it is dangerous. Many of the radiations can kill or damage living cells. An overdose of radiation can cause sterility, blindness, sickness or even death. Some of the elements which produce radiation, for example plutonium, are very poisonous. Great care has to be taken when handling or using radiations. Long tongs or robot arms are used to avoid touching the radioactive substances.

When exposure to radiation is part of a person's job, special precautions are taken. Radioactive sources in hospitals or nuclear installations are surrounded with lead. Workers in danger areas wear film badges which record the amount of radiation they receive. The employees also have frequent medical checks. However some people still think that the risks are too great. An accidental release of radioactive dust or liquid from a nuclear power station may increase the risk of cancer in people living nearby. It is difficult to prove this link because cancers caused by radiation may not appear until 20 or 30 years after exposure.

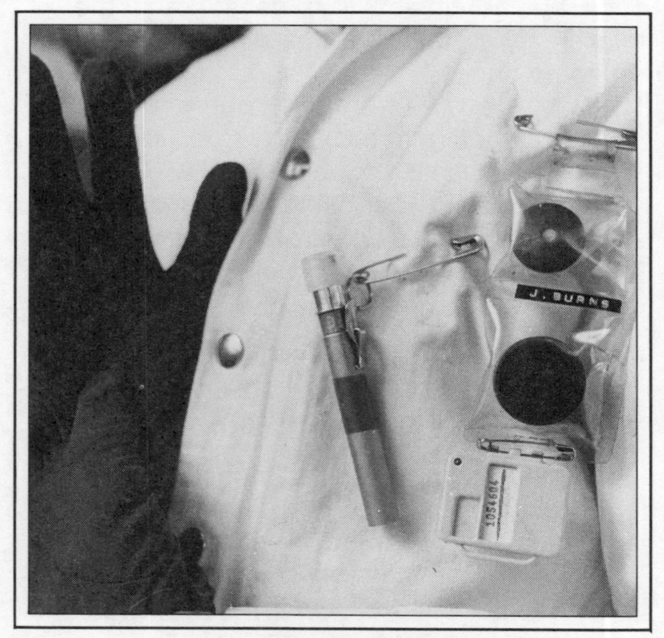

Fig. 5.33 An array of devices that monitor how much radiation a radiation worker is exposed to

RADIOACTIVE ELEMENTS

There are many radioactive elements. In nature, they are mostly heavy elements, heavier than lead. Uranium, plutonium and radium are examples. These elements are unstable. They may lose some nuclear energy and emit one of the three types of radiation. At the same time, the element itself may change into a different element.

Elements that emit alpha radiation become lighter elements:

$${}^{226}_{88}\mathrm{Ra} \longrightarrow {}^{222}_{86}\mathrm{Rn} + \alpha$$

radium → radon + alpha particle

Elements that emit beta radiation become a new element, but with the same mass:

$${}^{234}_{90}\mathrm{Th} \longrightarrow {}^{234}_{91}\mathrm{Pa} + \beta$$

thorium → protactinium + beta particle

An element that emits a gamma particle loses energy but does not become a different substance. Gamma radiation usually happens at the same time as alpha or beta radiation.

When an element gives out an alpha or a beta particle, it *decays* to become a new element. if the new element also decays, there will be a series of elements made in turn. This is called a decay series. Here is part of the decay series for uranium-238:

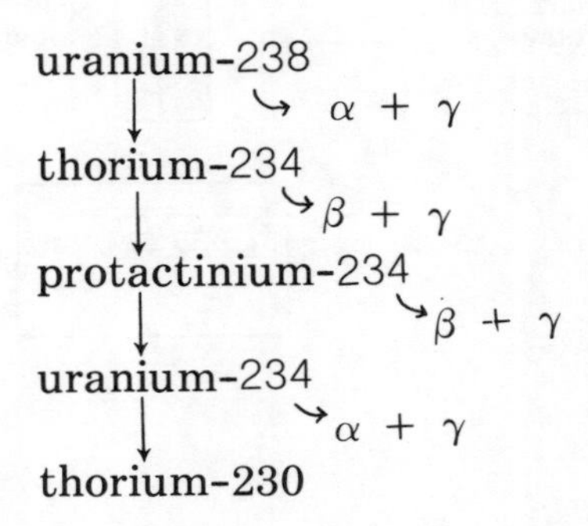

The series continues through ten more stages until it reaches lead-206 which is a stable element.

How fast do elements decay?

Radioactive elements do not give out particles in a steady flow. Some give them out quickly, others do it more slowly. The particles are emitted *at random*. Random means unpredictable. It is rather like your chance of winning on the Premium Bonds with only one bond. You cannot say when it will happen, if it will happen at all.

Because it is random, it is impossible to say how long it will take for all the atoms to decay. Instead, the average time for half of the atoms to decay is measured. This is called the HALF-LIFE. It takes one half-life, in years, hours or seconds, for half the atoms in a radioactive element to decay. Here is an example: thorium-234 decays to protactinium-234. It takes 24 days for half the thorium to decay (Fig. 5.34).

Of course, it does not really look like this! You would still see a 'ball' of elements. But some of the ball would be protactinium, and the protactinium would also be decaying.

It is possible to do experiments to measure short half-lives. The next assignment describes an experiment to measure the half-life of protactinium-234.

Knowing an element's half-life can be very useful. For example, it can be used to tell the age of fossils or old furniture. There is a small amount of the carbon isotope carbon-14 in the air. All living things absorb carbon, by photosynthesis or feeding. Even you contain a small amount of carbon-14 in your cells. When an organism dies, it stops taking in carbon. The carbon-14 in the organism gradually decays. As it has a half-life of about 6000 years, it remains present for a long time.

To work out the age of an object like the Dead Sea scrolls, scientists have to measure the radioactivity that the paper produces. They find the total amount of carbon in the paper, and then estimate how much carbon-14 was there to start with. The amount of radioactivity they find tells them how much carbon-14 is left. They can then work out for how long the carbon-14 has been decaying, and so estimate the age of the paper.

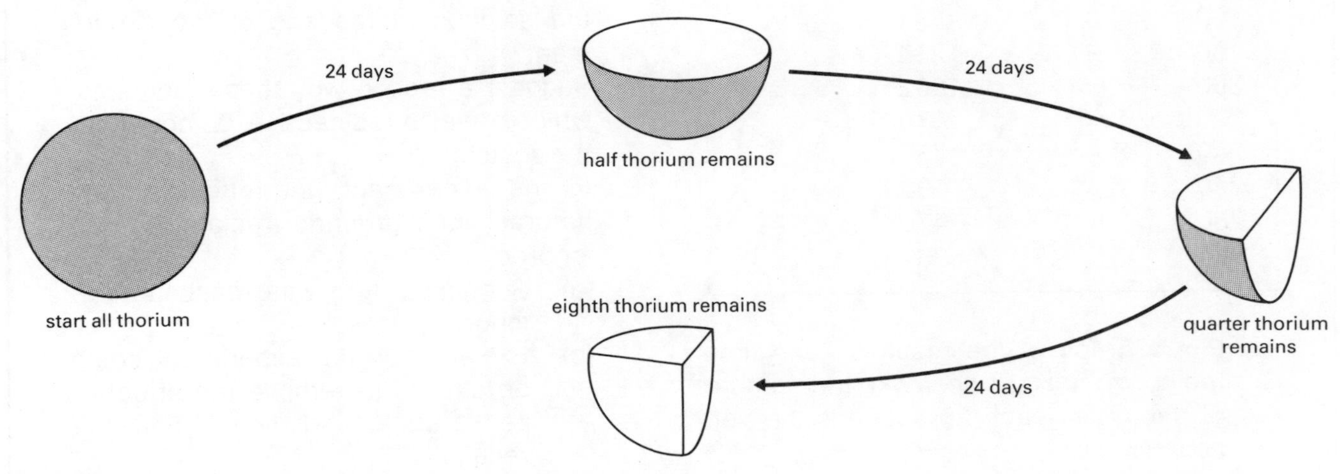

Fig. 5.34

Assignment – Measuring a half-life

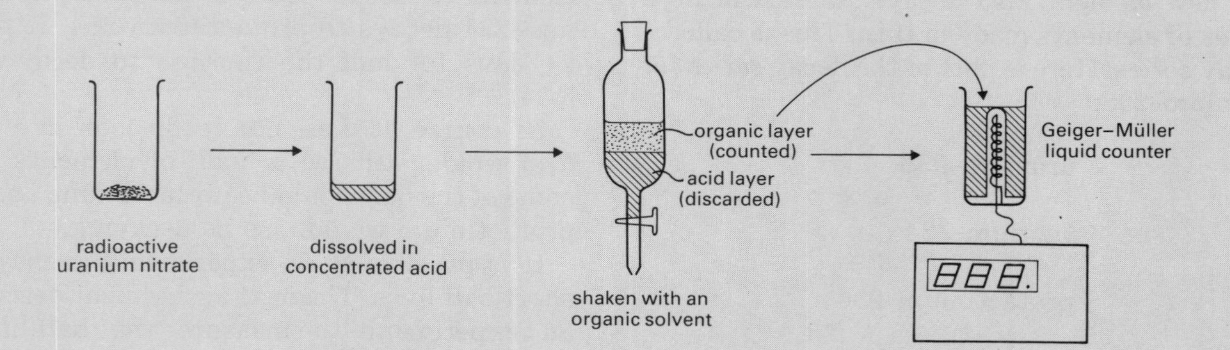

Some uranium nitrate is dissolved in concentrated hydrochloric acid. This gives a solution which contains uranium-238. An organic solvent is added. This is like a dry-cleaning liquid. It dissolves protactinium, one of the decay products of uranium. The dissolved protactinium is separated from the insoluble radioactive elements, and the radiations from it are measured. The flow diagram shows the method.

Results

Time/s	No. of decays since last reading
20	6300
40	5130
60	4270
80	3470
100	2820
120	2340
140	1910
160	1585
180	1230
200	1002
220	870
240	753
260	590
280	470
300	384

1 Plot a graph of the results. Make time the horizontal (*x*) axis. Make the number of decays in each 20 seconds the vertical (*y*) axis.

2 From your graph, when is the number of decays exactly half of the value at 20 seconds? Take 20 seconds from this time, and you have the half-life.

3 (a) How many decays are there between 120 and 140 seconds?
(b) From the graph, when is the rate exactly half this number?
(c) How many seconds are there between your answer to (b) and 140 seconds?
(d) What is the relation between your answer to question 2, and your answer to 3 (c)? Can you explain it?

4 Why did the protactinium have to be removed from the uranium solution before it was counted?

5 From the decay series on page 265 what particles does protactinium emit?

6 What other elements will be present in the solution at the end of the experiment?

7 Suggest a reason why these elements, which are also radioactive, do not affect the results.

8 In the experiment, the effect of any natural background radiation was ignored.
(a) Suggest why it was reasonable to ignore it.
(b) Explain how the experiment could be altered to remove the effect of background radiation from the results.

NUCLEAR POWER

To get useful energy from the nucleus, nuclear energy has to be converted into heat. It would be impossible to do this with natural radiations. The amount of energy produced is not enough to make much heat.

In the early part of this century, two brilliant scientists found the key to using nuclear energy. Albert Einstein (1879-1955) suggested that energy and mass were related by the equation

$$E = mc^2$$

E represents energy, m is mass and c^2 is the velocity of light squared. As light travels at 300 million metres per second, c^2 is very large. If you could convert all the nuclear energy in a bag of sugar into heat, you would get 10^{17} joules (about the same amount of energy as the whole world uses in one day).

Shortly after Einstein, Ernest Rutherford (see page 244) fired alpha particles at nitrogen atoms. He found that some oxygen atoms were produced, and a proton:

$^{14}_{7}N$	+ $^{4}_{2}He$	$\longrightarrow$	$^{1}_{1}H$	+ $^{17}_{8}O$
nitrogen	alpha particle		proton	oxygen isotope

This meant that the helium nucleus had been split. It was later called 'splitting the atom'. In the process, quite a lot of mass was converted into energy.

Einstein and Rutherford began the train of events that led to the atomic bomb, the hydrogen bomb and nuclear power stations.

The atomic bomb

In an atomic bomb, uranium-235 is bombarded with neutrons. Neutrons work better than Rutherford's alpha particles because they have no charge. This means that they are not repelled by the positive charges in the nuclei of the uranium atoms. The neutrons produce FISSION in the uranium-235. It splits into other, lighter, elements, and also produces more neutrons. These neutrons do the same to other uranium atoms, and a chain reaction follows. In each atom that splits, a small amount of mass is converted into energy. As the chain reaction happens very fast, and involves many atoms, a lot of energy is produced. There is a massive explosion. Heat and radiation are produced at an incredible rate.

Fortunately, atomic bombs are not easy to make. Natural uranium contains only a small amount of uranium-235. This has to be concentrated before a weapon can be made.

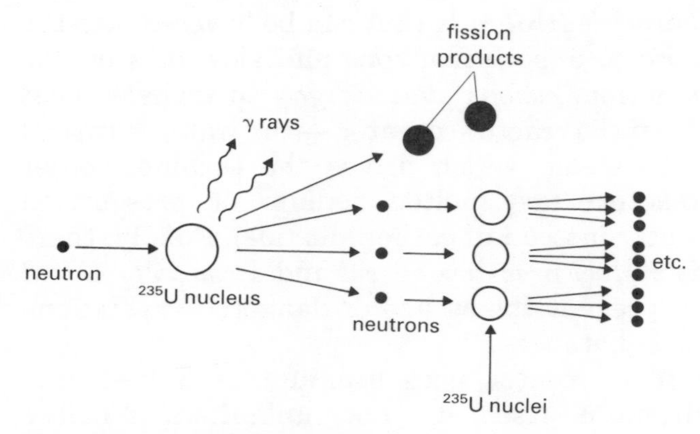

Fig. 5.35 A chain reaction

Fig. 5.36 The mushroom cloud of an atomic weapons test at Maralinga, Australia

Nuclear power stations

Most nuclear power stations rely on the same idea as an atomic bomb. Uranium atoms split, producing heat and neutrons. The difference is that the reaction is kept under control. An uncontrolled chain reaction is not allowed to start.

The reactor shown in the picture has *uranium* in fuel rods; a *graphite moderator* which slows the neutrons down and makes the reaction more efficient (some reactors use special water called heavy water for this); *boron control rods* that can be lowered into the core to absorb neutrons and slow or stop the reaction; *carbon dioxide gas* to transfer heat from the reactor to water — the water is turned into steam which drives the turbines (other reactors use molten sodium or pressurised water instead of carbon dioxide). Finally, there is a *steel pressure vessel* and a *concrete shield* to prevent the escape of dangerous radiations or substances.

This reactor uses uranium as a fuel. But uranium itself is not unlimited. Nuclear scientists are now looking for ways to produce power without using uranium. One answer is the fast breeder reactor. This is much more efficient than the reactor shown above, but produces some dangerous by-products. There is a small reactor at Dounreay in Scotland that works on this principle.

Another answer is to use FUSION instead of fission. Fusion means firing particles together with enough energy for them to stick to each other, or fuse. At the same time, some of their mass is converted into energy. This is how the sun produces energy. It is also the principle behind the hydrogen bomb. Unfortunately, it is very hard to control nuclear fusion. The atoms have to be at a very high temperature to fuse, about 100 million °C. In a hydrogen bomb this is done by using an atomic bomb to start the fusion reaction.

In a fusion reactor, the fuel, which is a mixture of hydrogen isotopes, is heated up. When it is hot enough the gas ionises; the nuclei and the electrons separate. It is now called a plasma, and can be heated further by passing electricity through it. Plasmas are not easy to handle. They tend to expand, so a large magnetic field is used to contain the plasma. All this is so difficult that no-one has yet achieved controlled nuclear fusion, let alone built a fusion reactor. There is a great deal of interest in fusion reactions as they may give us almost unlimited energy. But that is a long way off yet.

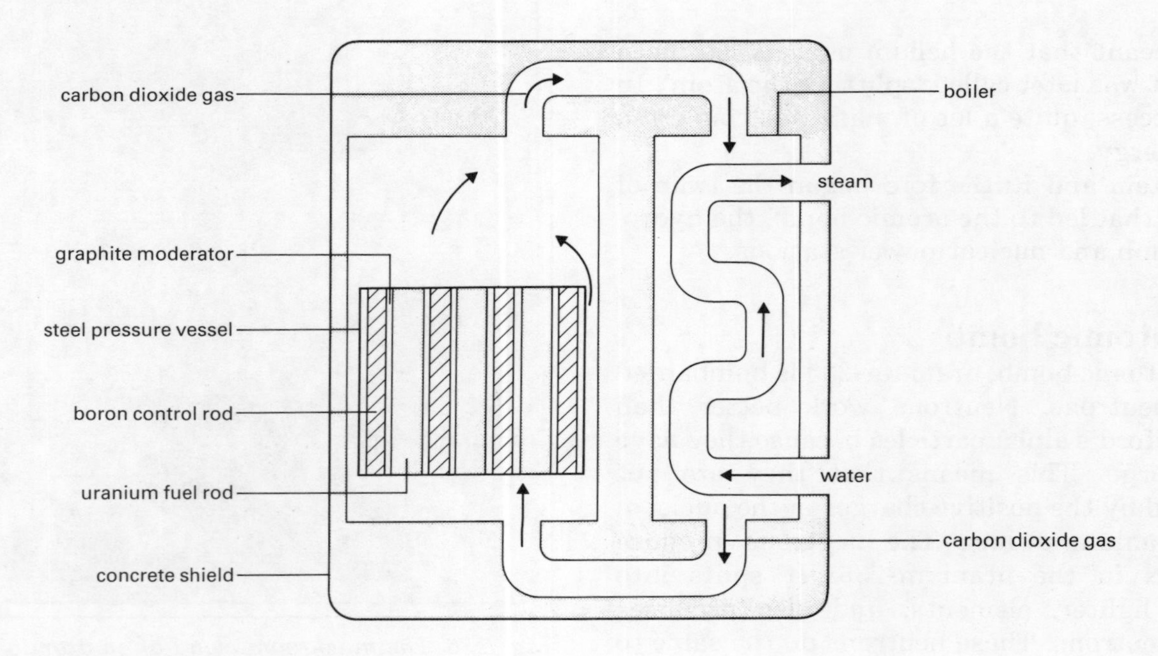

Fig. 5.37 A thermal nuclear reactor

Assignment – Nuclear weapons: yes or no?

Strong feelings exist about nuclear weapons. Some people argue that we need them to keep a balance of power. Others say that they are so dangerous that we cannot afford the risk of keeping them. The two quotations below reflect some of the views of the two sides.

From a Ministry of Defence fact sheet, 1983:

'The present position is a military stalemate. Both sides are able to survive a surprise nuclear attack and to launch a "second strike" nuclear retaliation. But the situation is not static. If one side makes a significant advance in the design or deployment of weapons the other side may have to counter it to ensure that the gap does not grow so wide that deterrence is undermined. As long as we do this war can be prevented - as it has been for over 30 years. Neither side can see advantage in starting a war or in threatening a nuclear attack.

'Deterrence is not an attractive way of ensuring peace. But at least it has worked. There is no alternative as yet which can keep us safe and free. And it gives us time and incentive to get international agreement on disarmament measures which everyone agrees are the best way of guaranteeing world peace.'

From *Questions and Answers about Nuclear Weapons*, published by CND in 1981

'We believe that nuclear arms reduction is right for reasons of survival, morality and finance. The possession of nuclear arms and bases makes Britain a number one priority target for retaliation if war breaks out. New missiles increase the danger for Britain. There are a growing number of people who agree with the mother who said on television: "If the use of nuclear bombs means causing the death of millions of men, women, and children in another country, then I want no part in it, under any circumstances."

'Britain's annual arms expenditure has already risen to nearly £13 000 000 000 and this is planned to be increased by 3% per year. It costs a family of four no less than £16 per week, paid in various taxes. The money would be better spent on housing, health, education, pensions and social services.

'These are immediate aims. The ultimate aim of CND is general and complete disarmament. It is opposed to the policies of any country that makes war more likely.'

1 Make a list of the arguments in favour of nuclear weapons. The passage above is only a starting point.
2 Make a list of the arguments against nuclear weapons. Again, the excerpt from the pamphlet does not give all the reasons.
3 Write a letter to your local newspaper arguing either in favour of nuclear weapons or against them. Remember that style and the use of humour, together with reasonable brevity, will make it more likely that your letter will be published.

Assignment – Nuclear power stations: do we need them?

THE GUARDIAN, Monday January 10 1983

Sizewell plan generates questions of safety and economics

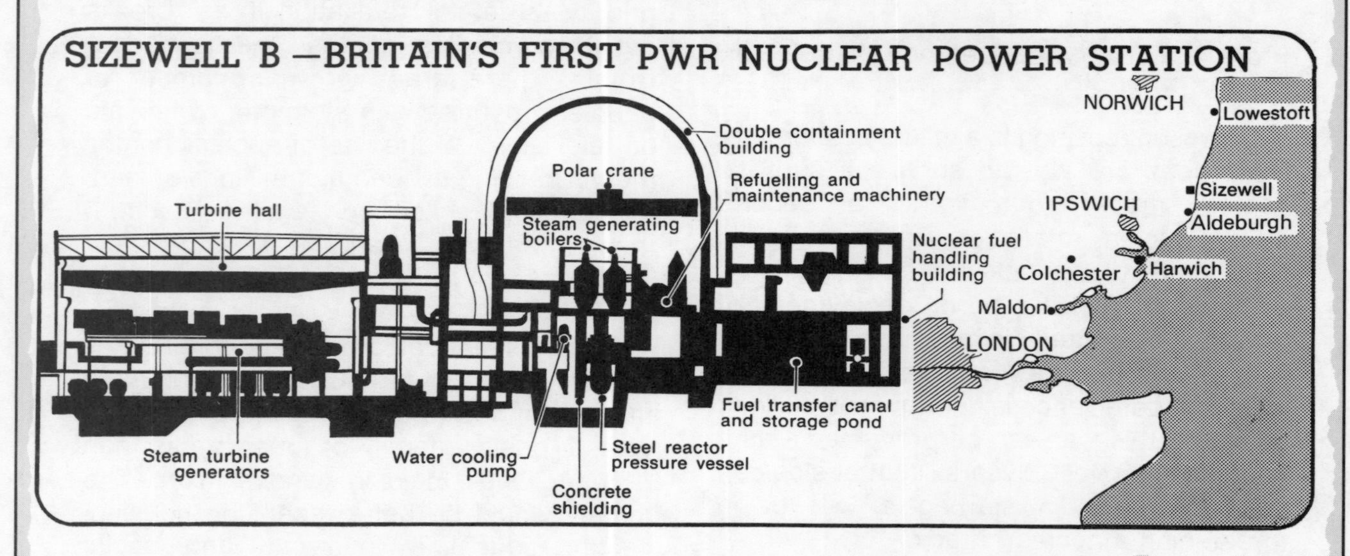

A public inquiry which could determine Britain's nuclear power programme for the rest of this century, opens at the Snape Maltings, Suffolk, tomorrow.

The Central Electricity Generating Board wants permission to build its first American-style pressurised water (PWR) nuclear power station behind the coastal sand dunes at Sizewell, alongside an existing Magnox nuclear station.

The PWR, cooled by water rather than the gas used in British designs, is a new venture for the nuclear industry and raises complex issues of safety — highlighted in the near-catastrophic accident to the American PWR at Three Mile Island in 1979 — and economics.

The inquiry will be chaired by Sir Frank Layfield QC, who took part in the Windscale inquiry into plans for a nuclear fuel processing plant. It is expected to last well into the autumn.

The Sizewell B station would be built to the US Westinghouse/Bechtel SNUPPS design, modified by the National Nuclear Corporation to meet British safety standards, which are more stringent than in the US.

These modifications will add about £100 million to the construction cost, estimated at £1,147 million in March 1982. The inspectorate has already said that, on the basis of the preliminary design, its safety requirements can be met.

Among the changes introduced to meet British standards are additional emergency cooling systems, pumps driven by steam rather than electricity, a second containment shell for the main reactor building and automatic equipment to lower the radiation exposure of the operating staff.

The PWR concept raises at least two basic safety problems. If the cooling water is depressurised — for example through a broken pipe — it turns to an unpredictable mixture of steam and water which may prevent the correct operation of crucial valves.

The reactor core is enclosed in a single large steel pressure vessel which has to be manufactured to an extremely high standard — in this case by the French firm Framatome — to avoid dangerous cracks.

The fact that a nuclear physicist and advocate of the PWR is now running the electricity supply industry will lead the inquiry to the broader question of whether the CEGB's nuclear programme is part of a Government strategy not to become too dependent on coalminers who provide more than 80 per cent of the fuel for power stations.

The board admits that it does not need Sizewell B simply for its capacity — "to keep the lights on," as board member John Baker puts it — until three or four years after it should be completed in 1991. But this, it will argue, is irrelevant, because the new plant will generate electricity at a lower cost than alternative coal or oil-fired plants.

Friends of the Earth and the Council for the Protection of Rural England have both indicated that they will try. They will also emphasise alternative energy strategies that would switch the heavy investment in nuclear power into conservation measures and "renewables" like solar power.

1 What is the fuel for this type of reactor? ('Nuclear' is not enough.)
2 Draw a simple labelled sketch to explain briefly how the energy from the fuel is converted into electrical energy.
3 What do the initials PWR stand for?
4 Why is this type of reactor called a PWR?
5 How does it differ from a reactor cooled by carbon dioxide?
6 Make a list of the properties you would look for in a reactor-cooling liquid or gas. Think first about what the coolant has to do, then list the properties.
7 List at least four safety devices to be incorporated in the design for this reactor.
8 Explain the function of each safety device that you have listed.
9 Why is it proposed that this reactor should be built at Sizewell?
10 Opponents of the reactor at Sizewell believe that they have a strong case. Write a summary of their arguments.
11 Why does the Electricity Generating Board want to build a PWR?

Energy from the nucleus

WHAT YOU SHOULD KNOW

1 The nucleus of an atom contains energy which can be converted into heat.
2 There are three main types of natural radiation emitted by radioactive substances: alpha, beta and gamma.
3 Some elements are naturally unstable and decay, emitting radiations.
4 Radiations from decaying elements are given out at random.
5 The rate at which the radiations are given out is measured by the half-life. This is the time taken for half the atoms in an element to decay.
6 Radioactive substances are very dangerous, but can also be very useful.
7 Energy to make electricity, or nuclear explosions, can be obtained from a nucleus that splits ('fissions').
8 The sun's energy is provided by atoms that fuse or join together ('fusion'). This may provide a source of energy on earth in the future.

QUESTIONS

1 Carbon-14 ($^{14}_{6}C$) is a radioactive isotope of carbon which decays by beta decay. It is present in living matter with carbon-12 ($^{12}_{6}C$).

(a) Copy the table below and show the number of protons, neutrons and electrons in the two isotopes of carbon.

	Protons	Neutrons	Electrons
Carbon-12 Carbon-14			

(b) The elements of the second period of the Periodic Table are shown below:

Symbol:	Li	Be	B	C	N	O	F	Ne
Atomic number:	3	4	5	6	7	8	9	10

Write a word equation which represents the decay of carbon-14.

(c) In freshly cut wood 1.1% of the carbon present is carbon-14. The graph below shows the percentage of carbon-14 in samples of wood of different ages.

(i) What is the half-life of carbon-14?

(ii) A wooden axe handle was found by radioactivity measurements to contain 0.4% of carbon-14. Approximately how old is the axe handle?

(d) Carbon-14 burns in oxygen to form carbon dioxide according to the following equation:

$$^{14}_{6}C(s) + O_2(g) \longrightarrow {}^{14}_{6}CO_2(g).$$

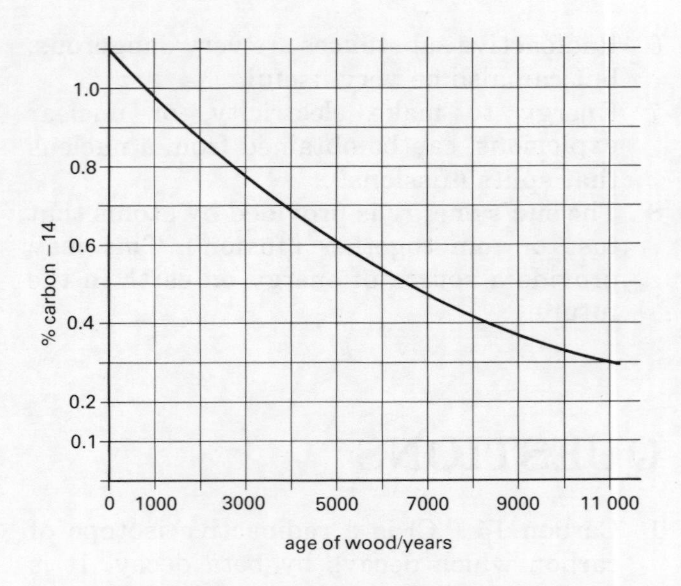

(i) What mass of carbon dioxide would be produced by burning 1 mole of ${}^{14}_{6}C$ atoms? (A_r (O) = 16) Show your working.

(ii) How would the density of carbon dioxide made from ${}^{12}_{6}C$ compare that from ${}^{12}_{6}C$? Explain your answer.

2 Below is shown a simplified schematic diagram for a nuclear powered electricity generating station.

One of the problems associated with nuclear powered generating stations is that of getting as much heat as possible from the reactor to the boiler in order to raise steam. Careful choice of a suitable fluid to be pumped around the heat exchanger system is essential. The following data are for three substances that could be used in the heat exchanger system:

Substance	Boiling point	Ability to absorb heat	Cost
Water	100 °C	good	very low
Carbon dioxide	– 78 °C	fair	low
Sodium	892 °C	very good	fairly expensive

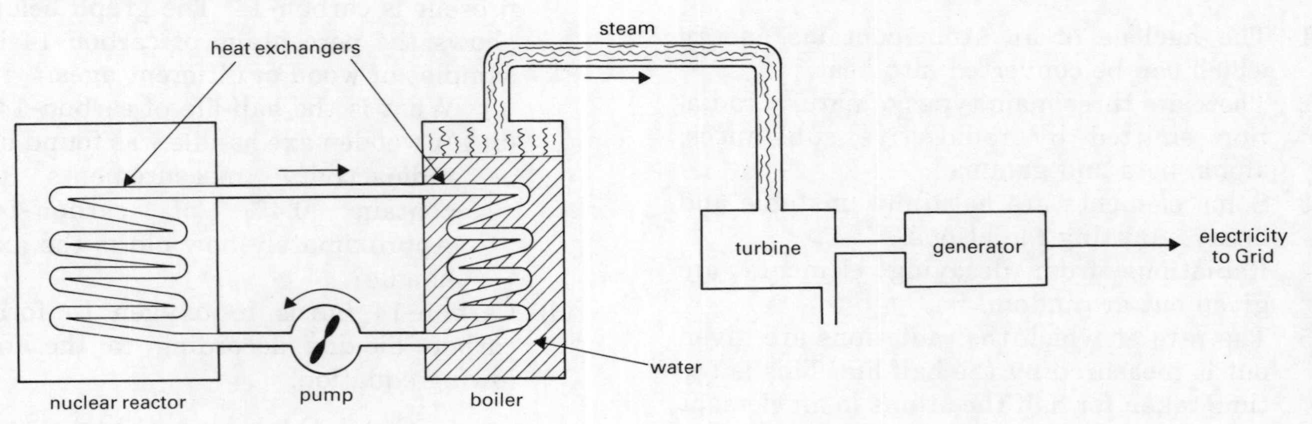

(a) Which chemical substance would you choose for the heat exchanger fluid?

(b) Give two reasons for choosing this substance.

(c) Give two reasons against using this substance.

(d) Explain how heat is generated by the reactor.

(e) What is done to stop the reactor generating too much heat?

(f) What precautions are taken to prevent radiation leakage?

THE PERIODIC TABLE

I	II											III	IV	V	VI	VII	0
1 H 1																	2 He 4
3 Li 7	4 Be 9											5 B 11	6 C 12	7 N 14	8 O 16	9 F 19	10 Ne 20
11 Na 23	12 Mg 24											13 Al 27	14 Si 28	15 P 31	16 S 32	17 Cl 35.5	18 Ar 40
19 K 39	20 Ca 40	21 Sc	22 Ti	23 V	24 Cr	25 Mn	26 Fe 56	27 Co	28 Ni	29 Cu 63.5	30 Zn 65	31 Ga	32 Ge	33 As	34 Se	35 Br 80	36 Kr
37 Rb	38 Sr	39 Y	40 Zr	41 Nb	42 Mo	43 Tc	44 Ru	45 Rh	46 Pd	47 Ag 108	48 Cd	49 In	50 Sn	51 Sb	52 Te	53 I 127	54 Xe
55 Cs	56 Ba	57 La	72 Hf	73 Ta	74 W	75 Re	76 Os	77 Ir	78 Pt	79 Au	80 Hg 200.5	81 Tl	82 Pb 207	83 Bi	84 Po	85 At	86 Rn
87 Fr	88 Ra	89 Ac	104 Rf*	105 Ha	106												

58 Ce	59 Pr	60 Nd	61 Pm	62 Sm	63 Eu	64 Gd	65 Tb	66 Dy	67 Ho	68 Er	69 Tm	70 Yb	71 Lu
90 Th	91 Pa	92 U	93 Np	94 Pu	95 Am	96 Cm	97 Bk	98 Cf	99 Es	100 Fm	101 Md	102 No	103 Lr

*Or Ku.

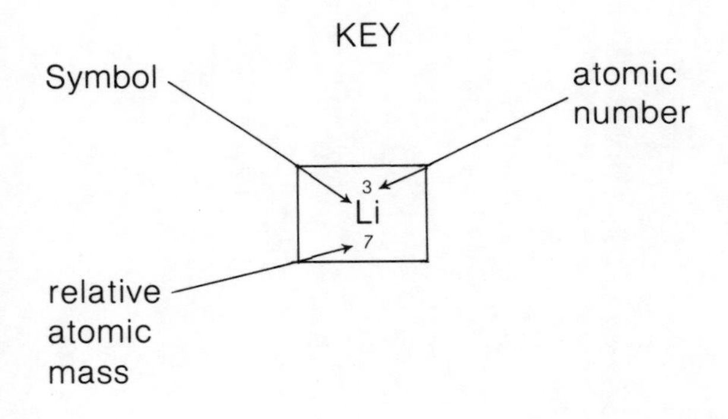

INDEX

Page numbers printed in **bold type** show the pages where the word is explained.